# Efficient Management of Wastewater: Its Treatment and Reuse

# Efficient Management of Wastewater: Its Treatment and Reuse

Edited by Gabriel Craig

STATES
ACADEMIC PRESS
www.statesacademicpress.com

**States Academic Press,**
109 South 5th Street,
Brooklyn, NY 11249, USA

Visit us on the World Wide Web at:
www.statesacademicpress.com



ISBN: 978-1-63989-753-7



**Cataloging-in-Publication Data**

Efficient management of wastewater : its treatment and reuse / edited by Gabriel Craig.
p. cm.
Includes bibliographical references and index.
ISBN 978-1-63989-753-7
1. Sewage disposal. 2. Water reuse. 3. Water--Waste. 4. Water--Purification.
5. Water quality management. I. Craig, Gabriel.
TD741 .E34 2023
628.3--dc23

# Table of Contents

# Preface

Wastewater treatment aims at reducing pollutants by using different treatment methods. To achieve desired performance, it is important to ensure the proper operation and maintenance of the water treatment plant. A number of factors determine the wastewater reuse and include potential revenues, operating costs, resource value and public acceptance of wastewater resources. Membrane filtration, automated variable filtration (AVF) technology, new urban sanitation technology, and urine separating process are some of the advanced technologies used for wastewater treatment. The new urban sanitation technology integrates electro flocculation (Elflox) and anaerobic digestion technologies for treating wastewater. Natural treatment systems are helpful in removing several contaminants such as nutrients, pathogens, micro-constituents, and endocrine disrupting chemicals. This treatment process is extremely effective for water reclamation (or water reuse). This book provides a detailed explanation of the wastewater treatment and reuse technologies. It aims to equip students and experts with the advanced topics and upcoming concepts in the area of wastewater treatment and management.

This book is a comprehensive compilation of works of different researchers from varied parts of the world. It includes valuable experiences of the researchers with the sole objective of providing the readers (learners) with a proper knowledge of the concerned field. This book will be beneficial in evoking inspiration and enhancing the knowledge of the interested readers.

In the end, I would like to extend my heartiest thanks to the authors who worked with great determination on their chapters. I also appreciate the publisher's support in the course of the book. I would also like to deeply acknowledge my family who stood by me as a source of inspiration during the project.

**Editor**

# 1

# Energy Issues in Sustainable Urban Wastewater Management: Use, Demand Reduction and Recovery in the Urban Water Cycle

**Andrea G. Capodaglio** [1,*] **and Gustaf Olsson** [2]

1 Department of Civil Engineering and Architecture, University of Pavia, 27100 Pavia, Italy
2 Department of Biomedical Engineering, Division of Industrial Electrical Engineering and Automation, Lund University, Box 117, 221 00 Lund, Sweden; gustaf.olsson@iea.lth.se
* Correspondence: capo@unipv.it

**Abstract:** Urban water systems and, in particular, wastewater treatment facilities are among the major energy consumers at municipal level worldwide. Estimates indicate that on average these facilities alone may require about 1% to 3% of the total electric energy output of a country, representing a significant fraction of municipal energy bills. Specific power consumption of state-of-the-art facilities should range between 20 and 45 kWh per population-equivalent served, per year, even though older plants may have even higher demands. This figure does not include wastewater conveyance (pumping) and residues post-processing. On the other hand, wastewater and its byproducts contain energy in different forms: chemical, thermal and potential. Until very recently, the only form of energy recovery from most facilities consisted of anaerobic post-digestion of process residuals (waste sludge), by which chemical energy methane is obtained as biogas, in amounts generally sufficient to cover about half of plant requirements. Implementation of new technologies may allow more efficient strategies of energy savings and recovery from sewage treatment. Besides wastewater valorization by exploitation of its chemical and thermal energy contents, closure of the wastewater cycle by recovery of the energy content of process residuals could allow significant additional energy recovery and increased greenhouse emissions abatement.

**Keywords:** water–energy nexus; urban used water cycle; wastewater treatment plant; waste sludge; energy demand; energy recovery; GHGs reduction

---

## 1. Introduction

The water–energy nexus has been long neglected but became recently a high-priority issue from which better understanding and development of new paradigms of water cycle sustainability may arise [1,2]. After decades of unchanging practices, in fact, new strategic and technological approaches to urban water cycle management are being proposed [3]. Urban water systems and, in particular, wastewater treatment plants (WWTPs) are one of the major energy consumers at the municipal level worldwide. It has been estimated that on the average these facilities alone may require about 1% to 3% of the total electrical energy output of a country [4] and that over 20% of public utilities' electrical energy consumption by municipalities is required for their operation [5,6]. Furthermore, energy use by utilities is expected to grow significantly in the next decade in some areas. In Australia, for example, due to an expected population growth of 25% by 2030, and to the need to access new, more energy-intensive water sources mostly by reuse of treated wastewater, energy use by water utilities for enhanced effluent processing is estimated to grow between 130% and 200% above existing levels to fulfil a steady per-capita consumption [7].

According to current technology levels, specific power consumption of state-of-the-art WWTPs should range between 20 and 45 kWh per PE (population equivalent) per year, even though some older plants may double these energy demands [8]. Other data indicate power consumption of between 0.3–2.1 kWh/m$^3$ of treated wastewater in the EU, and between 0.41 to 0.87 kWh/m$^3$, in the U.S., depending on type of treatment, plant size and topography, etc. [9]. Some assessments may include energy for pumping wastewater to the plant, while others do not, hence the wide range of reported values. According to the North Rhine Westphalia (Germany) Ministry of Environment's energy WWTP manual, the optimal target for overall electricity consumption at treatment plants is determined as 20 kWh/PE-year, with guide value at 26 kWh/PE-year [10]. Generally, the smaller the facility, the higher its specific power consumption. Power consumption depends not only on plant size but also on design and technology. The previous values pertain to modern facilities, which include nutrient (N and P) removal and anaerobic sludge digestion. Lack of nutrient removal processes implies lower energy requirements, lack of anaerobic digestion, on the other hand, will increase consumption (e.g., for aerobic sludge stabilization). The cited values do not include the upstream water supply cycle, wastewater conveyance (mostly energy for pumping) and residues post-processing and disposal (chemical and biological sludges). Consequently, these facilities may not only be considered as highly energy demanding, with high operational cost, but also significant sources of greenhouse gas (GHG) emissions, whose reduction has been recently mandated by European Union and other countries' policies.

On the other hand, wastewater contains energy in different forms: chemical, thermal and kinetic. The chemical energy contained in wastewater has been estimated by different studies as ranging from approximately 10 to 14 kJ/g COD (1.67–2.33 kWh/m$^3$ even assuming a diluted COD concentration of 600 mg/L), while thermal energy could yield about 21 MJ/m$^3$ (5.8 kWh/m$^3$) for a drop of 5 °C in wastewater temperature, and potential energy would yield just 30 kJ/m$^3$ (0.008 kWh/m$^3$) over a 10 m drop [11–13]. The first two numbers clearly illustrate the energy-saving potential of innovative wastewater management, highlighting the fact that the energy stored in wastewater is about six to nine times higher than the electric energy needed for treatment. Assuming this energy could be practically exploited, WWTPs could be transformed into energy-neutral or even in net energy producers [14,15].

Until very recently, the almost unique form of energy recovery from these facilities consisted of the anaerobic post-digestion of process residuals (waste activated sludge, WAS) by which chemical energy in the form of biogas (mostly methane) is obtained. This is still one of the main energy recovery options applied, generally sufficient to cover about half of total plants requirements, but with low (30% to 60%) conversion efficiency of the organically embedded chemical energy into a readily usable source [16]. However, implementation of new process technologies is nowadays making possible more efficient strategies of energy recovery from sewage [17]. Besides wastewater valorization by exploitation of its chemical and thermal energy contents, closure of the wastewater cycle by taking advantage of the energy contents of process residuals could provide significant additional energy recovery and increased GHGs abatement.

Wastewater and its residual products can therefore be considered renewable energy sources if addressed by proper technological solutions.

This paper summarizes the main items of energy consumption in the wastewater treatment cycle and discusses the most promising state-of-the-art technologies currently available for energy recovery from both wastewater and its residual byproducts.

## 2. Wastewater Embedded Energy

According to current paradigms, wastewater is still typically viewed as an energy-demanding problem requiring expensive solutions, rather than a resource. Despite the fact that most countries spend substantial amounts of energy treating sewage to release it as harmlessly as possible into receiving waters, wastewater represents a mostly-untapped, potentially huge source of energy, including intrinsic energy embedded within wastewater organics; thermal energy (5.8 kWh/m$^3$ per 5 °C) recoverable through heat extraction, e.g., by a heat pump; and an external fossil-fuel energy equivalent required for the production of the same amounts of fertilizing elements N and P contained in wastewater (19.3 kWh/kg N and 2.11 kWh/kg P) [18].

Owen [19] theoretically calculated the chemical energy of wastewater as 3.86 kWh/kg COD oxidized to $CO_2$ and $H_2O$. Several other studies attempted to determine experimentally the energy content of raw municipal wastewater by bomb calorimeter tests, finding energy values ranging around 3.2 kJ/g dry sample, which translates into organic-content-based value of 14.7 kJ/g COD (4.1 kWh/kg COD) [11]. Subsequent studies postulated that this value would in fact be a substantial underestimation of real wastewater internal energy content, given that the methodology previously used had some faulty premises and that the true value may in fact be significantly greater. By comparison, freeze-drying based wastewater energy estimation enabled a greater fraction of that energy to be measured, up to almost 30 kJ/g COD (8.3 kWh/kg COD), matching more closely with internal chemical energy calculations based on thermodynamic calculations from First Principles (reaction enthalpy) [20].

Shizas and Bagley [11] determined for the North Toronto WWTP a wastewater total energy content of 62.8 MWh/d, according to their embedded energy estimates, compared to a measured electrical consumption to run the entire treatment plant of 6.8 MWh/day, with ratio between embedded and required energy greater than 9.2. If by conservative calculations, a median energy content of 12 kJ/g COD (3.3 kWh/kg COD) (according to the earlier estimates) with a world population of 7.7 billion people (as of 2019), with each one producing at least 60 g COD/day, is assumed, wastewater-embedded energy would thus correspond to over 562 TWh/year. If all the energy contained in wastewater could be harnessed, water industries could become energy self-sufficient or even net producers, and wastewater could become also a needed source of energy in developing countries, which often lack an affordable local energy supply, through decentralized, local wastewater treatment systems [21]. To generate this large amount of energy, furthermore, almost 50 MTOEs (millions of tons of oil equivalent) would be required, whose combustion would produce GHG emissions of approximately $100 \times 10^6$ t $CO_2$ (about 1% of current worldwide $CO_2$-equivalent emission estimate).

*WAS Embedded Energy*

A residual of wastewater treatment, biological waste sludge is a concentrate (literally) of chemical energy. Due to higher organic solids concentration (raw wastewater has a Volatile to Total solids ratio of around 0.10, about 5 times lower than the one observed in primary sludge), sludge contains a higher embedded chemical energy than wastewater (in fact, about 5 times as much). It was estimated that the energy content of primary domestic sludge is between 15 and 16 kJ/g (d.w.) (4.2–4.4 kWh/kg COD) while mixed (primary plus secondary) sludge sent to digestion may contain between 11.5 and 13 kJ/g (3.2–3.6 kWh/g COD) [11,22]. According to the determination methodology of Heidrich et al. [20], these values may actually be substantially higher.

## 3. Wastewater Treatment Energy Demand

Energy demand (mostly electric) of wastewater treatment to required standards is a significant component of the urban water cycle overall costs. It was estimated that 30% to 35% of total cost of wastewater treatment facilities in the U.S.A. is due to electric energy supply [23]. More stringent limits for nutrient removal or mandatory removal of presently-unregulated contaminants, such as contaminants of emerging concern (CEC) and pharmaceuticals and personal care products (PPCP), with introduction of additional process steps, might imply a significant increase in energy demand of treatment facilities, therefore energy reduction and recovery represent an important sustainability issue for maintaining required standards.

To date, numerous efforts to adapt mainstream processes to make them less energy intensive have resulted in alternative outcomes, mainly due to the lack of standardized energy monitoring procedures and regular energy audits in these facilities, essential for identifying any potential for improvement [24,25]. A common practice is to compare energy demand in terms of kWh/$m^3$ treated or kWh/PE; however, this can be highly misleading as it is assumed that pollutant concentration in the influent (BOD or COD, TSS, N and P) would not differ significantly between facilities. A more sensible approach would be to compare energy consumption in terms of units of pollutant removed, for example, kWh/kg removed [26], or in terms of "electrical energy per order" (EEO), which is defined as the amount of kWh's required to reduce a pollutants' concentration by one order of magnitude (i.e. one log or 90%) in 1 $m^3$ of solution [27]. EEO values (expressed as $kWh/m^3$-log) therefore measure actual process efficiency and not absolute pollutant removal. They are normally used for advanced oxidation processes (AOPs) and other processes working on direct electric energy input to compare technology efficiency for a given reaction [28] but could easily be adopted for any process driven directly or indirectly (in this case also accounting for energy conversion efficiency) by an electric power supply.

One frequent reason for high energy demand in WWTPs is that, conventionally, they were designed to process wastewater using aerobic activated sludge processes (ASP), requiring pumping of large flows of compressed air or, less often, oxygen into the biological tank. This component may represent more than 50%, and up to 70%, of the entire energy demand of a conventional facility [29,30]. Although ASP is quite simple to handle, its operation is rather energy intensive.

Figure 1 shows typical figures [29] and ranges of energy consumption breakdowns in ASP WWTPs found in the literature. These values may vary significantly as WWTPs may have many possible configuration options, and no standardized method for energy breakdown among process units is established. Figures on aeration energy demand largely agree, although in a fairly wide range, but some of the ranges indicated will strongly depend on the specific type of technology adopted and on the desired end product of the process, varying within up to one order of magnitude: sludge dewatering with belt press may require less than 4% of the total plant energy, while achievement of lower final humidity could significantly boost this consumption. Disinfection requirements will also vary significantly depending on technology: Simple chlorination may require no more than 0.3% of total plant energy, while UV disinfection may account for up to 25% of total electrical energy use at a municipal WWTPs [31].

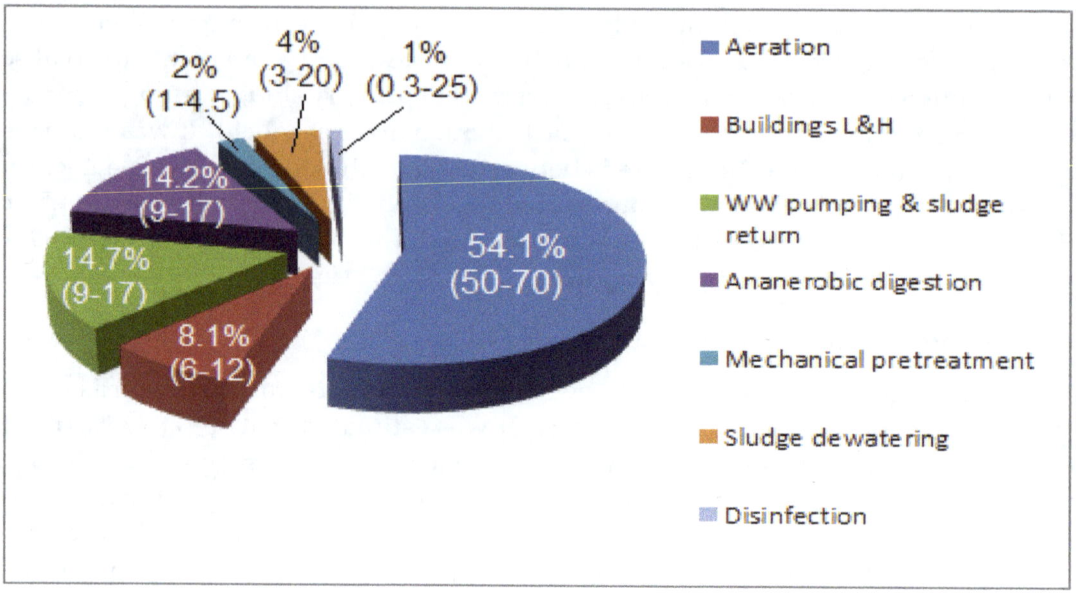


**Figure 1.** Wastewater treatment plant energy consumption breakdown.

Often, in-plant energy estimates neglect external energy embedded in chemicals contributing to the processes. The chemicals consumed at the plant result in both upstream and on-site energy use and emissions; their energy impact on treatment facilities has been estimated as 7% of total demand [32]. However, chemicals' costs are usually assessed as financial rather than energy costs (these also depend on indirect factors, such as their transport) and as such may be greater than pure energy costs at some facilities [33].

### 3.1. WAS Treatment Energy Demand

Finally, pre-disposal excess waste activated sludge (WAS) treatment (or post-treatment as it is usually called in the WWTP process chain) is costly both in financial and energy terms: The former component was estimated in up to 50% to 60% of the total operational costs of WWTPs [34]. These processes are needed to reduce excess sludge volume and stabilize organics and pathogenic bacterial contents that were generated during the treatment process. WAS in fact still contains large quantities of water (primary sludge usually has about 93% to 97% residual humidity, secondary sludge around 98% to 99%), and a significant organic load, making its disposal unsanitary and overly expensive. Various approaches, relying on a combination of physical, chemical and biological methods are employed to WAS post-processing [35]. These processes are usually quite energy demanding as, due to sludge particles properties, only some of the water present within may be removed mechanically, in proportion depending on actual sludge composition and characteristics [36].

Small or older facilities may stabilize WAS by extended aeration or aerobic digestion. This process is a modification of conventional ASP, providing organic matter stabilization without significant volume (water-content) reduction. Aerobically stabilized sludge usually has humidity content of 98% to 99% and thus requires subsequent drying. Extended aeration is typically designed for small WWTPs, where wastewater is aerated for longer than in conventional ASP. As a consequence, extended aeration WWTPs use more energy than other common small plant configurations at an average energy intensity of between 0.21–5.5 kWh/m$^3$ [37,38]. Aerobic sludge digestion is quite similar to ASP: As organic matter supply depletes; bacterial cells undergo an endogenous phase. Spent cell tissue is hence oxidized into carbon dioxide, water, and ammonia. Aerobic digestion is less capital and operationally demanding than other WAS post-processing technologies; however, it is more energy intensive, as oxygen supply must be continuously poured into the process. Aerobic digestion has been rated as the greatest consumer of energy out of all sludge stabilization methods, with median energy intensity of 0.52 kWh/m$^3$ in small plants [38]. While usually adopted in small WWTPs for its simplicity, aerobic stabilization has sometimes been adopted also in large facilities [39].

Energy demand for WAS post-processing can be benchmarked by thermal technologies, which may be considered a standardized baseline. Minimum thermodynamic requirement for sludge drying can in fact be calculated based on sensible and latent heat that is necessary to evaporate water, which is approximately 0.63 kWh/kg. In practice, such requirements (depending on technology) range from 0.82 to 1.1 kWh/kg evaporated water, and total sludge drying energy demand can be calculated at approximately 2500 kWh/ton dry sludge, on average [40]. Table 1 shows rated energy requirements of some commercially available waste activated sludge drying technologies.

**Table 1.** Rated energy requirements of commercially available WAS drying technologies.

| Technology | Total Energy * [kWh/kg] |
|---|---|
| Band dryer ™ | 0.82 |
| Paddle dryer ™ | 0.95 |
| Flash dryers | 1.05 |
| Rotary dryers | 1.05 |
| Drum | 1.07 |
| Fluid bed | 1.10 |
| Greenhouse | 1.53 |

* Sum of thermal (between 95% to 99%) and electric (1% to 5%).

WAS however, has a residual chemical energy content that is attractive for energy recovery purposes. In current practice, therefore, it is common to include in post-treatment WAS anaerobic digestion (a more appealing alternative to aerobic stabilization), with the purpose of recovering a fraction of the chemical energy still contained in the sludge as methane-containing biogas. This will be addressed in Section 5.

### 3.2. *Treatment Residuals Disposal Energy Demand*

Disposal of wastewater processes residuals (processed waste biological sludge) generally makes up a relevant share of the financial and energy costs of a wastewater treatment facility. Wastewater residues disposal may occur according to alternative technologies and disposal routes, with variable environmental impacts. Estimating the true "cost" of each is quite difficult due to the number and variety of the parameters involved. With equal adopted technology capital, operating and energy cost will vary not just from country to country, but even between different sites in the same country. Table 2 shows the range of observed direct financial costs for common final sludge disposal alternatives [41]. Transportation of sludge to its destination may have significant impact also on energy requirements for its disposal.

**Table 2.** Range of sludge treatment and disposal cost in EU countries.

| Disposal Type | Cost Range (Euro) | |
|---|---|---|
| | **Minimum** | **Maximum** |
| Land application | 25 | 210 |
| Landfilling (where allowed) | 125 | 255 |
| Composting | 150 | 310 |
| Thermal drying | 80 | 210 |
| Incineration | 80 | 438 |

As seen in Table 2, even actual observed costs for the most common disposal practices are distributed in a wide range of values. The lack of uniform analytical indicators implies that direct comparisons between the sparsely available data is practically meaningless for making reliable estimates, making general energy requirements assessments unfeasible. It should be noted however that, among the disposal options listed in Table 2, some could allow recovery of materials (e.g., nutrients and carbon) or energy (e.g., incineration), or both, as discussed later, with some cost recovery.

## 4. Towards More Energy-Efficient WWTPs

As mentioned, WWTPs energy audits can be a very effective tool to identify process units that are inefficient and overly energy hungry. In addition to equipment energy efficiency check-up and replacement, design verification of existing facilities should also be performed: One of the most overlooked issues in plants' energy requirements is pumping. Without considering losses due to pumps efficiency (typically between 65% and 80%), theoretical pumping energy requirements are 9.81 kJ/m$^3$-m, or 2.725 kWh/10$^3$m$^3$-m. Typical sewage systems pumping energy requirements have been estimated to 69 kWh/PE-year [42], or over twice the average amount of energy used for treatment of the wastewater to high standards. In-plant pumping requirements may also be significant (some estimates rate them at up to 50% of total energy) but they are often lumped together with treatment process needs in many surveys and are thus difficult to properly estimate. As most plants are constructed on flat terrain, pumping cannot be avoided, but the internal hydraulic profile should be designed to minimize head loss. Double pumping should be avoided: It is not uncommon to observe that flow is often pumped multiple times, e.g., from inlet sewer to headworks, then from headworks to primary sedimentation, and then from final clarifiers to disinfection and discharge. Every time flow drops into a sump well, total pumping head lost at each site could be as high as 5 m, requiring additional lifting energy of at least 0.014 kWh/m$^3$. Discharge pumping should be needed only when water level in the receiving stream is high, which may be only for short high flow periods. Minimization of in-plant static head losses may lead to significant reduction of energy consumption.

Implementation of low energy equipment and processes could result in significant energy savings [14]. Energy efficiency improvements may be obtained by implementing high efficiency aeration technology, including high efficiency blowers, fine or ultrafine bubble diffusers, oxygen sensor devices [43]. Oxygen requirements typically vary during the day by a factor of 5 to 7, depending on incoming loads. Online sensors and model-based control systems may enable operators to enact energy and performance efficient process control strategies [44]. Other measures to improve both energy use and process performance may consist in specific interventions such as use of low energy mixers in anaerobic process phases (e.g., biological denitrification and phosphorous removal). Many design manuals indicate energy requirements for mixing at around 8–16 W/m$^3$. Mixing energy can be reduced to 2 W/m$^3$, which by causing minimal surface turbulence reduce oxygen transfer into solution but still prevent solids stratification [45]. Therefore, in addition to direct power savings, such modifications may even improve process performance.

In order to identify appropriate process modifications in terms of energy and treatment efficiency, process simulation studies may also be conducted to test modified process schemes [46,47]. Sometimes, switching to a completely new process may prove more successful than attempting to update an old layout.

Some technological improvements in process treatment effectiveness may imply drawbacks of increased energy requirements. One example is given by membrane bioreactors (MBRs), a common upgrade of ASPs, which use more energy than the latter [48]. Significant potential reduction in MBRs energy use may derive from process modification, such as the use of anaerobic fluidized membrane bioreactor (AFMBR), combining an MBR with an anaerobic fluidized bed reactor (AFBR) [49], or other similar anaerobic schemes being developed.

Additional recent developments in process technology have perfected biological processes that have significantly lower energy requirements than traditional ASPs. For example, the NEREDA® Process [50], a patented process based on the aerobic granular sludge (AGS) formation phenomenon [51,52] and operated according to an optimized sequencing batch reactor (SBR) cycle, in addition to sporting a much reduced footprint, has demonstrated energy requirements of less than 60% of those of an equivalent ASP, resulting from lack of moving parts, mixers, or recycle pumps inside process units (Figure 2).

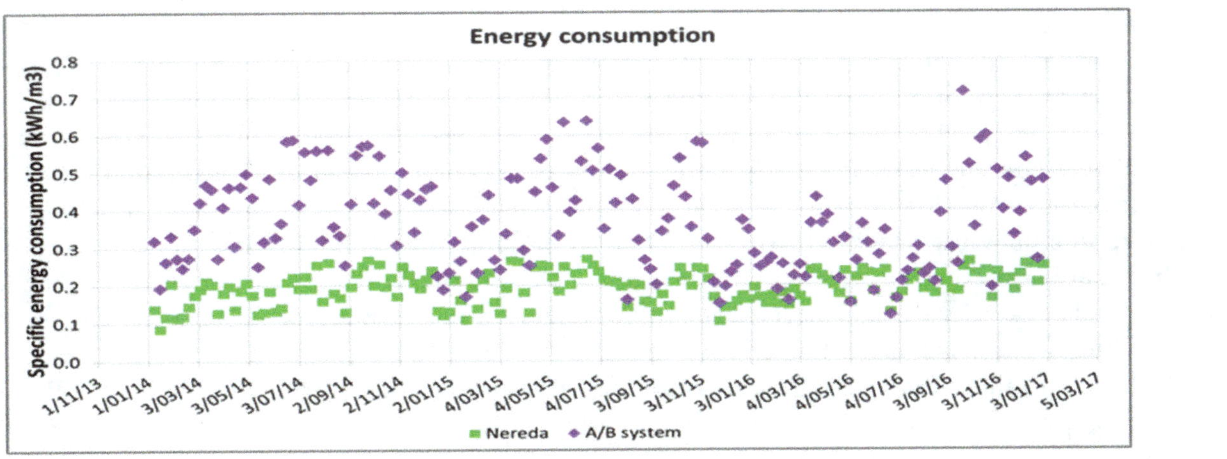


**Figure 2.** Energy consumption patterns for the Garmerwolde plant in Groningen, The Netherlands. Nereda® vs. traditional A/B process. (DHV Royal Haskoning).

The AGS process selectively develops biomass that forms granules, rather than flocs, which include phosphate-accumulating nitrifiers, denitrifiers and glycogen-accumulating organisms, allowing simultaneous anaerobic, aerobic and anoxic conditions to coexist, thus reducing the need for multiple tanks and recirculation and allowing simultaneous biological removal of organic, nitrogen and phosphorus components. Granular biomass also shows faster settling compared to flocs, disposing of the need for large clarifiers. In DHV's Garmerwolde facility in Groningen, The Netherlands, a Nereda® and an A/B (adsorption/bio-oxidation) process (two-stage N-removing modification of conventional ASP) are operated side-by-side, the former treating 40% of the total incoming wastewater flow, with an energy usage of 13.9 kWh/PE-year as opposed to the 25 to 28 kWh of the ASP process. Depending on wastewater flow and characteristics, a Nereda® plant could include multiple modular reactors (often 3) or a combination of buffer tank and one/two reactors. An advantage of AGS processes is that they can be implemented in traditional WWTPs relatively easily with minor modifications. Process configuration is flexible; therefore, Nereda® technology can be used to upgrade existing ASP or SBR plants, achieving approximately twice the active biomass concentration, with outstanding settling characteristics. As a result of retrofit to this technology, both biological and hydraulic capacity of existing plants would be significantly increased, with considerable improvement of effluent quality.

Disposing of aeration altogether, immediate consistent energy savings could be obtained, at least equal to the energy otherwise spent for air insufflation. This could be achieved by substituting ASPs (or other aerobic processes) with anaerobic biodegradation (anaerobic digestion, AD), that could in fact represent one of the principal core technologies in future urban sanitation realizations [53]. AD has existed as a technology for over 100 years and sports several advantages in addition to low energy requirements including low facilities' construction costs, low excess sludge production, plain operation and maintenance, energy generation in the form of biogas, robustness in terms of COD removal efficiency, pH stability and process recovery time.

The main problem preventing its generalized adoption is that domestic wastewater in developed countries is usually too diluted to render this process economically convenient as, being characterized by lower reaction rates, it would require much larger facilities than those needed by aerobic processes to reach the same COD removal objectives. Low temperatures are also mentioned as barriers to the feasibility of AD of domestic wastewaters; however, studies have shown that good performance of these systems could be attained even at temperatures as low as 5 °C [54]. Several researchers have remarked that if a technological shift in urban sewerage could be completed through the adoption of low-dilution collection systems, anaerobic processes could become the mainstream process for domestic wastewater treatment [55]. An ideal process technology in this sense could be represented by Upflow Anaerobic Sludge Blanket (UASB) reactors, already used to treat urban sewage in several large cities of tropical countries (e.g., Cali (Colombia), Belo Horizonte (Brazil)), due to low energy requirements, higher organic concentration in wastewater and extremely favorable climatic conditions [56]. UASB operation is based on the formation of granular sludge (1 to 3 mm in diameter) suspended in the tank within a thick layer ("blanket") by the effect of the upward influent current. Solids accumulation in the blanket (with thickness in the order of 1.5–2.5 m)enables disconnection between solid and liquid retention time in the reactor, increasing applicable organic loads. Wastewater organics are degraded by anaerobic microorganisms while flowing through the blanket; at the same time, methane-containing biogas is produced and may be collected as energy source. UASB operation is controlled by hydraulic loading rates. Upflow velocity of 0.7–1 m/h must be maintained to keep the sludge blanket in suspension and ensure that biomass is not washed out.

Enhancements of the UASB process include Expanded Granular Sludge Bed (EGSB) [57] and Anaerobic Baffled reactors (ABR), both designed to exploit more effectively the granular nature of UASB biomass under low temperature and low organics concentration conditions. The latter employ plug-flow staging of the various anaerobic treatment phases, this being a more efficient configuration than the complete mix scheme used by EGSBs [58]. In pilot scale applications, EGSB systems have shown energy consumption savings of up to 75%, compared to a conventional activated sludge process [59]. While EGSB applications are nowadays common, ABR use, showing promising results at pilot-scale, is still limited in full-scale installations.

If implemented under proper conditions, anaerobic wastewater treatment could become a net energy producer. Another non-negligible advantage of AD is that the amounts of digested sludge from this process are much lower (by 3–20-fold) than those produced by aerobic treatment, constituting another highly valuable energy (and economic) benefit in terms of reduced residue disposal demand.

Both direct and indirect energy recovery could derive from the optimization of nutrients removal and recovery strategies from wastewater. Currently, nutrients (N and P) removal before discharge is often required, in compliance to environmental protection regulations. In conventional ASP nitrification–denitrification, process energy consumption is 2.3 kWh/kg N removed [60]. A new, casually discovered, mechanism of autotrophic N-removal (Anammox) in which ammonium is oxidised under anaerobic conditions has recently been introduced in wastewater treatment practice, with energy requirements of just 0.9 kWh/kg N removed [61–63]. It was also shown that autotrophic N-removal determines lower sludge production by up to one order of magnitude [64], reducing disposal costs for residues. The Anammox process, however, has shown so far better results in the presence of high N concentrations and relatively high temperatures. Although recent studies have shown that Anammox may perform well even at low temperature (13 °C), such applications rely on the successful selection of the only "cold-tolerant" anammox bacterial species so far identified (*Ca. Kuenenia*) in the system's microbiodome [65]. It remains to be demonstrated that Anammox could be widely used in mainstream applications. If successful, conversion from conventional nitrification–denitrification to Anammox process in a WWTP could imply over 60% energy savings for nitrogen control. It should also be considered that recovery of fertilizing elements N and P in usable form would imply indirect energy savings equivalent to those required for their production less those required by the recovery process [66].

Liquid-phase P recovery technologies could reduce, in general, the cumulative energy demand of WWTPs, by reducing energy requirements for sludge handling [67]. On average, the application of P-recovery technologies could reduce cumulative WWTP energy demand by 17.7%, and it was suggested that up to 27% overall energy advantage could exist between a biological P-recovery facility and a conventional chemical P-removing facility [68]. Implementing P recovery technologies as struvite or other P-containing fertilizer materials could reduce a facility's energy requirements compared to conventional P removal by ferric chloride or ferric sulfate salts [66,69].

Finally, in a larger frame of reference, treatment facilities decentralization may be a long-term winning strategy for improving the energy efficiency of the Urban Water Cycle. Decentralized wastewater treatment will require less pumping and shorter pipes, implying less energy than centralized ones, therefore decreasing infrastructural energy requirements and cost. About 80% to 90% of capital costs in such systems are related to the collection system itself, and pumping energy is typically much larger than energy for actual treatment [42]. Almost all current wastewater treatment technologies could, in principle, be applied to such systems including "natural" solutions that are notoriously low-energy demanding, although they are area-extensive. Decentralized (cluster) wastewater treatment operating at local scale not only can have beneficial effects on wastewater disposal in environmental and public health terms but could reduce energy demand and increase reuse of wastewater [17]. Recent developments in renewable energy, in particular solar PV and wind, will make decentralized wastewater treatment even more attractive from an energy point of view, thus meeting the challenges related to the climate variability issue [70].

Less diluted sewage could be achieved by generalized adoption of vacuum, rather than gravity sewers. In these systems, the energy that moves waste along pipes is the difference between atmospheric pressure and vacuum mains negative pressure, therefore requiring less water for transport of waste through the system. Vacuum sewer systems have some notable advantages over gravity ones, including lower (by 30% to 35%) construction costs, lower O&M costs, including lower overall energy (by a similar percentage) for their operation [71]. Hence, from the viewpoint of sustainability, significant reduction of energy and GHGs emissions could be achieved within their life cycle [72]. Similar considerations extend to wastewater treatment, since the more energy-efficient anaerobic technology is ideally suited to these conditions. This technological paradigm switch could significantly improve the overall economics and sustainability of an urban water system, not just of the wastewater treatment processes.

With respect to the removal of emerging contaminants (CECs—contaminants of emerging concern) including pharmaceuticals, while current regulations worldwide do not generally foresee their mandatory removal from wastewater (with some notable exceptions such as Switzerland), stricter effluent standards are under discussion [73]. This would almost certainly increase energy requirements for wastewater treatment significantly, as their removal is currently based mostly on the combination of MBRs and AOPs [74], both highly energy intensive in absolute and specific, i.e., per unit of pollutant removed (EEO), terms [75]. A new class of processes indicated as AO(R)Ps (Advanced Oxidation and Reduction Processes) based on water radiolysis has been proposed for such purpose for its high removal efficiency yield and low EEO indexes [76]. These new technologies have been used for some time in different production sectors, but their adoption in water treatment is still poorly diffused, although showing very promising experimental results in a few existing applications. It has been shown that, in combination to conventional processes, they can achieve significant overall efficiency improvement not just towards removal of CECs but also of conventional contaminants [77]. Hence, Electron Beam and Plasma technologies could have a primary role in future strategies addressing emerging contaminants, contributing to optimized process energy demand [78].

## 5. Technological Energy Recovery Opportunities

Once established that wastewater and WAS contain considerable energy in the organic material, the major challenge consists in capturing that potential and doing so with minimal losses and costs. Unfortunately, under conventional approaches (anaerobic treatment of WAS) only about $\frac{1}{4}$–$\frac{1}{2}$ of a plants' energy needs can be satisfied by using biogas produced by digestion of organics. These limits are due to several factors, including partial degradability of some organics originally present in wastewater, biological process efficiency and thermodynamic efficiency of recovery processes technology. Thermal, chemical, radiolytic or electrical processes may be used to condition refractory organics to increase biodegradability (and biogas production), but the related direct or indirect energy cost may often offset gains.

Due to universal thermodynamic laws, some energy is always lost in any conversion processes and biological processes are no exception. In anaerobic processes, about 8% of the potential embedded energy is lost in the conversion of higher carbohydrates into lower ones (methane). Additionally, about 7% is lost due to metabolic requirements to form new microorganism cells that carry out the process. Further energy losses of about 5% can be attributed to wastewater process inefficiency, with combined losses totaling about 19% of the initial content. This means that biogas produced would contain under the best scenario only about 81% of the initial estimated energy potential [79].

Under the AD approach, biogas produced (about 70% methane) is generally used in combined heat and power (CHP) engines capable of converting only about up to 40% of the methane energy into electricity, while the remainder may be in part recovered as heat, which may or may not be useful, depending on local conditions. A significant drawback of biogas is that it contains trace components of corrosive agents (e.g., $H_2S$) that make it unsuitable for general uses other than CHP feeding. In order to allow its wider utilization, biogas should be converted into biomethane (a fuel containing >97% $CH_4$) with higher heat value, lower corrosion potential and therefore more valuable. This requires further, expensive, processing [80]. A different approach to produce electricity from biogas may be realized using chemical fuel cells application, which may increase electric conversion efficiency to 50% [81].

Advantages of the substitution of aerobic processes with anaerobic ones for wastewater treatment were illustrated in the previous section. However, this would require a major paradigmatic shift in urban sewerage planning and design to achieve maximum efficiency and sustainability, as the effectiveness of these processes is enhanced by concentrated substrates (COD > 4000 mg/L) and high process temperatures. A possible strategy to achieve this condition is through adoption of low-dilution sewerage, for example, vacuum sewers. If combined with source separation, these systems could result in much higher in-sewer COD concentration. Based on per capita organic matter contribution and modified wastewater discharge flows, COD values of approximately 8000 mg/L, with possible range extending up to 20,000 mg /L, depending on local conditions, could be achieved. Objections to this approach are mostly based on practitioners' and public agencies' unfamiliarity with this technology and on the cost for existing systems replacement. It should be considered, however, that in most of the developed world, urban water infrastructure is nowadays close or past its useful design lifespan (usually, 50–75 years) and thus due to undergo substantial rehabilitation/refurbishment in the near future. Gradually switching to decentralized facilities served by source-separated low dilution system could not only save energy but increase its recovered amount. It has in fact been reported that UASBs operated for low-strength (COD = 100–2600 mg/L) domestic wastewater treatment could generate 130–420 L $CH_4/m^3$ (at 97% COD removal) [82].

Other novel methods have been proposed, among them wastewater processing with bioelectrochemical methods, i.e., microbial fuel cells (MFCs), which could accomplish the direct conversion of the embedded chemical (organics) energy into electricity [83]. The theoretical electrochemical potential (voltage) can be estimated at about 1.1 V for wastewater organics [84]. Even though MFCs have shown considerable efficiency in organics removal from domestic and industrial wastewaters [85], their performance in terms of actual energy recovery is still quite unsatisfactory, achieving in optimal conditions power densities of 25–60A/$m^3$ against expected values exceeding 1000 A/$m^3$ indicated by other researchers [86,87]. This is due to high internal losses that can be attributed to operational factors, construction materials and process scale, with current values that at this stage of technological development may be much greater than 50% [88].

*Energy Recovery From Process Residuals*

Final wastewater processes residuals require safe and appropriate disposal that may require significant energy input and costs, as seen in Section 3.2. In addition to disposal through incineration, other technologies exist for useful recovery of energy and materials from these residuals. A severe technological limitation of these options is given by the water content of sludges, which unless it is well below 30%, will require more energy evaporation than that produced by the residual's combustion [41].

While incineration can recover energy only as heat (it may be used for centralized district heating of private and public buildings), other thermal processes (e.g., pyrolysis, hydrothermal carbonization) can extract energy from sludge in solid or liquid form, with more ample application possibilities [89]. Pyrolysis may be driven by conventional thermal energy (e.g., electric or from primary fuels), by convection or by microwave-assisted devices, which are more energetically efficient [90]. Still, effective sludge pyrolysis requires sludge humidity not exceeding 10%. WAS pyrolysis may produce different types of storable energy (e.g., py-oil, biochar, py-gas) and recoverable residuals that may feed local Circular Economy circuits [91]. Recovered products characteristics depend strongly on sludge properties [92]. Lately, combined ASP/microalgae wastewater treatment facilities were designed, built and operated to achieve more sustainable combined removal of carbon and nutrients in a single unit [93]. It was shown that pyrolysis of combined WAS and microalgae may improve the caloric value of these residuals [94].

The hydrothermal carbonization (HTC) process is a thermal conversion process that converts high-moisture biomass into hydrochar [95]. Unlike pyrolysis, HTC generates a process liquid residual that may be further processed for nutrients recovery and biogas production. This technology, operating at relatively low temperature (180–250 °C), short process times (1–12 h) and pressure up to 30 bar, generates a product (hydrochar) used not only as solid fuel and energy storage but also, similarly to biochar, as soil amendment, absorbent and $CO_2$ sequestrator [96]. Both biochar and hydrochar could be significant fossil fuel replacements in combustion processes.

## 6. Discussion

At the moment, the most efficient and technologically robust solutions in term of energy demand reduction in WWTPs seem to lie in AGS processes, as far as aerobic technologies are concerned, or, as an alternative, UASB/EGSB anaerobic treatment of domestic wastewater. In addition to reducing net treatment energy demand, AD at the moment has the best practical potential for capturing wastewater's embedded energy content and should be considered a preferred option in future design of domestic WWTPs [53]. However, retrofitting conventional aerobic plants to anaerobic processes could be costly and would require more concentrated sewage to achieve optimal efficiency. As an alternative, in addition to upgrading aerobic facilities to more energy-efficient equipment, introducing automatic system control for performance optimization and reducing solids retention in aeration tanks could result in lower energy expenditure and in higher fractions of organics in the biosolids sent to WAS digestion, with increased biogas production.

While literature has shown that anaerobic technology could be applied to diluted domestic sewage with low biogas recovery, AD is better suited to systems based on source separation and vacuum collection, as shown by the practical showcase application of Noorderhoek, Sneek (The Netherlands). In this experimental district system, counting 232 homes, organic waste and toilet water (blackwater) are collected by a vacuum system with low dilution ratio of 1:7 compared to conventional systems. Wastewater, treated by a UASB process, produces biogas which satisfies 12% of the heating demand in the district [97].

Pyrolysis or HTC processes, in addition to providing a safe and sustainable final disposal option for wastewater process residuals, could significantly improve the overall energy footprint of a WWTP, as WAS constitutes an ideal feedstock for the production of certain renewable biofuels [98].

Bioelectrochemical systems show promising perspectives, but the current technological progress involves high construction costs and is not yet mature for full-scale, significant energy recovery.

## 7. Conclusions

Energy represents a significant part of wastewater treatment operational requirements and costs worldwide, with pumping and aeration among the primary energy consumers within the treatment cycle. Residual sludge disposal also requires significant amounts of energy. Consequently, efficient pumping and aeration are crucial for high total energy efficiency. The switch to anaerobic treatment technology, which has increasingly been shown as a possible solution, would drastically cut process energy requirements and the amounts of process residual to be disposed of, while recovering part of the chemical energy embedded in wastewater organics. New technologies are being proposed to improve the efficiency and energy footprint of specific pollutant removal, such as pharmaceuticals and other contaminants of emerging concern.

Considering the costs of sewer construction and operation, decentralized systems, and particularly those based on low sewage dilution, such as vacuum sewers that favor the adoption of energy-recovery technologies like anaerobic digestion, will be increasingly interesting from an energy point of view. Other different forms of wastewater-contained energy include thermal and kinetic energy. Either, although less relevant than the chemical form, could be locally recovered with existing technologies. Bioelectrochemical processes show interesting potential, even though they still need to prove their applicability and energy recovery potential at large scale.

Energy recovery from wastewater process residuals may significantly contribute to improving the energy balance of WWTPs. In addition to traditional systems such as incineration, technologies such as pyrolysis and hydrothermal carbonization allow recovery of liquid and solid phases with energy values that may be transported more easily than biogas for off-site use. These may also find applications in the chemical synthesis industry for extraction of other valuable components and originate Circular Economy cycles at the local level.

Urban water systems need the supply of relevant amounts of energy for their operation. Future, more stringent emissions regulations on nutrients or emerging compounds may further increase these requirements. Improving internal energy efficiency by process modification and technological upgrade and exploiting available incoming energy wisely could lead in the next future to the "zero energy" wastewater treatment plant concept implementation, which not only reduces facilities' energy footprint but allows recovery of wastewater-embedded resources for reuse.

**Author Contributions:** Both authors equally contributed to the research effort and manuscript preparation. Both authors have read and agreed to the published version of the manuscript.

## References

1. Olsson, G. *Water and Energy: Threats and Opportunities*, 2nd ed.; IWA Publishing: London, UK, 2015; p. 496.
2. Lee, M.; Keller, A.A.; Chiang, P.C.; Den, W. Water-energy nexus for urban water systems: A comparative review on energy intensity and environmental impacts in relation to global water risks. *Appl. Energy* **2017**, *205*, 589–601. [CrossRef]
3. Capodaglio, A.G.; Ghilardi, P.; Boguniewicz-Zablocka, J. New paradigms in urban water management for conservation and sustainability. *Water Pract. Technol.* **2016**, *11*, 176–186. [CrossRef]
4. U.S. Department of Energy. *The Water-Energy Nexus: Challenges and Opportunities*; US DOE: Washington, DC, USA, 2014.
5. Means, E. *Water and Wastewater Industry Energy Efficiency: A Research Roadmap*; AWWA Research Foundation: Denver, CO, USA, 2004.
6. Kenway, S.J.; Lam, K.L.; Stokes-Draut, J.; Twomey, K.S.; Binks, A.N.; Bors, J.; Head, B.; Olsson, G.; McMahon, J.E. Defining water-related energy for global comparison, clearer communication, and sharper policy. *J. Clean. Prod.* **2019**, *236*, 117502. [CrossRef]
7. Kenway, S.; Lant, P.; Priestly, A.; Daniels, P. The connection between water and energy in cities: A review. *Water Sci. Technol.* **2011**, *63*, 1983–1990. [CrossRef] [PubMed]
8. Zessner, M.; Lampert, C.; Kroiss, H.; Lindtner, S. Cost comparison of wastewater treatment in Danubian countries. *Water Sci. Technol.* **2010**, *62*, 223–230. [CrossRef] [PubMed]
9. Gandiglio, M.; Lanzini, A.; Soto, A.; Leone, P.; Santarelli, M. Enhancing the Energy Efficiency of Wastewater Treatment Plants through Co-digestion and Fuel Cell Systems. *Front. Environ. Sci.* **2017**, *5*, 70. [CrossRef]
10. MURL. *Energy in WWTPs (in German)*; Ministry for Environment, Nature Protection, Agriculture and Consumer Protection in the German State of North Rhine Westphalia: Düsselforf, Germany, 1999.
11. Shizas, I.; Bagley, D.M. Experimental Determination of Energy Content of Unknown Organics in Municipal Wastewater Streams. *J. Energy Eng.* **2004**, *130*, 45–53. [CrossRef]
12. Korth, B.; Maskow, T.; Günther, S.; Harnisch, F. Estimating the Energy Content of Wastewater Using Combustion Calorimetry and Different Drying Processes. *Front. Energy Res.* **2017**, *5*, 23. [CrossRef]
13. Sohail, U.; Kwiatek, C.; Fung, A.S.; Joksimovic, D. Techno-Economic Feasibility of Wastewater Heat Recovery for A Large Hospital in Toronto, Canada. *Proceedings* **2019**, *23*, 1. [CrossRef]
14. Maktabifard, M.; Zaborowska, E.; Makinia, J. Achieving energy neutrality in wastewater treatment plants through energy savings and enhancing renewable energy production. *Rev. Environ. Sci. Biotechnol.* **2018**, *17*, 655–689. [CrossRef]
15. Remy, C.; Boulestreau, M.; Lesjean, B. Proof of concept for a new energy-positive wastewater treatment scheme. *Water Sci. Technol.* **2014**, *70*, 1709–1716. [CrossRef] [PubMed]
16. Oladejo, J.; Shi, K.; Luo, X.; Yang, G.; Wu, T. A Review of Sludge-to-Energy Recovery Methods. *Energies* **2019**, *12*, 60. [CrossRef]
17. Capodaglio, A.G. Integrated, decentralized wastewater management for resource recovery in rural and peri-urban areas. *Resources* **2017**, *6*, 22. [CrossRef]
18. Gellings, C.W.; Parmenter, K.E. Energy Efficiency in Fertilizer Production and Use. In *Knowledge for Sustainable Development*; Gellings, B., Ed.; Encyclopedia of Life Support Systems (EOLSS); EOLLS Publishers: Oxford, UK, 2004.
19. Owen, W.F. *Energy in Wastewater Treatment*; Prentice-Hall, Inc.: Englewood Cliffs, NJ, USA, 1982.

20. Heidrich, E.S.; Curtis, T.P.; Dolfing, J. Determination of the Internal Chemical Energy of Wastewater. *Environ. Sci. Technol.* **2011**, *45*, 827–832. [CrossRef]
21. Capodaglio, A.G.; Callegari, A.; Cecconet, D.; Molognoni, D. Sustainability of decentralized wastewater treatment technologies. *Water Pract. Technol.* **2017**, *12*, 463–477. [CrossRef]
22. Zanoni, A.E.; Mueller, D.L. Calorific value of wastewater plant sludges. *J. Environ. Eng. Div. ASCE.* **1982**, *108*, 187–195.
23. WERF. *Energy Efficiency in Wastewater Treatment in North America Compendium of Best Practices and Case Studies of Novel Approaches*; WERF Report OWSO4R07c; Water Environment Research Foundation: Alexandria, VA, USA, 2010.
24. Marner, S.T.; Schroter, D.; Jardin, N. Towards energy neutrality by optimising the activated sludge process of the WWTP Bochum-Olbachtal. *Water Sci. Technol.* **2016**, *73*, 3057–3063. [CrossRef]
25. Longo, S.; d'Antoni, B.M.; Bongards, M.; Chaparro, A.; Cronrath, A.; Fatone, F.; Lema, J.M.; Mauricio-Iglesias, M.; Soares, A.; Hospido, A. Monitoring and diagnosis of energy consumption in wastewater treatment plants. A state of the art and proposals for improvement. *Appl. Energy* **2016**, *179*, 1251–1268. [CrossRef]
26. Bodik, I.; Kubaská, M. Energy and sustainability of operation of a wastewater treatment plant. *Environ. Prot. Eng.* **2013**, *39*, 15–24.
27. Bolton, J.R.; Bircher, K.G.; Tumas, W.; Tolman, C.A. Figures-of-merit for the technical development and application of advanced oxidation technologies for both electric- and solar-driven systems (IUPAC Technical Report). *Pure Appl. Chem.* **2001**, *73*, 627–637. [CrossRef]
28. Trojanowicz, M.; Bojanowska-Czajka, A.; Capodaglio, A.G. Can radiation chemistry supply a highly efficient AO(R)P process for organics removal from drinking and waste water? A review. *Environ. Sci. Pollut. Res.* **2017**, *24*, 20187–20208. [CrossRef] [PubMed]
29. Energy Resources Center, University of Illinois at Chicago. Wastewater Treatment Facilities Program—Reducing Energy Usage in Wastewater Treatment. Available online: http://www.erc.uic.edu/energy-efficiency/illinois-energy-now-programs/waste-water-treatment-facilities-program (accessed on 30 October 2019).
30. WSUEP. Ultraviolet Disinfection of Water and Wastewater. Water Disinfectant: Ultraviolet vs. Chemical or Ozone. Washington State University Energy Program. Available online: http://e3tnw.org/ItemDetail.aspx?id=13 (accessed on 23 December 2019).
31. Nouri, J.; Naddafi, K.; Nabizadeh, R.; Jafarinia, M. Energy recovery from wastewater treatment plant. *Pak. J. Biol. Sci.* **2006**, *9*, 3–6.
32. Singh, P.; Carliell-Marquet, C.; Kansa, A. Energy pattern analysis of a wastewater treatment plant. *Appl. Water Sci.* **2012**, *2*, 221–226. [CrossRef]
33. Venkatesh, G.; Brattebø, H. Analysis of chemicals and energy consumption in water and wastewater treatment, as cost components: Case study of Oslo, Norway. *Urban Water J.* **2011**, *8*, 189–202. [CrossRef]
34. Campos, J.L.; Otero, L.; Franco, A.; Mosquera-Corral, A.; Roca, E. Ozonation strategies to reduce sludge production of a seafood industry WWTP. *Bioresour. Technol.* **2009**, *100*, 1069–1073. [CrossRef]
35. Liu, Y.; Tay, J.H. Strategy for minimization of excess sludge production from the activated sludge process. *Biotechnol. Adv.* **2001**, *19*, 97–107. [CrossRef]
36. Vesilind, P.A. The Role of Water in Sludge Dewatering. *Water Environ. Res.* **1994**, *66*, 4–11. [CrossRef]
37. Yang, L.; Zeng, S.; Chen, J.; He, M.; Yang, W. Operational energy performance assessment system of municipal wastewater treatment plants. *Water Sci. Technol.* **2010**, *62*, 1361–1370. [CrossRef]
38. ENERWATER. Deliverable 2.1 Study of Published Energy Data. European Commission, Horizon 2020 ENERWATER Project. Available online: http://www.enerwater.eu/wpcontent/uploads (accessed on 25 October 2019).
39. Daneshgar, S.; Buttafava, A.; Callegari, A.; Capodaglio, A.G. Simulations and Laboratory Tests for Assessing Phosphorus Recovery Efficiency from Sewage Sludge. *Resources* **2018**, *7*, 54. [CrossRef]
40. Treviño Arjona, B.; Rodríguez Cisneros, J. Analysis of Drying Technologies for Wastewater Treatment Plant Sludge as an Alternative Source of Energy. In Proceedings of the Third LACCEI International Latin American and Caribbean Conference for Engineering and Technology (LACCET'2005) "Advances in Engineering and Technology: A Global Perspective", Cartagena de Indias, Colombia, 8–10 June 2005.

41. Kacprzak, M.; Neczaj, E.; Fijałkowski, K.; Grobelak, A.; Grosser, A.; Worwag, M.; Rorat, A.; Brattebo, H.; Almås, A.; RamSingh, B. Sewage sludge disposal strategies for sustainable development. *Environ. Res.* **2017**, *156*, 39–46. [CrossRef]
42. NAWCA. Utilities of the Future Energy Findings. Available online: http://www.nacwa.org/news-publications/white-papers-publications (accessed on 30 October 2019).
43. Rosso, D. (Ed.) *Aeration, Mixing and Energy: Bubbles and Sparks*; IWA Publishing: London, UK, 2019.
44. Olsson, G.; Nielsen, M.; Yuan, Z.; Lynggaard-Jensen, A.; Steyer, J.P. *Instrumentation, Control and Automation in Wastewater Treatment Plants*; Scientific and Technical Report No. 15; IWA Publishing: London, UK, 2005.
45. Cavanaugh, L.; Carson, K.; Lynch, C.; Phillips, H.; Barnard, J.; McQuarrie, J. A Small Footprint Approach for Enhanced Biological Phosphorus Removal: Results from a 106 mgd Full-Scale Demonstration. In Proceedings of the 85th Annual Water Environment Federation Technical Exhibition and Conference, New Orleans, LA, USA, 29 september–3 October 2012.
46. Callegari, A.; Boguniewicz-Zablocka, J.; Capodaglio, A.G. Energy recovery and efficiency improvement for an activated sludge, agro-food WWTP upgrade. *Water Pract. Technol.* **2018**, *13*, 909–921. [CrossRef]
47. Tomei, M.C.; Stazi, V.; Daneshgar, S.; Capodaglio, A.G. Holistic approach to phosphorus recovery from urban wastewater: Enhanced biological removal combined with precipitation. *Sustainability* **2019**, submitted.
48. Sun, J.; Liang, P.; Yan, X.; Zuo, K.; Xiao, K.; Xia, J.; Qiu, Y.; Wu, Q.; Wu, S.; Huang, X.; et al. Reducing Process Aeration Energy Consumption in MBRs. The MBR Site 2017. Available online: http://www.thembrsite.com/features/reducing-process-aeration-energy-consumption-in-membrane-bioreactors/ (accessed on 30 October 2019).
49. Kim, J.; Kim, K.; Ye, H.; Lee, E.; Shin, C.; McCarty, P.L.; Bae, J. Anaerobic fluidized bed membrane bioreactor for wastewater treatment. *Environ. Sci. Technol.* **2011**, *45*, 576–581. [CrossRef] [PubMed]
50. DHV Royalskoning. About Nereda®Technology. Available online: http://www.royalhaskoningdhv.com/en-gb/nereda/about-nereda-technology (accessed on 30 October 2019).
51. Barnard, J.L. A review of biological phosphorous removal in activated sludge process. *Water Sa* **1976**, *2*, 136–144.
52. Bassin, J.P.; Winkler, M.K.H.; Kleerebezem, R.; Dezotti, M.; Van Loosdrecht, M.C.M. Improved phosphate removal by selective sludge discharge in aerobic granular sludge reactors. *Biotechnol. Bioeng.* **2012**, *109*, 1919–1928. [CrossRef] [PubMed]
53. Zeeman, G.; Kujawa, K.; de Mes, T.; Hernandez, L.; de Graaff, M.; Abu-Ghunmi, L.; Mels, A.; Meulman, B.; Temmink, H.; Buisman, C.; et al. Anaerobic treatment as a core technology for energy, nutrients and water recovery from source-separated domestic waste(water). *Water Sci. Technol.* **2008**, *57*, 1207–1212. [CrossRef] [PubMed]
54. Dague, R.R.; Banik, G.C.; Ellis, T.G. Anaerobic sequencing batch reactor treatment of dilute wastewater at psychrophilic temperatures. *Water Environ. Res.* **1998**, *70*, 155–160. [CrossRef]
55. Zeeman, G.; Kujawa-Roeleveld, K. Resource recovery from source separated domestic waste(water) streams; full scale results. *Water Sci. Technol.* **2011**, *64*, 1987–1992. [CrossRef]
56. Lew, B.; Lustig, I.; Beliavski, M.; Tarre, S.; Green, M. An integrated UASB-sludge digester system for raw domestic wastewater treatment in temperate climates. *Bioresourc. Technol.* **2011**, *102*, 4921–4924. [CrossRef]
57. Jeison, D.; Chamy, R. Comparison of the behaviour of expanded granular sludge bed (EGSB) and upflow anaerobic sludge blanket (UASB) reactors in dilute and concentrated wastewater treatment. *Water Sci. Technol.* **1999**, *40*, 91–97. [CrossRef]
58. van Lier, J.B.; van der Zee, F.P.; Frijters, C.T.M.J.; Ersahin, M.E. Celebrating 40 years anaerobic sludge bed reactors for industrial wastewater treatment. *Rev. Environ. Sci. Biotechnol.* **2015**, *14*, 681. [CrossRef]
59. Ratanatamskul, C.; Siritiewsri, T. A compact on-site UASB–EGSB system for organic and suspended solid digestion and biogas recovery from department store wastewater. *Int. Biodeter. Biodegrad.* **2015**, *102*, 24–30. [CrossRef]
60. Henze, M.; Harremoës, P.; La Cour Jansen, J.; Arvin, E. *Wastewater Treatment: Biological and Chemical Processes*, 2nd ed.; Springer: Heidelberg, Germany, 1996.
61. Mulder, A.; Van der Graaf, A.A.; Robertson, L.A.; Kuenen, J.G. Anaerobic ammonium oxidation discovered in a denitrifying fluidized bed reactor. *FEMS Microbiol. Ecol.* **1995**, *16*, 177–184. [CrossRef]

62. Jetten, M.S.; Wagner, M.; Fuerst, J.; van Loosdrecht, M.C.M.; Kuenen, G.; Strous, M. Microbiology and application of the anaerobic ammonium oxidation ('anammox') process. *Curr. Opin. Biotechnol.* **2001**, *12*, 283–288. [CrossRef]
63. Capodaglio, A.G.; Hlavínek, P.; Raboni, M. Advances in wastewater nitrogen removal by biological processes: State of the art review. *Rev. Ambiente Agua* **2016**, *11*, 250–267. [CrossRef]
64. Mulder, A. The quest for sustainable nitrogen removal technologies. *Water Sci. Technol.* **2003**, *48*, 67–75. [CrossRef] [PubMed]
65. He, S.; Chen, Y.; Qin, M.; Mao, Z.; Yuan, L.; Niu, Q.; Tan, X. Effects of temperature on anammox performance and community structure. *Bioresour. Technol.* **2018**, *260*, 186–195. [CrossRef]
66. Daneshgar, S.; Buttafava, A.; Callegari, A.; Capodaglio, A.G. Economic and energetic assessment of different phosphorus recovery options from aerobic sludge. *J. Clean. Prod.* **2019**, *223*, 729–738. [CrossRef]
67. Muhmood, A.; Wu, S.; Lu, J.; Ajmal, Z.; Luo, H.; Dong, R. Nutrient recovery from anaerobically digested chicken slurry via struvite: Performance optimization and interactions with heavy metals and pathogens. *Sci. Total Environ.* **2018**, *635*, 1–9. [CrossRef]
68. Amann, A.; Zoboli, O.; Krampe, J.; Rechberger, H.; Zessner, M.; Egle, L. Environmental impacts of phosphorus recovery from municipal wastewater. *Resour. Conserv. Recycl.* **2018**, *130*, 127–139. [CrossRef]
69. Levlin, E.; Hultman, B. *Phosphorus Recovery from Phosphate Rich Side-Streams in Wastewater Treatment Plants*; Polish-Swedish Seminar: Gdansk, Poland, 2003; pp. 47–56.
70. Olsson, G. *Clean Water Using Solar and Wind: Outside the Power Grid*; IWA Publications: London, UK, 2018; Available online: https://iwaponline.com/ebooks/book/738/Clean-Water-Using-Solar-and-Wind-Outside-the-Power (accessed on 25 November 2019).
71. Shafiqul Islam, M. Comparative evaluation of vacuum sewer and gravity sewer systems. *Int. J. Syst. Assur. Eng. Manag.* **2017**, *8*, 37–53. [CrossRef]
72. Panfil, C.; Mirel, I.; Szigyarto, I.; Isacu, M. Technical, economical, social and ecological characteristics of vacuum sewage system. *Environ. Eng. Manag. J.* **2013**, *12*, 1017–1022. [CrossRef]
73. Eggen, R.I.L.; Hollender, J.; Joss, A.; Schärer, M.; Stamm, C. Reducing the discharge of micropollutants in the aquatic environment: The benefits of upgrading wastewater treatment plants. *Environ. Sci. Technol.* **2014**, *48*, 7683–7689. [CrossRef] [PubMed]
74. Cecconet, D.; Molognoni, D.; Callegari, A.; Capodaglio, A.G. Biological combination processes for efficient removal of pharmaceutically active compounds from wastewater: A review and future perspectives. *J. Environ. Chem. Eng.* **2017**, *5*, 3590–3603. [CrossRef]
75. Capodaglio, A.G.; Bojanowska-Czajka, A.; Trojanowicz, M. Comparison of different advanced degradation processes for the removal of the pharmaceutical compounds diclofenac and carbamazepine from liquid solutions. *Environ. Sci. Pollut. Res.* **2018**, *25*, 27704. [CrossRef] [PubMed]
76. Capodaglio, A.G. High-energy oxidation process: An efficient alternative for wastewater organic contaminants removal. *Clean Technol. Environ Policy* **2017**, *19*, 1995–2006. [CrossRef]
77. Capodaglio, A.G. Could eb irradiation be the simplest solution for removing emerging contaminants from water and wastewater? *Water Pract. Technol.* **2018**, *13*, 172–183. [CrossRef]
78. Capodaglio, A.G. Contaminants of Emerging Concern Removal by High-Energy Oxidation-Reduction Processes: State of the Art. *Appl. Sci.* **2019**, *9*, 4562. [CrossRef]
79. McCarty, P.L.; Bae, J.; Kim, J. Domestic Wastewater Treatment as a Net Energy ProducerCan This be Achieved? *Environ. Sci. Technol.* **2011**, *45*, 7100–7106. [CrossRef]
80. Capodaglio, A.G.; Callegari, A.; Lopez, M.V. European Framework for the Diffusion of Biogas Uses: Emerging Technologies, Acceptance, Incentive Strategies, and Institutional-Regulatory Support. *Sustainability* **2016**, *8*, 298. [CrossRef]
81. US EPA. *Opportunities for and Benefits of Combined Heat and Power at Wastewater Treatment Facilities*; EPA-430-R-07-003; U.S. Environmental Protection Agency: Washington, DC, USA, 2007; p. 42.
82. Stazi, V.; Tomei, M.C. Enhancing anaerobic treatment of domestic wastewater: State of the art, innovative technologies and future perspectives. *Sci. Total Environ.* **2018**, *635*, 78–91. [CrossRef] [PubMed]
83. Capodaglio, A.G.; Molognoni, D.; Dallago, E.; Liberale, A.; Cella, R.; Longoni, P.; Pantaleoni, L. Microbial Fuel Cells for Direct Electrical Energy Recovery from Urban Wastewaters. *Sci. World J.* **2013**, *2013*, 8. [CrossRef] [PubMed]

84. Logan, B.E.; Hamelers, B.; Rozendal, R.; Schroder, U.; Keller, J.; Freguia, S.; Aelterman, P.; Verstraete, W.; Abaey, K. Microbial fuel cells: Methodology and technology. *Environ. Sci. Technol.* **2006**, *40*, 5181–5192. [CrossRef] [PubMed]
85. Molognoni, D.; Chiarolla, S.; Cecconet, D.; Callegari, A.; Capodaglio, A.G. Industrial wastewater treatment with a bioelectrochemical process: Assessment of depuration efficiency and energy production. *Water Sci. Technol.* **2018**, *77*, 134–144. [CrossRef]
86. Callegari, A.; Cecconet, D.; Molognoni, D.; Capodaglio, A.G. Sustainable processing of dairy wastewater: Long-term pilot application of a bio-electrochemical system. *J. Clean. Prod.* **2018**, *189*, 563–569. [CrossRef]
87. Rozendal, R.A.; Hamelers, H.M.V.; Rabaey, K.; Keller, J.; Buisman, C.J.N. Towards practical implementation of bioelectrochemical wastewater treatment. *Trends Biotechnol.* **2008**, *26*, 450–459. [CrossRef]
88. Capodaglio, A.G.; Molognoni, D.; Pons, A.V. A multi-perspective review of microbial fuel-cells for wastewater treatment: Bio-electro-chemical, microbiologic and modeling aspects. *AIP Conf. Proc.* **2016**, *1758*, 030032. [CrossRef]
89. Callegari, A.; Hlavinek, P.; Capodaglio, A.G. Production of energy (biodiesel) and recovery of materials (biochar) from pyrolysis of urban waste sludge. *Rev. Ambiente Agua* **2018**, *13*, e2128. [CrossRef]
90. Capodaglio, A.G.; Callegari, A.; Dondi, D. Microwave-Induced Pyrolysis for Production of Sustainable Biodiesel from Waste Sludges. *Waste Biomass Valoriz.* **2016**, *7*, 703–709. [CrossRef]
91. Callegari, A.; Capodaglio, A.G. Properties and beneficial uses of (bio)chars, with special attention to products from sewage sludge pyrolysis. *Resources* **2018**, *7*, 20. [CrossRef]
92. Capodaglio, A.G.; Callegari, A. Feedstock and process influence on biodiesel produced from waste sewage sludge. *J. Environ. Manag.* **2018**, *216*, 176–182. [CrossRef] [PubMed]
93. Arbib, Z.; de Godos, I.; Fernandez, M.; Lara, E.; Llamas, B.; Rogalla, F. Wastewater as Biofuel: Nutrient and Energy Recovery with Algae Treatment. Available online: http://www.researchgate.net/publication/314461731_Wastewater_as_Biofuel_Nutrient_and_Energy_Recovery_with_Algae_Treatment/citations (accessed on 7 December 2019).
94. Bolognesi, S.; Bernardi, G.; Callegari, A.; Dondi, D.; Capodaglio, A.G. Biochar production from sewage sludge and microalgae mixtures: Properties, sustainability and possible role in circular economy. *Biomass. Conv. Bioref.* **2019**. [CrossRef]
95. Wang, T.; Zhai, Y.; Zhu, Y.; Lia, C.; Zeng, G. A review of the hydrothermal carbonization of biomass waste for hydrochar formation: Process conditions, fundamentals, and physicochemical properties. *Renew. Sustain. Energy Rev.* **2018**, *90*, 223–247. [CrossRef]
96. Koottatep, T.; Fakkaew, K.; Tajai, N.; Pradeep, S.V.; Polprasert, C. Sludge stabilization and energy recovery by hydrothermal carbonization process. *Renew. Energy* **2016**, *99*, 978–985. [CrossRef]
97. de Graaf, R.; van Hell, A.J. *New Sanitation Noorderhoek, Sneek*; STOWA (Dutch Foundation for Applied Water Research): Amersfoort, The Netherlands, 2014.
98. Callegari, A.; Bolognesi, S.; Cecconet, D.; Capodaglio, A.G. Production technologies, current role, and future prospects of biofuels feedstocks: A state-of-the-art review. *Crit. Rev. Environ. Sci. Technol.* **2020**, *50*, 384–436. [CrossRef]

2

# Livestock Wastewater Treatment in Constructed Wetlands for Agriculture Reuse

**Sofia Dias [1,2], Ana P. Mucha [1,3], Rute Duarte Crespo [3], Pedro Rodrigues [2] and C. Marisa R. Almeida [1,2,*]**

1 CIIMAR—Interdisciplinary Centre of Marine and Environmental Research of the University of Porto, Terminal de Cruzeiros do Porto de Leixões, Avenida General Norton de Matos, 4450-208 Matosinhos, Portugal; sofiamdias5@gmail.com (S.D.); amucha@ciimar.up.pt (A.P.M.)

2 Chemistry and Biochemistry Department, Faculty of Sciences, University of Porto, Rua do Campo Alegre 687, 4150-171 Porto, Portugal; up201709923@fc.up.pt

3 Biology Department, Faculty of Sciences, University of Porto, Rua do Campo Alegre 790, 4150-171 Porto, Portugal; up201603331@fc.up.pt

* Correspondence: calmeida@ciimar.up.pt

**Abstract:** The aim of this study focused on the evaluation of constructed wetlands (CWs) microcosms, on a laboratory scale, for the removal of metals from a pig industry effluent while maintaining effluent organic matter and nutrients levels for its later used as a fertilizer. CWs with different macrophytes (*Phragmites australis* and *Typha latifolia*) and different substrates (light expanded clay aggregate and lava rock) were tested. Results showed high removals of metals during CWs treatment, with removal rates reaching >80% for Cd, Cr, Cu, Fe, Mn, and Zn after 2 days of treatment in CWs planted with *T. latifolia* and >60% in CWs planted with *P. australis*. Significant differences were only found between substrates for Fe and Mn in CWs with *P. australis*. Removal of organic matter (through chemical oxygen demand (COD)) was >77%, with no significant differences between substrates or plants. Removals of ammonium and phosphate ions ranged between 59–84% and 32–92%, respectively, in CWs with *P. australis* and 62–75% and 7–68% in CWs with *T. latifolia*, with no significant differences between substrates. Overall, CWs showed potential to be efficient in removing toxic contaminants, as metals, while maintaining moderated levels of nutrients, allowing the use of reclaimed water in agriculture, namely as fertilizer. If one aims for a short CW treatment, CW planted with *T. latifolia* and expanded clay as substrate could be the more suitable choice.

**Keywords:** phytoremediation; constructed wetland; wastewater; fertilizer

---

## 1. Introduction

Continued worldwide industrialization and exponential growth of the human population have led to an increase in wastewater generated. Wastewaters may contain different chemical compounds some of which can cause environmental and health risks [1]. Therefore, wastewater treatment is needed before its reuse.

Reusing treated wastewater represents an important part of sustainable water resource management and circular economy, reducing the use of freshwater [1]. Reclaimed water can be reused for different purposes depending on its quality and legislation in force. The higher use of reclaimed water is agricultural irrigation, because of the additional value of water due to the presence of fertilizers, nutrients, and macro elements [2–7]. An estimation indicates that water resources used for irrigation purposes is about 70% of that available in the world [6]. Therefore, the reuse of treated wastewater in the agricultural sector is of particular interest. However, as mentioned, wastewaters need to be adequately treated before its reuse to remove a wide variety of pollutants such as pesticides, pharmaceuticals, or metals [8]. In the case of livestock, metals and pharmaceuticals are the main contaminants of concern [9]. For instance, Zn and Cu are often used in high doses to enhance animals' immune systems [10]. In fact, Zn salts have been used as an alternative to infeed antibiotics, which is believed to contribute to the emergence of methicillin-resistant *Staphylococcus aureus* [10,11]. The EU has banned the inclusion of pharmacological levels of ZnO after 2022. In Portugal, maximum levels of metals are legislated for fertilizers [12] and for wastewater discharges [13]. Recommend values for metals in the Portuguese legislation for irrigation water are identical or lower to those established in international legislation regarding water quality for irrigation [14,15]. Therefore, there is an emergence in remove/reduce these pollutants from livestock wastewaters.

A possible methodology to treat livestock wastewater is phytoremediation. Based on natural processes, phytoremediation is an efficient and extensively employed environmental cleanup biotechnology, that uses plants and associated microorganisms to remove, accumulate, metabolize, absorb and/or degrade organic and inorganic pollutants [9,16]. Phytoremediation has been shown to be an efficient process for removing metals from wastewater and this technology can have a key role when integrated in constructed wetlands (CWs) [9].

CWs have been reported to be efficient for the treatment of different types of wastewaters such as urban, agricultural, industrial, among others [17–20]. Livestock wastewaters tend to be more difficult to treat than urban wastewaters, due to their complex matrix and higher organic contents [21]. But several studies have shown the potential of CWs for the removal of different contaminants, including not only organic matter and nutrients but also biological contaminants, metals, or emergent pollutants (e.g., pharmaceuticals), from livestock wastewaters [8,10,21–29]. CWs are low-cost engineered systems designed to mimic natural wetland processes, using wetland vegetation, soils/substrate, and associated microorganisms to treat wastewater [8]. Pollutant removal is reached through a combination of physical, chemical, and biological processes in which all the elements of CWs, plants, microorganisms, and substrates, can have a significant influence [24]. Plants play an important role in CWs by uptaking pollutants, namely metals, by moving oxygen through their roots to the bottom of treatment wetlands, and by providing a medium beneath the water surface for the attachment of microorganisms that contributes to pollutant removal [30]. For instance, the use of microalgae or floating aquatic plants, such as duckweed, to phytoremediate livestock wastewater has been widely studied, because these plants can absorb organic matter and nutrients from water for their own growth needs [9]. Different macrophytes are also known to remove pollutants from water, such as *Juncus maritimus, Phragmites australis, Typha latifolia, Chrysopogon zizanioides, Colocasia esculenta*, among others [10,27,31–33]. *T. latifolia* and *P. australis* are two of the most commonly used plants in CWs, having properties of hyperaccumulation for different metals [30]. Substrates also have a significant role in CWs. Substrates support plants and microorganisms, promote biofilms formation, and improve the efficiency of CWs systems being involved in the adsorption, absorption, sedimentation, and filtration processes [30]. So, suitable substrates should be selected, as different layers and porous materials can have a positive impact on the removal of contaminants. Studies have shown that, for instance, light expanded clay aggregate and lava rock are appropriate for the development of plants and microorganisms in CWs [34–36].

In general, studies involving CWs aim the removal of all types of contaminants present in the wastewater to be treated. However, livestock wastewaters have a fertilizer potential, so removing toxic contaminants, such as metals, maintaining the nutritional value is a challenge. To our knowledge, there are no applications of CWs with this purpose. CWs offer a more sustainable and low-cost alternative to water treatment, so its potential should be explored.

The aim of this study focused on the evaluation of CWs microcosms simulating a vertical subsurface flow (VSSF) CW, on a laboratory scale, for the removal of metals from a pig industry effluent while maintaining effluent organic matter and nutrients levels to be later used as a fertilizer. To optimize the operation of the CWs, two types of porous substrate, lava rock and expanded clay, and two species of emerging macrophytes, *T. latifolia* and *P. australis*, which are among the most used plants in built wetlands [37], were tested. The final objective was to evaluate metals removal and nutrients permanence, so that nutrients final concentrations in the final effluents are suitable for fertilization, after the removal of the toxic contaminants.

## 2. Materials and Methods

Experiments were carried out in controlled conditions in six microcosms simulating CWs.

### 2.1. Sampling

Livestock wastewater was collected on a pig farm on the 14th and 29th of January and 13th of February of 2020. The wastewater was collected from a decantation lagoon used to store the liquid part of pig residues and was used as collected. An initial characterization was done, showing high levels of metals and nutrients (see Section 3).

*P. australis* plants were collected in River Lima Estuary (41.689822, −8.816289) in December of 2019 in Viana do Castelo, NW Portugal, and *T. latifolia* plants were collected in a riverside area near the coast in Leça da Palmeira, Matosinhos (41.205082, −8.715423) in February 2020. In the laboratory, plants were washed and replanted in each CW, ca. of 30 individual plants per microcosm for *P. australis* and ca. of 10 plants per microcosm for *T. latifolia*, as the latter have higher individual biomass.

### 2.2. Microcosms Assembly

Six CWs microcosms were assembled, as described previously [29], using plastic containers (40 cm × 30 cm × 30 cm) all with a first layer of gravel (3 cm) (Supplementary Material Figure S1). Three of the systems were filled with a second layer of lava rock (3 cm) (L systems: L1, L2, L3) and the other three with a second layer of lightweight expanded clay aggregate (A systems: A1, A2, A3). Finally, a third layer of sand (20 cm) was added to all systems, into which plants were transplanted. The first experiment was carried out with *P. australis*. At the end of this experiment, plants and sand were removed and the second experiment was initiated by transplanting *T. latifolia* into the same systems with a new sand layer.

All microcosms were wrapped in aluminum foil to prevent light penetration at a substrate level and possible photodegradation of the compounds. The wastewater was poured on top, to percolate through the different layers of the system, and drained out through a tap at the bottom when necessary, simulating a VSSF CW. Throughout the whole experience, wastewater was daily recirculated to prevent oxygen reduction in the system and avoid the development of anoxic microbiota. Microcosms were kept under greenhouse conditions, under normal light:dark regime and with a temperature variation between 9.3 °C and 32.3 °C.

### 2.3. Microcosms Operation

#### 2.3.1. Phragmites Australis

CW microcosms were firstly acclimatized for 6 weeks with a Hoagland nutrient solution in which solution was changed every two/three days, to reduce plant stress and allow the creation of rhizospheric microbiomes and biofilms in the systems. Afterwards, two stabilization weeks were performed (SW1 and SW2) in which livestock wastewater was added to the systems, to adapt plants to the load of organic and inorganic matter. In SW1, 2.5 L of wastewater plus 0.5 L of deionized water were added to each system. After one week, all water was removed from the CWs systems, and for SW2, 1 L of new wastewater plus 1.5 L of deionized water was added.

After stabilization, all water was removed from the CWs systems and the experiments started in which new livestock wastewater was added for a 15 days-cycle treatment (1 L of wastewater + 1.5 L of deionized water). Only 1 L of wastewater was added as SW1 showed that higher volumes of wastewater were not being properly treated with the wastewater still maintaining most of its smell and organic load, probably due to the small dimensions of the microcosms. This volume (2.5 L) was established to maintain the water level just above the substrate surface, allowing a complete saturation of the substrate (saturation rate of ca. 100%) [29]. During the first 7 days, samples were collected along the week for analysis of metals and nutrients (t1, t2, t3, t5, t6, and t7). At the end of this first cycle (t15), treated wastewater was removed and samples were collected for analysis of organic matter (through chemical oxygen demand (COD)), nutrients, and metal levels. New livestock wastewater was added and a second 15 days-cycle was initiated. Samples were collected after 7 days (t7.2) and 15 days (t15.2) and analyzed for the same parameters as the previously treated wastewater. All samples were collected by draining the liquid phase of each CW microcosm and adding deionized water to achieve the same volume as the initial added. So, the wastewater was changed every 15 days.

Water samples collected were stored at 4°C or −20°C depending on the parameter to be analyzed. Initial wastewater was also stored for analysis before being added to the systems. At the end of the experiments, all plants were removed from the systems, washed, and put to dry at room temperature until constant weight. Then, plants were separated into roots, rhizomes, leaves, and stems. Samples of sand in contact with the plant roots were also collected and dried at room temperature until constant weight.

#### 2.3.2. Typha Latifolia

CW microcosms were firstly acclimatized for one week with a Hoagland nutrient solution in which solution was changed every weekday, for plant establishment. Afterwards, CW microcosms were stabilized for 15 days (St) in which 950 mL of livestock wastewater plus 1.5 L of deionized water were added to each system.

After the stabilization, the wastewater was treated during two 15 day-cycles (1 L of wastewater + 1.5 L of deionized water). During the first 7 days of the first cycle, samples were collected along the week (t1, t2, t3, t6, and t7). At the end of the first cycle (t15), treated wastewater was removed and samples were collected for analysis of organic matter (through COD), metals, and nutrients. A new 15 days-cycle started, adding 1 L of new livestock wastewater plus 1.5 L of deionized water to each CW. During the first five days, water was not recirculated daily due to logistic constraints. Samples were collected after seven (t7.2) and fifteen days (t15.2) and analyzed for the same parameters as the previously treated wastewater. All samples were collected by draining the liquid phase of each CW microcosm and adding deionized water to achieve the same volume as the initial added. So, the wastewater was changed every 15 days.

Water samples were stored at 4 °C or −20 °C depending on the parameter to be analyzed. Initial wastewater was also stored for analysis before being added to the systems. At the end of the experiments, all plants were removed from the systems washed, and put to dry at room temperature until constant weight. Then, plants were separated into roots, rhizomes, and leaves. Samples of sand in contact with the plant roots were also collected and dried at room temperature until constant weight.

### 2.4. *Samples Analysis*

All reagents used were pro analysis or equivalent. All material was washed with deionized water (conductivity < 0.1 $\mu S\ cm^{-1}$), immersed in nitric or chloric acid solution (20% *v/v*) for 24 h, washed again with deionized water, and dried in a clean oven.

For metals, initial livestock wastewaters, sand samples, and plant tissues (dry samples) were digested at high pressure with concentrated nitric acid ($HNO_3$) and hydrogen peroxide ($H_2O_2$) solution. Treated wastewater samples were only acidified (with 1% $HNO_3$) before direct analysis. Digestions were carried out in a high-pressure microwave system (Ethos, Milestone, Sorisole, Italy) in closed Teflon vessels. Metals (Cd, Cr, Cu, Fe, Mn, Ni, Pb, and Zn) were measured as described in previous studies [27], by atomic absorption spectrophotometry with flame atomization (AAnalyst 200 PerkinElmer, PerkinElmer Inc., Waltham, MA, USA) or electrothermal atomization (4100ZL, PerkinElmer, PerkinElmer Inc., Waltham, MA, USA) depending on metal levels. A calibration curve obtained with aqueous standard solutions of different metal concentrations (0–3 mg/L) was used for metal quantification. These standard solutions were prepared from 1000 mg/L stock standard solutions of each metal. Each individual sample was analyzed in triplicate with relative standard deviation always below 5%. For quality control, doped samples were prepared and analyzed with recovery values always between 80–120%. The detection limits (LOD) were 0.05 mg/L for Cu, Zn and Mn, 0.3 mg/L for Fe, 6.6 μg/L for Pb, 0.6 μg/L for Cd, 2.5 μg/L for Ni and 1 μg/L for Cr.

COD was measured using the Kits HI93754B-25 (Hanna Instruments, Padua, Italy), MR from 0 to 1500 mg/L and following provided protocol.

For nutrients, all samples were filtered (nitrate cellulose filters, 0.45 μm porosity). Dissolved ammonium, nitrate, nitrite, and phosphate ions concentrations were determined following the methods described in [8]. Nitrate was quantified by an adaptation of the spongy cadmium reduction technique [38], evaluating the reduction percentage to nitrites. All the analyses were performed in triplicate. The detection limits were 0.05 mg/L for ammonium and phosphate ions, 0.1 mg/L for nitrate ions and 0.01 mg/L for nitrite ions. For quality control, doped samples were prepared and analyzed with recovery values always between 80–120%.

### 2.5. *Statistical Analysis*

All parameters in wastewaters were measured in triplicate (three independent microcosms) and means and standard deviations were calculated for each treatment ($n = 3$). To evaluate statistically significant differences ($p < 0.05$), t-student tests and a parametric one-way analysis of variance (ANOVA) were applied. The presence of significant differences was detected by a multiple Tukey comparison test.

## 3. Results

To assess the functionality of CWs over time, analyzes of livestock effluent quality parameters (metals, nutrients, and organic matter content) were performed before and after CW treatment.

During the experiment, some changes in the visual aspect of the plants were observed. Initially, they were well-nourished and alive; however, at the end of the experiment, they were visually weak and malnourished indicating a negative impact of the wastewater.

### 3.1. *Metals*

#### 3.1.1. Water

Phragmites Australis

In the experiments with *P. australis*, concentrations of each metal in the initial livestock wastewater were in the order Fe > Zn > Mn > Cu > Ni > Pb > Cr > Cd (Supplementary Material—Table S1). Initial concentrations of Cr, Ni, and Pb were the only ones already lower than the legislated recommended value for irrigation water and legislated limit for wastewater discharge (Supplementary Material—Table S6) [13,14]. Initial concentrations of Cd, Cu, Mn, and Zn were lower than the legislated maximum allowable value for irrigation water [13], but Fe concentrations were higher. Cd initial concentration was also lower than the legislated limit for wastewater discharge [13], but Cu, Mn, and Fe were higher (no legislated values for Zn).

Figure 1 shows the removal rates of the different analyzed metals during the first week of treatment in systems with clay as the porous substrate (A systems). For Cu and Pb no significant differences among removals were observed. For Pb, removals tended to increase but with a high variability among microcosm replicates. Removal rates ranged between 83–88% during the week for Cu and between 40–69% for Pb. Zn, Fe, Mn, and Cr had significant increasing removals during the week, achieving removals of 79%, 64%, 78%, and 74% after 7 days of treatment. Cd showed a different behavior, with removals increasing and decreasing during the week, ranging between 46–74%. For Ni, in general, no removal during the week was observed. Concentrations of Fe and Zn after two days of treatment were already lower than legislated recommended value for irrigation water [13,14]. However, after 7 days, concentrations of Cd, Cu, and Mn were still higher than the recommended value. Regarding wastewater discharge legislation [13], after one day of treatment, concentrations of Cu and Mn were already lower, but concentrations of Fe were still higher than legislated value after seven days of treatment.

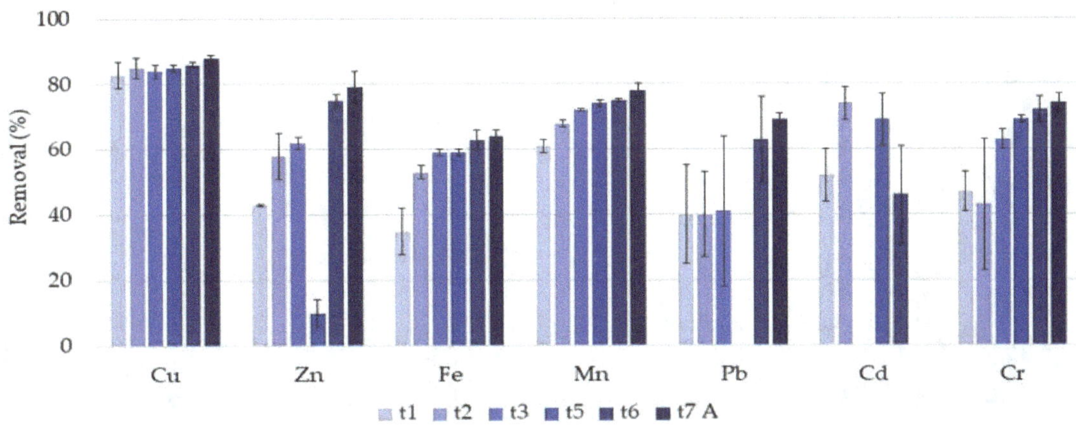


**Figure 1.** Removal percentage and respective standard deviation ($n$ = 3) for each metal, in constructed wetlands (CWs) with expanded clay (A) and *P. australis* during the first week of treatment. Samples taken after one day (t1), two days (t2), three days (t3), five days (t5), six days (t6) and seven days (t7) of treatment. At t5 there was no removal for Pb. For Cd no removal was observed at t3.

Figure 2 shows the removal rates of each metal during the first week of treatment in systems with lava rock as the porous substrate (L systems). For Cu, Mn, Pb, and Cd no significant differences along the week were observed. During the week, removal rates ranged between 83–88% for Cu, 26–49% for Mn, 32–58% for Pb, and 67–85% for Cd, although for Cd it varied greatly during the week. Zn, Fe, and Cr showed significant increasing removals over the week, achieving values of 77%, 80%, and 75%, respectively, at t7. Once again, no removal of Ni was observed. Concentrations of Fe, Cd, and Zn after one, two, and three days of treatment, respectively, were already lower than legislated recommended value for irrigation water [13]. Concentrations of Cu and Mn were still higher than the recommended value after seven days of treatment. Regarding wastewater discharge legislation, concentrations of Cu, Mn, and Fe, after one, three, and seven days of treatment, respectively, were already lower.

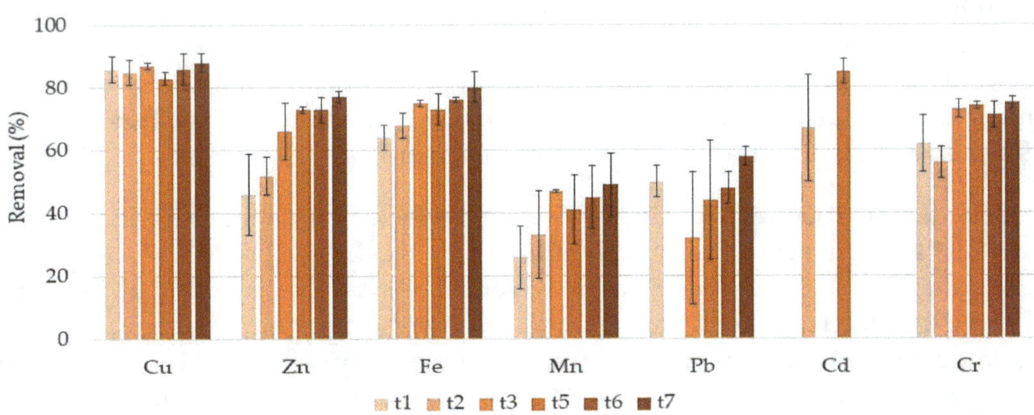


**Figure 2.** Removal percentage and respective standard deviation ($n$ = 3) for each metal, in CWs with lava rock (L) and *P. australis* during the first week of treatment. Samples taken after one day (t1), two days (t2), three days (t3), five days (t5), six days (t6), and seven days (t7) of treatment. For Pb, there was no removal at t2 and for Cd, there were no removals at t1, t3, t6, and t7.

Figure 3 shows a comparison in terms of removal of both substrates (A and L) in the two cycles of treatment. For Cu, Zn, and Pb no significant differences between substrates were observed. It is possible to notice, for these metals, an increase in removal in the second week of treatment (t15 A, t15.2 A, t15 L, t15.2 L) in both systems in comparison to the first week (t7). So, a 15-day cycle could be more appropriate to remove a higher metal level.

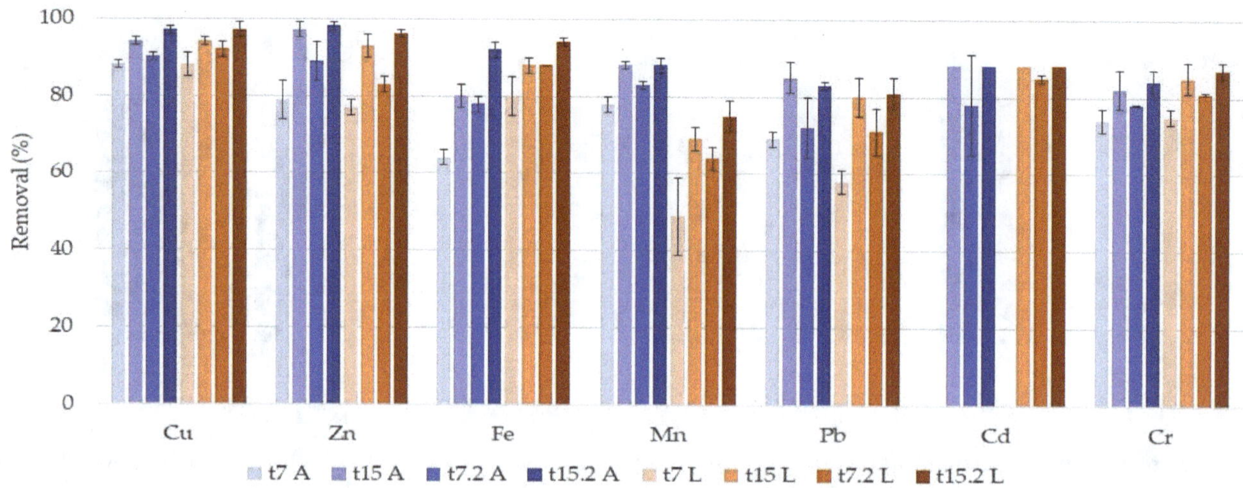


**Figure 3.** Removal percentage and respective standard deviation ($n = 3$) for each metal, in CWs with *P. australis* and expanded clay (t7 A, t15 A, t7.2 A, t15.2 A) and lava rock (t7 L, t15 L, t7.2 L, t15.2 L) over the two cycles of treatment. First cycle: samples taken after one week (t7) and at the end of the two-week cycle (t15). Second cycle: samples taken after one week (t7.2) and at the end of the two-week cycle (t15.2). For Cd, there was no removal at t15.2 A.

For Fe at t7 and t7.2 a significant difference between substrates was observed, with L systems removing more efficiently Fe after one week of treatment. However, at t15 and t15.2 no significant differences between substrates occurred. The second week of treatment had also higher removals than the first week, in both cycles for both substrates. Moreover, for both systems, a difference between cycles of treatment was observed. The second cycle showed higher removals than the first one, for both substrates. For Mn, significant differences between substrates were also observed for both cycles, but A systems were more efficient in removing Mn than L systems. In the second week of treatment higher removal values were achieved. For this metal, no differences between cycles were observed. For Cr, no significant differences between substrates, or between cycles, were observed. Besides, no significant differences between one and two weeks of treatment were also observed, which suggests that one-week treatment is efficient in removing this metal. At t7 no removals of Cd in either system occurred and at t15 and t15.2 Cd concentrations were below the limit of detection (LOD = 0.6 mg/L), so removals were >88%, without differences between substrates.

As mentioned, for Cd, Cu, Zn, Fe, Mn, and Pb a two-week treatment is needed to remove a higher amount of metals. Nevertheless, concentrations after one week (t7 A, t7.2 A, t7 L, t7.2 L) were lower than the maximum allowed levels for irrigation water for all analyzed metals [13], with the exception of Mn, Cu, and Cd in the first cycle that required two-week treatment. In general, both substrates were equally efficient, although there were some differences for Fe and Mn. For Ni, no removals were observed, initial concentration was 126 ± 23 μg/L and at the end of the first cycle, they were of 159 ± 7 μg/L and 143 ± 10 μg/L for A and L systems, respectively. Initial concentration in the second cycle was 124 ± 12 μg/L and at the end, they were 132 ± 14 μg/L and 113 ± 9 μg/L. These values were lower than legislated ones [13,14].

Typha Latifolia

In the experiment with *T. latifolia*, the measured concentrations of each metal, in the initial waters were in the order Fe > Zn > Mn > Cu > Ni > Cr > Pb > Cd (Supplementary Material Table S1). Initial concentrations of Cd, Cr, Ni, and Pb were the only ones with already lower than the legislated recommended value for irrigation water and legislated limit for wastewater discharge [13,14]. Initial concentrations of Cu, Mn, and Zn were lower than the legislated maximum allowable value for irrigation water [13], but Fe concentrations were in some cases higher. Cu and Mn initial concentrations were lower than the legislated limit for wastewater discharge [13], but Fe concentrations were higher (no legislated values for Zn).

Figure 4 shows the removal rates of the different analyzed metals during the first week of treatment in A systems. For Cu, Zn, Fe, Mn, Pb and Cd no significant differences among removals along the week were observed. During the week, removal rates ranged between 63–84% for Cu, 58–86% for Zn, 41–78% for Fe, 76–87% for Mn, 56–71% for Pb, and 61–87% for Cd, although once again there were variations for Cd. For Cr removals increase from the first day (t1) to the following days, achieving a removal of 87% at t7. For Ni, in general, no removal was observed. Concentrations of Fe and Zn after one day of treatment were already lower than legislated recommended value for irrigation water [13,14]. Concentrations of Cu and Mn were higher than the recommended value for irrigation water after seven-day treatment. In concern of wastewater discharge legislation, after three days of treatment concentrations of Fe were already lower.

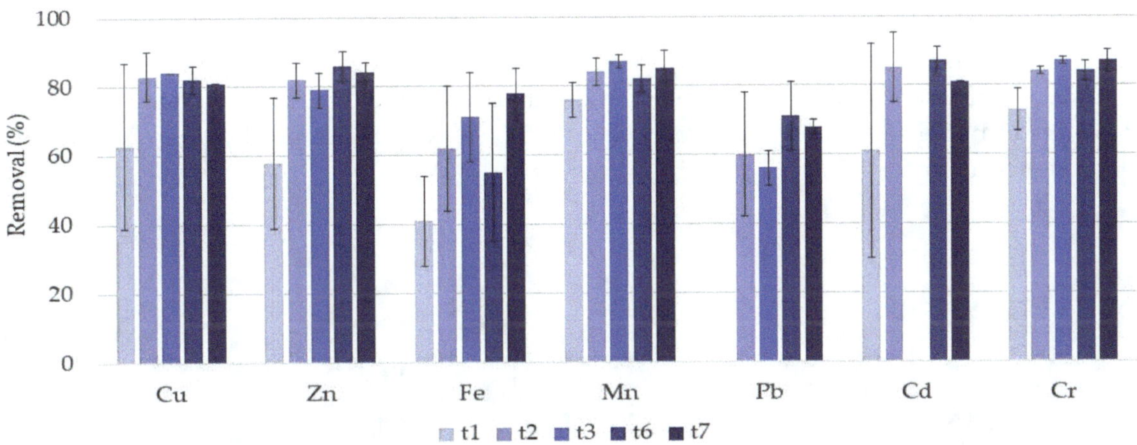


**Figure 4.** Removal percentage and respective standard deviation ($n = 3$) for each metal, in CWs with expanded clay (A) and *T. latifolia* during the first week of treatment. Samples taken after one day (t1), two days (t2), three days (t3), five days (t5), six days (t6), and seven days (t7) of treatment. There was no removal of Cd at t3.

Figure 5 shows the removal rates of each metal during the first week of treatment in L systems. For Cu and Cd, no significant differences among removals along the week were observed. Removal rates ranged between 72–86% during the week for Cu and between 80–91% for Cd. Zn, Fe, Mn, Pb, and Cr showed an increase between the first day (t1) and the following days, achieving removal rates of 93%, 95%, 80%, 77%, and 91%, respectively, at t7. Concentration of Fe at t7 was below the LOD of 0.3 mg/L, so removal was >95%. Once again, no removal of Ni was observed. Concentrations of Fe and Zn after one day of treatment were already lower than legislated recommended value for irrigation water [13,14]. Concentrations of Cu and Mn concentrations were still higher than the recommended value after seven-day treatment. In concern of wastewater discharge legislation, after one day of treatment, concentrations of Fe were already lower than the legislated limit.

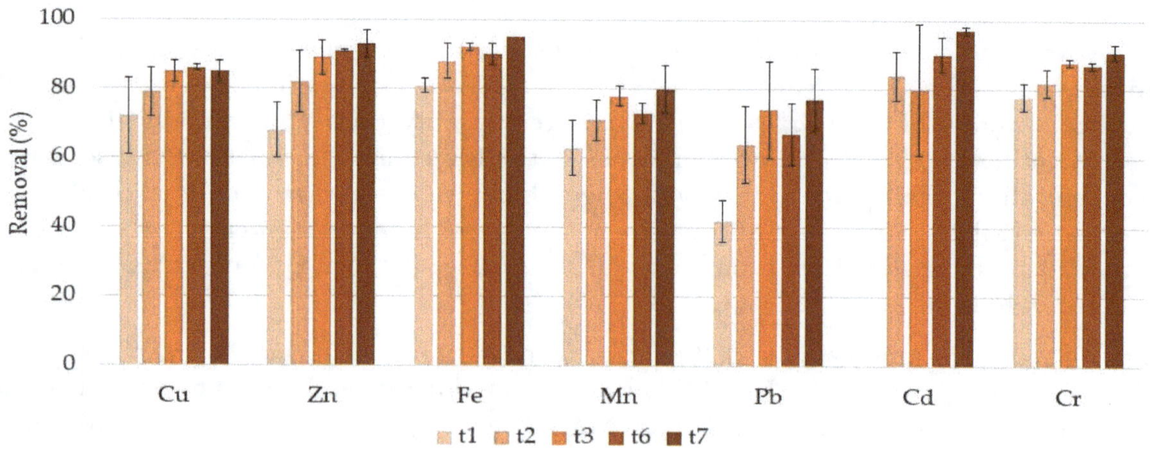


**Figure 5.** Removal percentage and respective standard deviation ($n = 3$) for each metal, in CWs with lava rock (L) and *T. latifolia* during the first week of treatment. Samples taken after one day (t1), two days (t2), three days (t3), five days (t5), six days (t6), and seven days (t7) of treatment. For Cd, there was no removal at t1.

Figure 6 shows a comparison in terms of metal removal of both substrates (A and L) in the two cycles of treatment. For Cu in A systems, no significant differences between cycles were observed, but in L systems removal rates decreased in the second cycle. In general, no significant differences between the first and second week of treatment in both cycles, or between substrates, were observed. Concentrations of Cu at t15 (A and L) and t15.2 L were below the LOD of 0.05 mg/L, so removals were > 90% for t15 and >79% for t15.2. For Zn and Pb, also no significant differences between the first and second week of treatment, or between cycles, were observed. No significant differences between substrates were either observed for these metals. At t15 L, concentrations of Pb were below the LOD of 6.6 mg/L so removal was >87%.

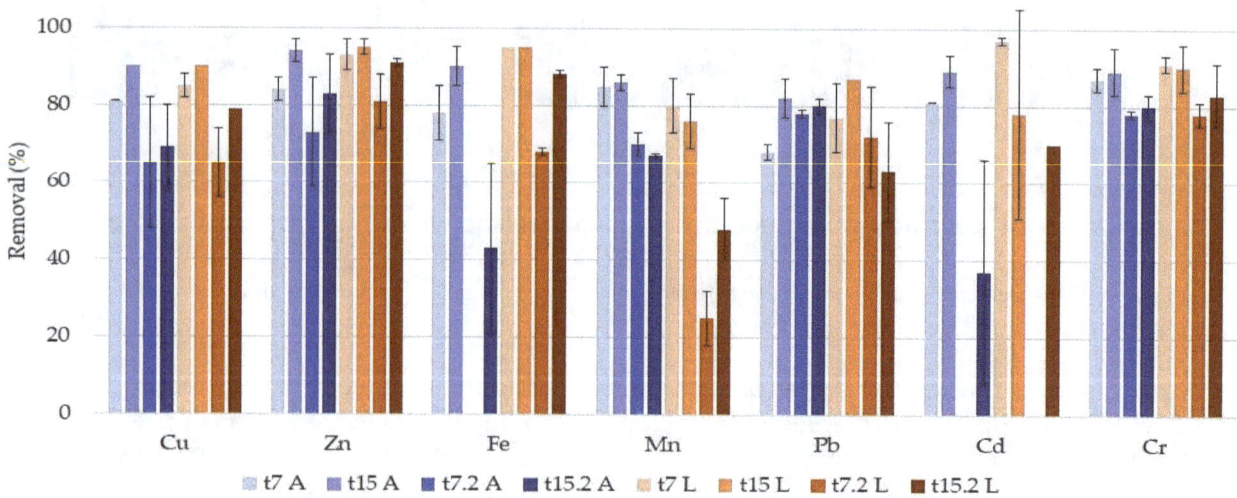


**Figure 6.** Removal percentage and respective standard deviation ($n = 3$) for each metal, in CWs with *T. latifolia* and expanded clay (t7 A, t15 A, t7.2 A, t15.2 A) and lava rock (t7 L, t15 L, t7.2 L, t15.2 L) over the two cycles of treatment. First cycle: samples taken after one week (t7) and at the end of the two-week cycle (t15). Second cycle: samples taken after one week (t7.2) and at the end of the two-week cycle (t15.2). At t7.2 A there was no removal of Fe and Cd, and at t7.2 L there was no removal of Cd.

For Fe differences between cycles occurred, with lower removals in the second cycle. Comparing the removals of the first and second weeks, there were differences between t7 A–t15 A and t7.2 L–t15.2 L. At t15 no differences were observed between substrates. No removal of Fe occurred at t7.2 A and concentrations at t7 L and t15 L were below the LOD of 0.3 mg/L, so removals were >95%. For Mn significant differences between cycles were observed, with the second cycle showing lower removals. In general, differences between the first and second weeks of treatment were not observed. In the first cycle, there were no significant differences between substrates, but in the second cycle, Mn removal rates were lower in L systems. For Cd, significant differences between cycles occurred, with the second cycle showing lower removal rates. At t7.2 no removals of Cd were observed in either system. In the first cycle, no significant differences were observed between the first and second weeks. Additionally, no significant differences were found between substrates. For Cr, no significant differences between weeks of treatment or differences between substrates were observed. Between cycles, differences were found at t7 and t7.2 in both substrates but at t15 and t15.2 no significant differences were found.

For this plant also, in general, both substrates were equally efficient although some differences for Fe and Mn occurred. Contrarily to what was observed for *P. australis*, in general, no significant increase on metal removal was observed in the second week of each cycle, with values being in general below the recommended value for irrigation water. But for Cu values were lower than the recommended value for irrigation water only after a two weeks treatment. For Mn, even after a two-week treatment values were still higher than legislated ones. For Ni, no significant removals were observed: initial concentration was 83 ± 22 µg/L and at the end of the first cycle they were of 80 ± 5 µg/L and 47 ± 6 µg/L for A and L systems, respectively. Initial concentration in the second cycle was 59 ± 5 µg/L and at the end, Ni concentrations were 84 ± 5 µg/L and 55 ± 9 µg/L. These values were lower than legislated ones [13,14].

### 3.1.2. Plant Tissues

For *P. australis*, in general, the initial metal concentrations in leaves, stems, roots, and rhizomes were higher than the concentrations measured at the end of the experiment except for Fe, Mn, and Zn in some plant tissues (Supplementary Material—Table S2). Mn was the only metal accumulated in all plant tissues, with accumulation values between 47% and 64%, depending on the plant tissue. Zn showed a high accumulation in stems, 76% and 75% for A and L system, respectively, and leaves, 60% and 68% for A and L system, respectively (Supplementary Material—Tables S3 and S4). No Zn accumulation in below-ground tissues was observed. Fe was only accumulated in plant leaves, 78% and 64% for A and L system, respectively.

In *T. latifolia*, the concentrations of metals at the end of the experiment were equal or lower than the initial values, indicating that metals were not accumulated in any of tissue plants (Supplementary Material—Tables S2–S4).

In general, no significant differences were observed between substrates.

Both plant species showed signs of toxicity due to the progressive appearance of dry leaves, with the aerial part of *P. australis* ending up dying at the end of the experiment.

### 3.1.3. Sand Substrate

In the initial sand, concentrations of Cu, Mn, Ni, and Zn were below the detection limit (LOD = 4 µg/g for Cu, Zn, and Mn and 0.2 µg/g for Ni), whereas Cd, Cr, and Pb were detected showing concentrations of 0.7 ± 0.1 µg/g, 0.9 ± 0.3 µg/g and 13 ± 8 µg/g, respectively. (Supplementary Material —Table S2). Metals levels in sand collected at the experiment with *P. australis* only showed accumulation of Mn in L systems, of 40% (for A system concentration was identical to initial ones). Concentrations of Cr, Cu, Fe, Ni, and Zn showed no significant differences comparing to the initial sand, whereas concentrations of Cd were below the detection limit (LOD = 0.05 µg/g) and of Pb were 1.0 ± 0.3 µg/g and 1.7 ± 0.5 µg/g for A and L systems, respectively.

Metals levels in sand collected at the experiment with *T. latifolia* only showed accumulation of Mn, ca. of 65% for A and 44% for L systems. Concentrations of Cu, Fe, Ni, and Zn showed no significant differences to the initial sand. Concentrations of Pb were below the detection limit (LOD = 0.5 µg/g). Concentrations of Cd and Cr showed no significant differences in A systems and were below the detection limit (LOD = 0.5 µg/g for Cd and 0.08 µg/g for Cr) in L systems.

### 3.2. Nutrients

The rates of nutrient removal from swine wastewater were calculated and evaluated.

#### 3.2.1. Phragmites Australis

Figures 7 and 8 show the removal rates of ammonium, nitrite, and phosphate ions in systems A and L, respectively.

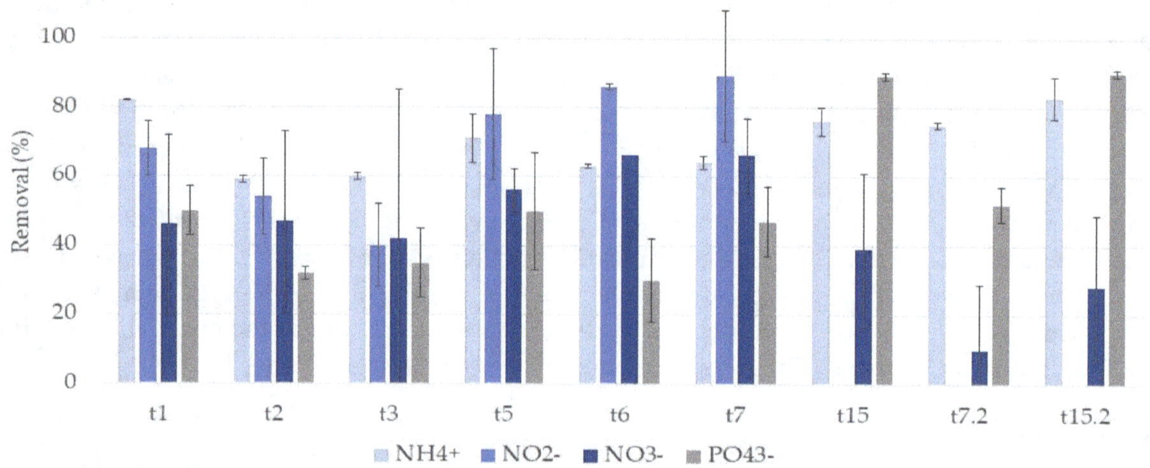


**Figure 7.** Removal percentage and respective standard deviation ($n$ = 3) for each nutrient ion, in CWs with expanded clay (A) and *P. australis* during the two cycles of treatment. First cycle: samples taken after one day (t1), two days (t2), three days (t3), five days (t5), six days (t6), seven days (t7), and at the end of the two-week cycle (t15). Second cycle: samples taken after one week (t7.2) and at the end of the two-week cycle (t15.2) of treatment. At t6, only one replicate was considered for nitrate ion.

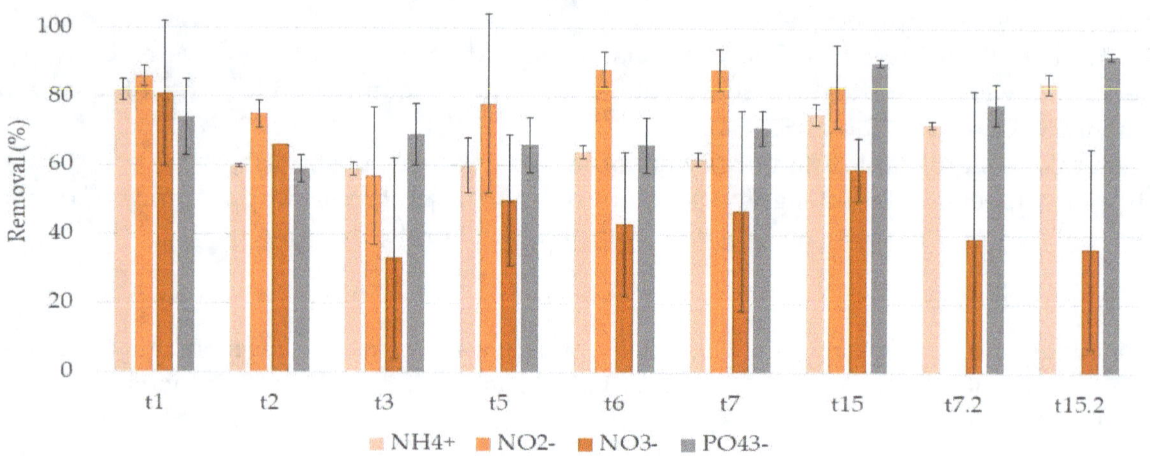


**Figure 8.** Removal percentage and respective standard deviation ($n$ = 3) for each nutrient, in CWs with lava rock (L) and *P. australis* during the two cycles of treatment. First cycle: samples taken after one day (t1), two days (t2), three days (t3), five days (t5), six days (t6), seven days (t7), and at the end of the two-week cycle (t15). Second cycle: samples taken after one week (t7.2) and at the end of the two-week cycle (t15.2) of treatment. At t2, only one replicate was considered for nitrate ion.

For ammonium ion, removals show no clear tendency to increase or decrease during the week. At t7 and t15 no significant differences were observed between substrates, with removal percentages of 64% and 76% respectively for A systems, and 62% and 75% for L systems. The second cycle showed higher removal rates, 75% and 83% at t7.2 and t15.2 for A systems and 72% and 84% for L systems.

Nitrite ion removals oscillated during the first week in both systems, with no significant differences between substrates. A systems achieved a removal of 89% at t7 and showed no removals at t15. L systems showed removals of 88% and 83% at t7 and t15, respectively. In the second cycle, no removals were observed in both systems, with nitrite ion concentration increasing.

Regarding nitrate ion, removals were very variable during the first week for both substrates. In general, L systems showed higher removals rates. For A systems, at t7 and t15 removals of 66% and 39% were observed and for L systems removals were 47% and 59%. The second cycle showed lower removal rates for both systems. For A systems removals were 10% and 28% at t7.2 and t15.2 for L and 39% and 36%. Concentrations of nitrate ion were very variable, even in replicate microcosm CWs.

For phosphate ion, removals vary throughout the first week, with no clear tendency to increase or decrease in both systems. During the first week (from t1 to t7), in general, L systems showed higher removal rates than A systems, achieving 71% at t7, while A systems presented a removal of 47%. At the end of the second week (t15), no differences between the substrates were observed, being 89% for A systems and 90% for L. The second cycle was similar to the first, showing removals of 52% and 90% at t7.2 and t15.2 for A systems and 78% and 92% for L systems.

So, in general, no influence of the substrates in nutrient ions removals was observed. Moreover, nutrient removal occurred mainly on the first day, and no significant increments were observed during the following days of treatment. In addition, no significant differences were observed between cycles.

### 3.2.2. Typha Latifolia

Figures 9 and 10 show the removal rates of ammonium, nitrate, and phosphate ions during both cycles of treatment for systems A and L, respectively.

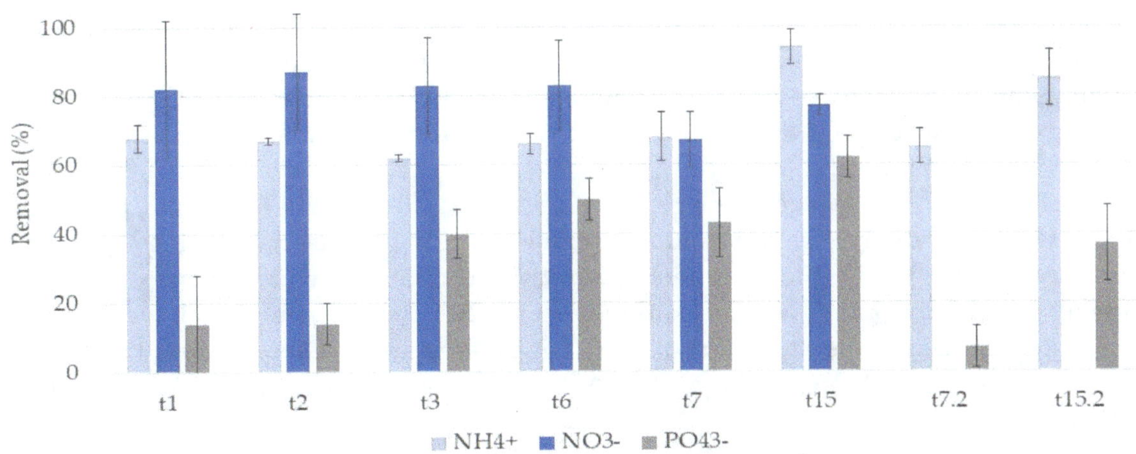


**Figure 9.** Removal percentage and respective standard deviation (n = 3) for each nutrient ion, in CWs with expanded clay (A) and *T. latifolia* during the two cycles of treatment. First cycle: samples taken after one day (t1), two days (t2), three days (t3), five days (t5), six days (t6), seven days (t7), and at the end of the two-week cycle (t15). Second cycle: samples taken after one week (t7.2) and at the end of the two-week cycle (t15.2) of treatment.

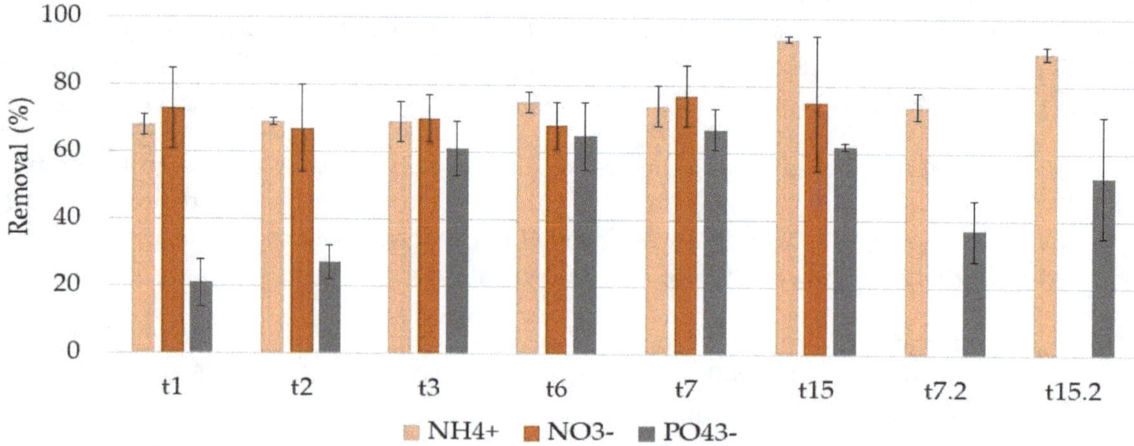


**Figure 10.** Removal percentage and respective standard deviation ($n$ = 3) for each nutrient ion, in CWs with lava rock (L) and *T. latifolia* during the two cycles of treatment. First cycle: samples taken after one day (t1), two days (t2), three days (t3), five days (t5), six days (t6), seven days (t7) and at the end of the two-week cycle (t15). Second cycle: samples taken after one week (t7.2) and at the end of the two-week cycle (t15.2) of treatment.

Regarding ammonium ion results, during the first week, removals rates ranged between 62–68% for A systems and 68–75% for L systems. At t15 removals increase with values of 94% for both systems. The second cycle was similar to the first, removals of 65% for A systems and 74% for L systems at t7.2 and of 85% and 90% at t15.2, respectively. No significant differences were found between substrates.

For nitrite ion, no removals were observed in both cycles. The initial concentration was 5.3 ± 0.4 mg/L (Supplementary Material—Table S5), and at the end of the first cycle, concentrations were 7 ± 5 mg/L and 11 ± 4 mg/L for A and L systems, respectively. In the second cycle, initial nitrate ion concentration was 1.3 ± 0.2 mg/L, and at the end of the cycle, they were 10 ± 3 mg/L and 10 ± 6 mg/L for A and L systems, respectively.

Nitrate ion concentrations were monitored showing positive removal results in the first cycle, while in the second cycle the results were inconclusive as they did not show significant removals over time. In the second cycle, initial nitrate ion concentration was 8 ± 1 mg/L, and at the end of the cycle were 24 ± 5 mg/L and 19 ± 5 mg/L for A and L systems, respectively. The initial concentration of nitrate ion in the second cycle was ca. ten times lower than the initial concentration in the first cycle, which may explain the differences between removals in the two cycles. In the first cycle, nitrate ion removals ranged between 67–87% for A systems and 67–88% for L systems. No significant differences were found between substrates.

For phosphate ion, in A systems, the removal increased, during the first cycle, being 44% and 62% at t7 and t15, respectively. For L systems, removals slightly increased, reaching 68% and 62% at t7 and t15, respectively. No significant differences between substrates were observed. In the second cycle, removals increased from the first week to the second in A systems, but with lower removal rates than in the first cycle. L systems showed no significant differences between t7.2 and t15.2. A systems showed 7% and 37% removal rates at t7.2 and t15.2, and L systems showed 37% and 54% removals at t7.2 and t15.2.

So, for this plant also no significant differences were observed when comparing the two substrates. For *T. latifolia*, nutrient ions removal in general increased over time.

### 3.3. Organic Matter

Organic matter removal was evaluated through COD analysis. In the CWs with *P. australis* (Figure 11), at the end of the first week (t7) there was a removal of 79% and 76% for A and L systems, respectively. After two weeks, at the end of the first cycle (t15) removals of 82% for both systems were observed. In the second cycle, significant differences were found between the first and second weeks, 74% and 70% at t7.2 and 89% and 88% at t15.2 for A and L, respectively. For both cycles, no significant differences between substrates were observed.

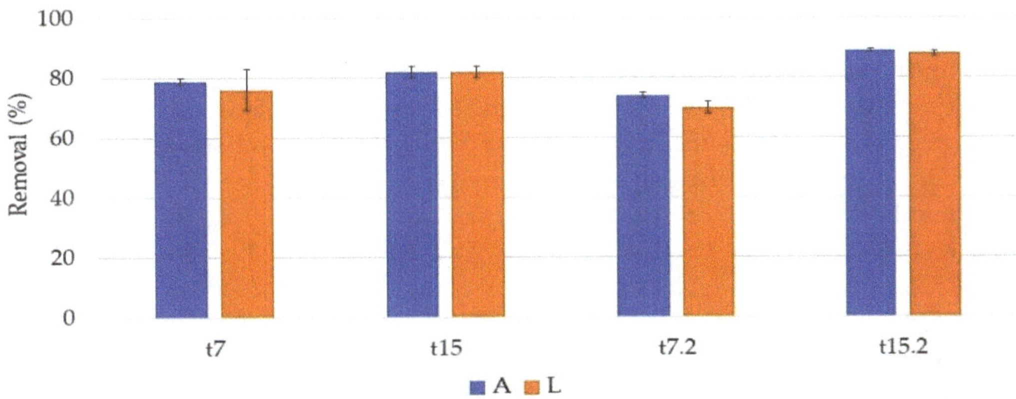


**Figure 11.** Removal percentage and respective standard deviation (n = 3) of COD, in CWs with *P. australis* and expanded clay (A) or lava stone (L) over the two cycles of treatment. First cycle: samples taken after one week (t7) and at the end of the two-week cycle (t15). Second cycle: samples taken after one week (t7.2) and at the end of the two-week cycle (t15.2).

In CWs with *T. latifolia* (Figure 12), at the end of the first week (t7) a removal of 60% and 74% for A and L systems, respectively, was observed. After one week, at the end of the first cycle (t15), a slight removal increase was observed for both systems, 79% and 83% for A and L, respectively. The second cycle showed a similar behavior but with lower removal values, 44% and 65% at t7.2 and 77%, and 80% at t15.2 for A and L, respectively. In general, no significant differences were found between substrates.

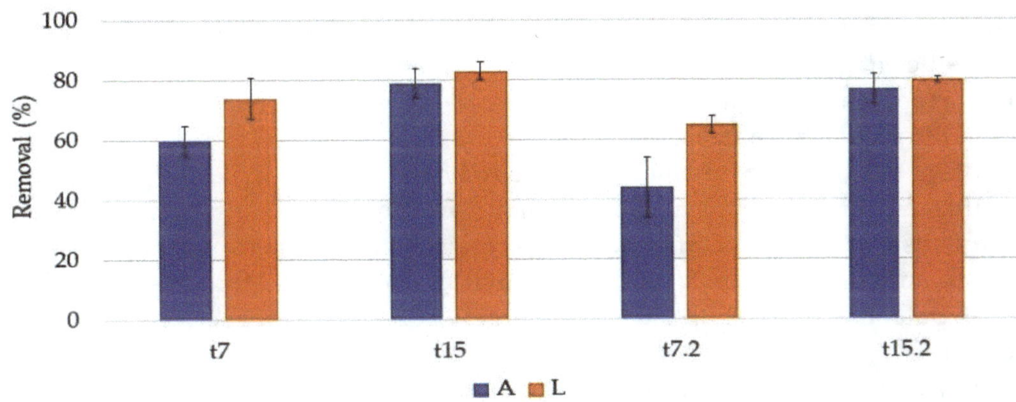


**Figure 12.** Removal percentage and respective standard deviation (n = 3) of COD, in CWs with *T. latifolia* and expanded clay (A) or lava rock (L) over the two cycles of treatment. First cycle: samples taken after one week (t7) and at the end of the two-week cycle (t15). Second cycle: samples taken after one week (t7.2) and at the end of the two-week cycle (t15.2).

## 4. Discussion

The use of reclaimed water is a sustainable alternative to reduce freshwater consumption, being an important part of the concept of the circular economy [39]. Livestock wastewaters are highly contaminated with metals, pharmaceuticals, and other contaminants that need to be removed during treatment if one aims to use this reclaimed water as fertilizer. CWs can be a suitable solution for the removal of toxic contaminants. But the challenge is to maintain the nutritional value of the reclaimed water. To our knowledge, there are no applications of CWs with this purpose. In the present study, high removal rates

(69–98%) of metals during livestock wastewater treatment by CWs were observed, being similar to those reported previously [27]. Moreover, metals concentrations after CWs treatment were lower than the maximum values reported in the Portuguese and international legislation for irrigation water [13–15]. Lower values than recommended values were achieved for all metals after the treatment, except for Mn, with values slightly higher.

Results indicated that different metals interact differently with the components of CWs, at different paces, resulting in different removal rates over time.

Removal of metals in CWs can occur through different processes, such as uptake by plants, bacteria, and algae, precipitation as insoluble salts, and adsorption to substrates [27]. Studies reported that substrate is considered the primary sink for metals and only a small fraction of the influent loading is removed by accumulation in the aboveground biomass of plants [27,40]. Metals have a positive charge, so are readily adsorbed, complexed, and bound with suspended particles, settling in the substrate [41]. Combinations of different layers of substrates and porous materials can have a positive impact on the removal of contaminants. Commonly used substrates in CWs are, for instance, clay, gravel, volcanic (lava) gravel, and sand [42]. In this study, sand was used as support media for the plant roots. Underneath, a layer of a more porous substrate was included, testing, in this case, two materials, expanded clay and lava rock. Mn was the only metal showing significant accumulation in sand by the end of the experiment in CWs of both plants. In fact, it has been reported that fine grains, with silt or clay, tend to favor metal adsorption [43]. According to ISO 14688-1:2002, the sand used in the CWs setup is considered a coarse sand (~0.7 mm), with no fine grains, which may explain the non-retention of metals in this substrate.

As mentioned, lava rock and expanded clay were used as substrates, both having high absorption capacity, porosity, and good ion exchange capacity. These substrates have been reported to be efficient in removing metals from contaminated water in CWs, so metals might have been adsorbed to these substrates. However, metals were not determined in lava rock and expanded clay as their recovery at the end of the experiment (by dismantling the systems) was not possible. Still, no significant differences were found in metal removals between substrates in CWs with *T. latifolia*. In CWs with *P. australis*, significant differences were only found for Fe and Mn, being A systems more efficient for Mn and L systems more efficient for Fe, showing the different interactions that different metals can have within CWs. But metals removals were high for either substrate. So, both substrates are suitable for metal removal from livestock wastewater.

Removal of metals from wastewaters in CWs may also occur by plant uptake. Several studies reported the efficiency of *P. australis* and *T. latifolia* to accumulate metals [10,27,44,45], being two of the most used plants in CWs [30]. Different factors can influence metal uptake by plants, such as clay content, cation exchange capacity, soil pH, organic matter content, and the presence of other ions [45]. Additionally, the potential of metal uptake varies largely between plant species, being important to do a proper selection of the plant species for specific target metals [45]. Therefore, in the present study removal efficiency of *P. australis* and *T. latifolia* were compared. Accumulated metal can be distributed through different plant tissues. Thus, metals were determined in the different tissues of the two plant species used: root, rhizome, stem, and leaves for *P. australis* and root, rhizome, and leaves for *T. latifolia* (this plant does not have a stem structure). In the present study, *P. australis* showed accumulation of Mn in all plant tissues, Zn in stems and leaves, and Fe only in leaves. For all other metals, levels by the end of the experiment were identical or lower to those initially detected. *T. latifolia* showed high initial concentrations of all metals, with levels decreasing by the end of the experiments, resulting in no accumulation of metals in any plant tissues. In fact, plants appear to have released some metals. Initial concentrations in *P. australis* were similar to previous studies, but initial concentrations of *T. latifolia* were higher than values reported previously [10,26,46]. So, considering the high concentration of metals already present in plants and also, considering the high biomass of each macrophyte in each bed system, the metal concentration removed from the wastewater represented a low value that did not result in significant accumulations in plant tissues. Nevertheless, systems with either plant removed high amounts of all metals, the only exception being Ni. So, either plant is suitable to be included in CWs for the removal of metals from livestock wastewater.

CWs systems were able to remove a high amount of metals from livestock wastewater without saturation of the system as no significant amounts of metals were accumulated in plant tissues or sand substrate. In addition, no major differences were observed between cycles of treatment, despite the usual cumulative effect in these systems, which also suggests that the systems were not saturated. CWs retention time usually varies between 4 and 15 days but studies showing a shorter retention time can also be effective [47], including in the removal of metals. For instance, removals of 63–93% of different metals have been reported in CWs in a retention time of 3 days [40]. In the present case, metal removals did not show a major difference between 7- and 15-days treatments. Moreover, high removal occurred already on the first days of CW treatment, increasing in most cases only slightly over time, indicating that metals were mostly retained in CW system components as soon as the wastewater contacted them. So, a shorter retention time might be enough for the removal of metals, which allows greater retention of nutrients and organic matter.

One of the main constituents of livestock wastewater is organic matter. Results obtained of COD showed high removal rates (44–89%) of organic matter from the wastewater during CWs treatment. Previous studies have shown the potential of CWs for organic matter removal with identical or slightly higher removal rates, including from livestock wastewater [8,23–25]. Removal of organic matter involves biological, chemical, and physical processes, such as separation and retention in substrate allowing a proliferation of microorganisms and biodegradation [8]. Plants also lead to higher removal of organic matter by increasing microbial activity [23]. So, different substrates and/or different plants can promote different microbial communities which can influence organic matter removal. However, in the present study, no differences were found between CWs with expanded clay and CWs with lava rock. Results also suggest that both CWs with *P. australis* and CWs with *T. latifolia* are efficient in removing organic matter, with no significant differences between them.

Other main constituents of livestock wastewater are the nutrients nitrogen and phosphorus. Nitrogen can be present in different forms, namely in the form of ammonium, nitrite, and nitrate ions. The major mechanisms of nitrogen removal from wastewater in CWs are microbial processes such as nitrification and denitrification. Nitrification is the biological oxidation of ammonium to nitrite and then nitrate in a sequential reaction. Denitrification is the process in which nitrate is transformed in dinitrogen via intermediates nitrite, nitric oxide, and nitrous oxide [8]. For phosphorus, major removal processes involve fixation by iron and aluminum in the CW's substrates [48]. Along with these, there are other processes involved in nutrient removal, such as ammonium volatilization, plant and microbial uptake, sorption, desorption, burial, and leaching among others [8]. So, both substrate type and plant species can influence nutrients removal.

In the present study, ammonium ion removal was slightly lower than in previous studies in similar CWs [8,21]. In CWs with *P. australis*, ammonium ion removals reached ca. 85%, showing high removals in the first days, without significant differences between substrates. In CWs with *T. latifolia*, ammonium ion removals were also similar during the first 7 days, increasing to ca. 94% after 15 days of treatment. Both substrates and plants can stimulate different microbial communities that could influence the removals of ammonium, but in the present study, no significant differences were found between systems with different plants or substrates. Removals of nitrites and nitrates ions showed high variability in their removals. This variation can be related to the balance among the different processes of nitrification, denitrification, and assimilation by plants. Daily recirculation of wastewater also enhances nitrification processes, and consequently removals of ammonium [49], but in the present study, recirculation might not be always efficient. Additionally, in some of the microcosms, some anaerobic zones might have occurred leading to variations in removal rates.

The main processes for phosphorus removal are precipitation, sorption, and plant uptake, namely through plant roots. Reported removal rates in similar CWs [8,21] were identical to the ones obtained in this study. In both experiments, removals of phosphate ion showed a tendency to increase over time.

In general, no significant differences were found between substrates. But CWs with *P. australis* showed higher removal rates than CWs with *T. lafitolia* (reaching 92% vs. 65%). Although the plant species could have influenced this result, substrate saturation should also be considered as *T. latifolia* were transplanted into CWs with expanded clay and lava rock from CWs with *P. australis*.

High removals of metals were already observed in the first days of CWs treatment, which suggests that a short retention time can be a good option. A shorter retention time can lead to lower removals of nutrients, promoting the use of reclaimed water as fertilizer, although in the present study nutrients removals were already high after only two or three days of CW treatment.

As mentioned, the choice of plants is an important issue in CWs once they mediate important processes. For example, plant metabolic activity releases oxygen into the rhizosphere, promoting the activity of bacteria involved in carbon and nitrogen cycles. The most widely used plants in CWs are cattails (*Typha* spp.) and reeds (such as *Phragmites australis*) [50], being these the plants selected for this study.

Different studies have already reported the suitability of the two plants for metal removal from contaminated wastewater. For instance, Hejna et al. [10] reported that *T. latifolia* accumulates and translocates Zn and Cu from contaminated wastewater into plant tissues, the increase of both metals in plants is related to a decrease of metals in water. Almeida et al. [27] observed also a high removal of different metals (Zn, Cu, Fe, and Mn) from swine wastewater during treatment in CW planted with *P. australis*.

Although the efficiency for metal removal varied with the plant species [51] as well as with the metal itself, removal rates as high as 90% have been reported (e.g., [52]. Similar removal rates were observed in the present study, with removal rates reaching in general >80% for Cd, Cr, Cu, Fe, Mn, Pb, and Zn after 7 or 15 days CW treatment planted with either plant. Previous studies have shown that both *P. australis* and *T. latifolia* are suitable species for phytoremediation, and CWs with these plants can be used for wastewater treatment and be an option for improving irrigation water quality [8,26,33].

Kumari et al. [44] reported that *P. australis* showed higher accumulative capacities for Cu, Cd, Cr, Ni, and Fe than *T. latifolia* but, in the present study, no significant differences in metal accumulation were observed and metal removal rates were identical between the two plants. However, regarding only two days of treatment, these removal rates are only attained in CWs planted with *T. latifolia* (with the exception of Pb with removal rates between 60 and 70%). In CWs planted with *P. australis*, after two treatment days, metal removals rates were ca. 60% (except for Pb with removal ca. 40%). So, CWs planted with *T. latifolia* are more suitable for the aim of this work, that nutrients final concentrations in reclaimed water are maintained for fertilization, after the removal of the toxic contaminants.

The efficiency of metal removal can also vary with the substrate composition. As mentioned, substrate is considered the primary sink for metals accumulation. The selection of a substrate with a high sorption capacity can be an important step in the optimization of CWs performance [53]. Most CWs employ crushed stones, sand, and gravel as substrates not only to support the plant growth but also to act as a filter; however, other substrates such as clay soil, zeolites, shells, and industrial wastes (furnace slag, steel slag, sludge from waste treatment plants) had also been found as efficient filter materials (Arivoli et al., 2015). In the present work, two commonly used substrates were selected, expanded clay and lava rock, to complement the sand used to give support to plant roots.

Calheiros et al. [35] reported that expanded clay aggregates were adequate for the establishment of *T. latifolia*, allowing a high plant propagation level. Moreover, Dordio et al. [53] observed high removals of TSS, COD, $NH_4$-N from swine wastewater when using CW with expanded clay substrate and *P. australis* plants and Almeida et al. [8] also observed a similar behavior when using CW with lava rock substrate and *P. australis* plants. In the later system, the CW system also showed high metal removal [27], similar to what was observed in the present study, removal >80%.

Looking at only two days of treatment, metal removal rates were, in general, identical between CWs with expanded clay or lava rock, although slightly higher removals were observed for Fe with lava rock whereas for Mn removals were slightly higher for expanded clay. But for nutrients, namely nitrate and phosphate ions a slightly lower removal was observed when CW had expanded clay. So, both porous substrates can be an adequate option for agricultural wastewater treatment, but expanded clay could be more suitable if one wants lower nutrient removal during a short CW treatment.

## 5. Conclusions

Present results demonstrate both microcosms simulating CWs planted with *P. australis* and planted with *T. latifolia* were able to remove significant amounts of toxic contaminants, such as metals, from livestock wastewater, showing high removal rates after 7 or 15 days of CW treatment. In fact, no difference between plant performance was observed indicating that *P. australis* and *T. latifolia* are appropriate species to be included in CWs for wastewater treatment. However, for a short CW treatment, *T. latifolia* could be a more suitable option as metal rates are already very high after two treatment days.

Results also show that either substrate selected, lava rock and expanded clay, combined with the sand used for plant roots support, contributed to pollutants removal without significant difference between them. So, both substrates can be suitable options for CWs assemblage, although in a short CW treatment expanded clay could be a more suitable choice as it showed slightly lower nutrients removals (nitrate and phosphate ions) in the first days of treatment.

Results also indicate that concentrations of nitrogen and phosphorus decreased during CWs treatment, while still maintaining levels that could be useful for fertilization.

In general, metal concentrations in wastewater treated by the CWs microcosms were below values established in Portuguese and international legislation for irrigation.

Thus, CWs seem to be a valuable and low-cost alternative to treat livestock wastewater so that it can be reused in agriculture, promoting a circular economy.

**Supplementary Materials:**
Figure S1: (a) Schematic representation of CW microcosm; (b) Picture of real CWs microcosms with *T. latifolia*, Tablez S1: Initial concentrations (mean and standard deviation, $n$ = 3) for each metal in livestock wastewater collected, for the first (t0) and second (t0.2) cycles, Table S2: Initial concentrations (mean and standard deviation, $n$ = 3) for each metal in sand and plant tissues, Table S3. Final concentrations at t15.2 and respective standard deviation ($n$ = 3) for each metal in sand and plant tissues for the CWs with expanded clay, Table S4. Final concentrations at t15.2 and respective standard deviation ($n$ = 3) for each metal in sand and plant tissues for the CWs with lava rock, Table S5: Initial concentrations (mean and standard deviation, $n$ = 3) for each nutrient ion in livestock wastewater collected, for the first (t0) and second (t0.2) cycles, Table S6: Legislated values of metals in irrigation wastewater and wastewater discharge in Portuguese and international legislation.

**Author Contributions:** Conceptualization, S.D., A.P.M. and C.M.R.A.; methodology, S.D., A.P.M. and C.M.R.A.; validation, S.D. and C.M.R.A.; formal analysis, S.D., R.D.C. and P.R.; investigation, S.D. and C.M.R.A.; resources, A.P.M. and C.M.R.A; writing—original draft preparation, S.D.; writing—review and editing, C.M.R.A. and A.P.M.; supervision, A.P.M. and C.M.R.A.; project administration, A.P.M. and C.M.R.A.; funding acquisition, A.P.M. and C.M.R.A. All authors have read and agreed to the published version of the manuscript.

## References

1. Shakir, E.; Zahraw, Z.; Al-Obaidy, A.H.M.J. Environmental and health risks associated with reuse of wastewater for irrigation. *Egypt. J. Pet.* **2017**, *26*, 95–102. [CrossRef]
2. Sgroi, M.; Vagliasindi, F.G.A.; Roccaro, P. Feasibility, sustainability and circular economy concepts in water reuse. *Curr. Opin. Environ. Sci. Heal.* **2018**, *2*, 20–25. [CrossRef]
3. Chen, W.; Lu, S.; Pan, N.; Wang, Y.; Wu, L. Impact of reclaimed water irrigation on soil health in urban green areas. *Chemosphere* **2015**, *119*, 654–661. [CrossRef] [PubMed]

4. Forslund, A.; Ensink, J.H.J.; Battilani, A.; Kljujev, I.; Gola, S.; Raicevic, V.; Jovanovic, Z.; Stikic, R.; Sandei, L.; Fletcher, T.; et al. Faecal contamination and hygiene aspect associated with the use of treated wastewater and canal water for irrigation of potatoes (Solanum tuberosum). *Agric. Water Manag.* **2010**, *98*, 440–450. [CrossRef]
5. Lopez, A.; Pollice, A.; Lonigro, A.; Masi, S.; Palese, A.M.; Cirelli, G.L.; Toscano, A.; Passino, R. Agricultural wastewater reuse in southern Italy. *Desalination* **2006**, *187*, 323–334. [CrossRef]
6. Maiolo, M.; Pantusa, D. A proposal for multiple reuse of urban wastewater. *J. Water Reuse Desalin.* **2018**, *8*, 468–478. [CrossRef]
7. Oron, G.; Gillerman, L.; Buriakovsky, N.; Bick, A.; Gargir, M.; Dolan, Y.; Manor, Y.; Katz, L.; Hagin, J. Membrane technology for advanced wastewater reclamation for sustainable agriculture production. *Desalination* **2008**, *218*, 170–180. [CrossRef]
8. Almeida, C.M.R.; Santos, F.; Ferreira, A.C.F.; Lourinha, I.; Basto, M.C.P.; Mucha, A.P. Can veterinary antibiotics affect constructed wetlands performance during treatment of livestock wastewater? *Ecol. Eng.* **2017**, *102*, 583–588. [CrossRef]
9. Hu, H.; Li, X.; Wu, S.; Yang, C. Sustainable livestock wastewater treatment via phytoremediation: Current status and future perspectives. *Bioresour. Technol.* **2020**, *315*, 123809. [CrossRef]
10. Hejna, M.; Moscatelli, A.; Stroppa, N.; Onelli, E.; Pilu, S.; Baldi, A.; Rossi, L. Bioaccumulation of heavy metals from wastewater through a Typha latifolia and Thelypteris palustris phytoremediation system. *Chemosphere* **2020**, *241*, 125018. [CrossRef]
11. van Alen, S.; Kaspar, U.; Idelevich, E.A.; Köck, R.; Becker, K. Increase of zinc resistance in German human derived livestock-associated MRSA between 2000 and 2014. *Vet. Microbiol.* **2018**, *214*, 7–12. [CrossRef] [PubMed]
12. Ministerio da Economia. Decreto-Lei n.º 103/2015. In *Diário da República n.º 114/2015*; Ministério da Economia: Lisboa, Portugal, 2015; pp. 3756–3788.
13. Ministerio do Ambiente. Decreto-Lei n.º 236/98. In *Diário da República nº 176/1998*; Ministério do Ambiente: Lisboa, Portugal, 1998; pp. 3676–3722.
14. U.S. Environmental Protection Agency. *Guidelines for Water Reuse*; U.S. Agency for International Development: Washington, DC, USA, 2004.
15. Norton-Brandão, D.; Scherrenberg, S.M.; van Lier, J.B. Reclamation of used urban waters for irrigation purposes—A review of treatment technologies. *J. Environ. Manag.* **2013**, *122*, 85–98. [CrossRef] [PubMed]
16. Dias, S.; Correia, B.; Fraga-Santiago, P.; Silva, C.; Baptista, P.C.; Gomes, C.R.; Almeida, C.M.R. Potential of an estuarine salt marsh plant (Phragmites australis (Cav.) Trin. Ex Steud10751) for phytoremediation of bezafibrate and paroxetine. *Hydrobiologia* **2020**, 1–14. [CrossRef]
17. Liang, Y.; Zhu, H.; Bañuelos, G.; Yan, B.; Zhou, Q.; Yu, X.; Cheng, X. Constructed wetlands for saline wastewater treatment: A review. *Ecol. Eng.* **2017**, *98*, 275–285. [CrossRef]
18. Gorito, A.M.; Ribeiro, A.R.; Almeida, C.M.R.; Silva, A.M.T. A review on the application of constructed wetlands for the removal of priority substances and contaminants of emerging concern listed in recently launched EU legislation. *Environ. Pollut.* **2017**, *227*, 428–443. [CrossRef] [PubMed]
19. Wang, M.; Zhang, D.; Dong, J.; Tan, S.K. Application of constructed wetlands for treating agricultural runoff and agro-industrial wastewater: A review. *Hydrobiologia* **2017**, *805*, 1–31. [CrossRef]
20. A, D.; Zhang, X.; Dai, Y.; Chen, C.; Yang, Y. Occurrence and removal of quinolone, tetracycline, and macrolide antibiotics from urban wastewater in constructed wetlands. *J. Clean. Prod.* **2020**, *252*, 119677. [CrossRef]
21. Santos, F.; Almeida, C.M.R.; Ribeiro, I.; Mucha, A.P. Potential of constructed wetland for the removal of antibiotics and antibiotic resistant bacteria from livestock wastewater. *Ecol. Eng.* **2019**, *129*, 45–53. [CrossRef]
22. Reddy, G.B.; Hunt, P.G.; Phillips, R.; Stone, K.; Grubbs, A. Treatment of swine wastewater in marsh-pond-marsh constructed wetlands. *Water Sci. Technol.* **2001**, *44*, 545–550. [CrossRef]
23. Lee, C.Y.; Lee, C.C.; Lee, F.Y.; Tseng, S.K.; Liao, C.J. Performance of subsurface flow constructed wetland taking pretreated swine effluent under heavy loads. *Bioresour. Technol.* **2004**, *92*, 173–179. [CrossRef]
24. Meers, E.; Rousseau, D.P.L.; Blomme, N.; Lesage, E.; Du Laing, G.; Tack, F.M.G.; Verloo, M.G. Tertiary treatment of the liquid fraction of pig manure with Phragmites australis. *Water Air Soil Pollut.* **2005**, *160*, 15–26. [CrossRef]
25. Meers, E.; Tack, F.M.G.; Tolpe, I.; Michels, E. Application of a full-scale constructed wetland for tertiary treatment of piggery manure: Monitoring results. *Water Air Soil Pollut.* **2008**, *193*, 15–24. [CrossRef]

26. Anning, A.K.; Korsah, P.E.; Addo-Fordjour, P. Phytoremediation of Wastewater with Limnocharis Flava, Thalia Geniculata and Typha Latifolia in Constructed Wetlands. *Int. J. Phytoremediation* **2013**, *15*, 452–464. [CrossRef] [PubMed]
27. Almeida, C.M.R.; Santos, F.; Ferreira, A.C.F.; Gomes, C.R.; Basto, M.C.P.; Mucha, A.P. Constructed wetlands for the removal of metals from livestock wastewater—Can the presence of veterinary antibiotics affect removals? *Ecotoxicol. Environ. Saf.* **2017**, *137*, 143–148. [CrossRef]
28. Ibekwe, A.; Murinda, S. Continuous Flow-Constructed Wetlands for the Treatment of Swine Waste Water. *Int. J. Environ. Res. Public Health* **2018**, *15*, 1369. [CrossRef] [PubMed]
29. Gorito, A.M.; Ribeiro, A.R.; Gomes, C.R.; Almeida, C.M.R.; Silva, A.M.T. Constructed wetland microcosms for the removal of organic micropollutants from freshwater aquaculture effluents. *Sci. Total Environ.* **2018**, *644*, 1171–1180. [CrossRef] [PubMed]
30. Batool, A.; Saleh, T.A. Removal of toxic metals from wastewater in constructed wetlands as a green technology; catalyst role of substrates and chelators. *Ecotoxicol. Environ. Saf.* **2020**, *189*, 109924. [CrossRef]
31. Mucha, A.P.; Teixeira, C.; Reis, I.; Magalhães, C.; Bordalo, A.A.; Almeida, C.M.R. Response of a salt marsh microbial community to metal contamination. *Estuar. Coast. Shelf Sci.* **2013**, *130*, 81–88. [CrossRef]
32. Anning, A.K.; Akoto, R. Assisted phytoremediation of heavy metal contaminated soil from a mined site with Typha latifolia and Chrysopogon zizanioides. *Ecotoxicol. Environ. Saf.* **2018**, *148*, 97–104. [CrossRef]
33. Rana, V.; Maiti, S.K. Municipal wastewater treatment potential and metal accumulation strategies of *Colocasia esculenta* (L.) Schott and Typha latifolia L. in a constructed wetland. *Environ. Monit. Assess.* **2018**, *190*, 1–15. [CrossRef]
34. Dordio, A.; Carvalho, A.J.P. Constructed wetlands with light expanded clay aggregates for agricultural wastewater treatment. *Sci. Total. Environ.* **2013**, *463*, 454–461. [CrossRef] [PubMed]
35. Calheiros, C.S.C.; Rangel, A.O.S.S.; Castro, P.M.L. Evaluation of different substrates to support the growth of Typha latifolia in constructed wetlands treating tannery wastewater over long-term operation. *Bioresour. Technol.* **2008**, *99*, 6866–6877. [CrossRef] [PubMed]
36. Collison, R.S.; Grismer, M.E. Nitrogen and Chemical Oxygen Demand Removal from Septic Tank Wastewater in Subsurface Flow Constructed Wetlands: Substrate (Cation Exchange Capacity) Effects. *Water Environ. Res.* **2013**, *86*, 314–323. [CrossRef]
37. Carvalho, P.N.; Basto, M.C.P.; Almeida, C.M.R. Potential of Phragmites australis for the removal of veterinary pharmaceuticals from aquatic media. *Bioresour. Technol.* **2012**, *116*, 497–501. [CrossRef]
38. Jones, M.N. Nitrate reduction by shaking with cadmium: Alternative to cadmium columns. *Water Res.* **1984**, *18*, 643–646. [CrossRef]
39. Villamar, C.-A.; Vera-Puerto, I.; Rivera, D.; De la Hoz, F. Reuse and Recycling of Livestock and Municipal Wastewater in Chilean Agriculture: A Preliminary Assessment. *Water* **2018**, *10*, 817. [CrossRef]
40. Yadav, A.K.; Abbassi, R.; Kumar, N.; Satya, S.; Sreekrishnan, T.R.; Mishra, B.K. The removal of heavy metals in wetland microcosms: Effects of bed depth, plant species, and metal mobility. *Chem. Eng. J.* **2012**, *211*, 501–507. [CrossRef]
41. Bhatia, M.; Goyal, D. Analyzing remediation potential of wastewater through wetland plants: A review. *Environ. Prog. Sustain. Energy* **2014**, *33*, 9–27. [CrossRef]
42. Sandoval-Herazo, L.; Alvarado-Lassman, A.; Marín-Muñiz, J.; Méndez-Contreras, J.; Zamora-Castro, S.A. Effects of the Use of Ornamental Plants and Different Substrates in the Removal of Wastewater Pollutants through Microcosms of Constructed Wetlands. *Sustainability* **2018**, *10*, 1594. [CrossRef]
43. Almeida, C.M.R.; Mucha, A.P.; Teresa Vasconcelos, M. Role of different salt marsh plants on metal retention in an urban estuary (Lima estuary, NW Portugal). *Estuar. Coast. Shelf Sci.* **2011**, *91*, 243–249. [CrossRef]
44. Kumari, M.; Tripathi, B.D. Efficiency of Phragmites australis and Typha latifolia for heavy metal removal from wastewater. *Ecotoxicol. Environ. Saf.* **2015**, *112*, 80–86. [CrossRef] [PubMed]
45. Bonanno, G. Comparative performance of trace element bioaccumulation and biomonitoring in the plant species Typha domingensis, Phragmites australis and Arundo donax. *Ecotoxicol. Environ. Saf.* **2013**, *97*, 124–130. [CrossRef] [PubMed]
46. Bonanno, G.; Cirelli, G.L. Comparative analysis of element concentrations and translocation in three wetland congener plants: Typha domingensis, Typha latifolia and Typha angustifolia. *Ecotoxicol. Environ. Saf.* **2017**, *143*, 92–101. [CrossRef] [PubMed]

47. Ghosh, D.; Gopal, B. Effect of hydraulic retention time on the treatment of secondary effluent in a subsurface flow constructed wetland. *Ecol. Eng.* **2010**, *36*, 1044–1051. [CrossRef]
48. Stottmeister, U.; Wießner, A.; Kuschk, P.; Kappelmeyer, U.; Kästner, M.; Bederski, O.; Müller, R.A.; Moormann, H. Effects of plants and microorganisms in constructed wetlands for wastewater treatment. *Biotechnol. Adv.* **2003**, *22*, 93–117. [CrossRef]
49. Saeed, T.; Sun, G. A review on nitrogen and organics removal mechanisms in subsurface flow constructed wetlands: Dependency on environmental parameters, operating conditions and supporting media. *J. Environ. Manag.* **2012**, *112*, 429–448. [CrossRef]
50. Adhikari, A.; Acharya, K.; Shanahan, S.; Zhou, X. Removal of nutrients and metals by constructed and naturally created wetlands in the Las Vegas Valley, Nevada. *Environ. Monit. Assess.* **2011**, *180*, 97–113. [CrossRef]
51. Vymazal, J.; Březinová, T. Heavy metals in plants in constructed and natural wetlands: Concentration, accumulation and seasonality. *Water Sci. Technol.* **2015**, *71*, 268–276. [CrossRef]
52. Vymazal, J.A.N. Removal of Heavy Metals in a Horizontal Sub-Surface Flow Constructed Wetland. *J. Environ. Sci. Health* **2005**, *40*, 1369–1379. [CrossRef]
53. Dordio, A.V.; Teimão, J.; Ramalho, I.; Carvalho, A.J.P.; Candeias, A. Selection of a support matrix for the removal of some phenoxyacetic compounds in constructed wetlands systems. *Sci. Total. Environ.* **2007**, *380*, 237–246. [CrossRef]

# 3

# Removal of Diclofenac in Wastewater using Biosorption and Advanced Oxidation Techniques: Comparative Results

**José M. Angosto, María J. Roca and José A. Fernández-López ***

Department of Chemical and Environmental Engineering, Technical University of Cartagena, Paseo Alfonso XIII, 52, 30203 Cartagena, Murcia, Spain; jm.angosto@upct.es (J.M.A.); mjose.roca@upct.es (M.J.R.)

* Correspondence: josea.fernandez@upct.es

**Abstract:** Wastewater treatment is a topic of primary interest with regard to the environment. Diclofenac is a common analgesic drug often detected in wastewater and surface water. In this paper, three commonly available agrifood waste types (artichoke agrowaste, olive-mill residues, and citrus waste) were reused as sorbents of diclofenac present in aqueous effluents. Citrus-waste biomass for a dose of 2 $g{\cdot}L^{-1}$ allowed for removing 99.7% of diclofenac present in the initial sample, with a sorption capacity of 9 mg of adsorbed diclofenac for each gram of used biomass. The respective values obtained for olive-mill residues and artichoke agrowaste were around 4.15 $mg{\cdot}g^{-1}$. Advanced oxidation processes with $UV/H_2O_2$ and UV/HOCl were shown to be effective treatments for the elimination of diclofenac. A significant reduction in chemical oxygen demand (COD; 40–48%) was also achieved with these oxidation treatments. Despite the lesser effectiveness of the sorption process, it should be considered that the reuse and valorization of these lignocellulosic agrifood residues would facilitate the fostering of a circular economy.

**Keywords:** agrowaste biomass; biosorption; diclofenac removal; advanced oxidation treatments; low-cost sorbents

---

## 1. Introduction

Water is a renewable but limited natural resource, which forces us to manage it, especially in arid and semiarid areas where water resources are limited. The presence of some pollutants can jeopardize water recovery, and the reuse of urban and industrial effluents [1]. In addition to pollution from municipal discharges, water bodies are also contaminated by industry, mining, and pollution from agrochemicals used on irrigated land, which in many cases are discharged directly into lakes and rivers without prior treatment [2].

The reuse of urban and industrial wastewater with some pollutants can contribute to soil contamination, as well as its passage into the tissue of cultivated plants. This can produce adverse effects on the ecosystem [3–6].

While the international scientific community continues to make considerable efforts to minimize or alleviate the problem of what we may call traditional pollutants (heavy metals, organic matter, inorganic nutrients, etc.), other environmental pollutants, called emerging pollutants (ECs), are of growing concern [7] and are leading us to adapt our strategies to respond to their threats [8,9].

Aqueous effluents that arrive at urban wastewater-treatment plants show a complex composition associated with current human and industrial development. Therefore, the presence of certain pollutants can, in many cases, interfere with the proper functioning of biological wastewater-treatment systems. Emerging contaminants are considered among these pollutants [9]. Emerging contaminants are compounds of different origin and chemical nature of which the presence in the environment has gone largely unnoticed [10].

The lack of international monitoring programs for ECs makes it impossible to fully know their effects, problems, and behavior from an ecotoxicological point of view [2]. They can be a danger to the environment and human health [11,12]. Another particularity of this type of pollutant is that, due to its high production and consumption, and the constant and continuous introduction of these pollutants into the environment, they do not need to be persistent to cause negative effects.

Emerging contaminants can be classified as pharmaceutical and personal-care products, plasticizers, food additives, wood preservatives, contrast media, detergents, surfactants, fire retardants, biocides, hormones, and some disinfection byproducts. Advances in analytical technologies have helped to identify this group of contaminants [13,14].

These compounds have high transformation/disposal rates that can compensate for their continued introduction into the environment. For this reason, it is necessary to raise awareness of their origin and transformation, and the effects that this new generation of pollutants can have in order to propose modifications or improvements in water-treatment mechanisms. Some authors suggested the use of toxicity tests with chemical analysis to evaluate the efficiency of treatment methods [15–18].

The ongoing processing of aqueous effluents in conventional wastewater-treatment systems does not provide an adequate approach for the removal of emerging contaminants. These can arrive in their original form and/or be metabolized, making it difficult to operate biological wastewater treatment, sludge treatment, and the subsequent reuse of wastewater; there is also a lack of knowledge about the possible impact of these pollutants on the environment [19]. In many cases, products resulting from the transformation of the original emerging contaminants produce compounds that turn out to be more toxic than the original ones are [20].

Diclofenac (2-(2,6-dichloroanilino) phenylacetic acid) is one of the best-selling nonsteroidal anti-inflammatory drugs [21]. The high consumption level of this compound means that it is one of the most detected pharmaceuticals in aqueous effluents, and according to reported data, the measured values of diclofenac in municipal wastewater can be up 7.1 $\mu g \cdot L^{-1}$ [22,23]. A distinctive feature of diclofenac is that it undergoes low biodegradability and high persistence in wastewater-treatment plants, leading to its bioaccumulation in surface waters, sediments, and sludges [24,25]. Therefore, it is necessary to implement specific treatments to reduce or eliminate its presence in effluents.

Different sorption studies of diclofenac are reported in the literature. The highest bioremoval efficacies were obtained with carbon nanotubes [26], clays [27], and biochars [28] as sorbents. Raw lignocellulosic biomass was also used as a low-cost biosorbent, but the achieved efficiency was not as high as that with previous sorbents [29,30]. Despite this, biosorption has lately been presented as a very interesting technology, in line with the concept of the circular economy [31,32]. Efficient biowaste management is among the most important challenges in agrifood industries. Suggested uses for these lignocellulosic residues include their utilization as sorbents of pollutants in aqueous effluents [33–35] since they present very interesting physicochemical properties, such as their high surface area and the presence of a variety of functional groups that facilitate the sorption process [36–38]. Bioadsorptive techniques are the best ecofriendly solutions for removing contaminants because they are economical, efficient, highly selective on pollutants, and easily operable.

The most commonly used treatments for the removal of emerging contaminants, such as diclofenac, are advanced oxidation processes [39–43], which utilize free radical reactions to directly degrade chemical contaminants as an alternative to traditional treatments. For these oxidative treatments, it is necessary to carry out research studies that examine what the most appropriate combination of oxidizing reagents is to degrade a specific pollutant.

The main objective of this work was to carry out a comparative study of diclofenac elimination in aqueous effluents through biosorption with agrifood residues and through different advanced oxidation techniques (ultraviolet radiation (UV), sodium hypochlorite, and combinations of ultraviolet radiation with hypochlorite and with hydrogen peroxide).

## 2. Materials and Methods

### 2.1. Preparation and Characterization of Agrowaste Biomass

In preliminary studies, different agricultural wastes residues were processed and examined for their ability to remove diclofenac from a test solution. Of these, the three most interesting agrifood waste types that are abundant in the Murcia region (Spain) were chosen to be investigated in this study: artichoke agrowaste, olive-mill residues, and citrus waste from agrifood industries. Biomass samples were washed with distilled water and dried at 70 °C for 24 h before being ground and passed through a sieve with a number 18 mesh (1 mm). The different surface morphologies of each biomass were explored, and a qualitative determination of each was made using a field-emission scanning electron microscope and an energy-dispersive X-ray analyzer (SEM) (Hitachi S-3500N). FTIR measurements of the sorbents were performed in the range of 4000 to 400 $cm^{-1}$ in a Thermo Nicolet 5700 (Karlsruhe, Germany). Moreover, the pH point of zero charge ($pH_{ZPC}$) was determined by the mass-titration method [44].

### 2.2. Preparation of Diclofenac Solutions

A diclofenac stock solution (1 $g{\cdot}L^{-1}$) was prepared using diclofenac sodium ($C_{14}H_{10}Cl_2NNaO_2$, ≥99%; Sigma-Aldrich, Madrid, Spain) in deionized water. The working solution (20 L) was achieved by diluting the stock solution in treated and filtered wastewater from a wastewater-treatment plant. The pH value was adjusted to 6.6 with a 0.1 $mol{\cdot}L^{-1}$ HCl or 0.1 $mol{\cdot}L^{-1}$ NaOH solution using a Metrohm 654 pH meter with a pH combination electrode. All chemicals were of analytical grade. The analytical characterization of the working solution was performed by American Public Health Association (APHA) methods [45].

### 2.3. Batch-Biomass Tests

Batch-biosorption experiments were performed at 25 °C under stirring in a reciprocal contact agitator with 100 mL of the diclofenac working solution at a fixed concentration ($C_o$ = 37.6 $mg{\cdot}L^{-1}$) and a known amount of sorbent for 24 h in a conical flask at a constant stirring speed (150 rpm). Lastly, the solutions were filtered and quantified obtaining the final concentration ($C_e$). The initial ($C_o$) and final ($C_e$) concentrations were quantified by high-performance liquid chromatography (HPLC).

The amounts of diclofenac adsorbed at equilibrium ($q_e$), also called diclofenac removal efficiency and expressed as mg diclofenac/g dry biomass, were determined by mass-balance equation (Equation (1)) on the basis of values of diclofenac concentration in the solution at the beginning ($C_o$) and end ($C_e$) of the test:

$$q_e = \frac{(C_o - C_e) \cdot V}{m}, \quad (1)$$

where $V$ is the solution volume (L) and m the sorbent dry weight (g).

The percentage of diclofenac elimination in each case was calculated by Equation (2):

$$\%\ Removal = \frac{(C_o - C_e)}{C_o} \times 100, \quad (2)$$

where $C_o$ is the initial concentration ($mg{\cdot}L^{-1}$) and $C_e$ is the equilibrium concentration ($mg{\cdot}L^{-1}$) of diclofenac.

### 2.4. Diclofenac Determination

Diclofenac concentration in the water samples was determined by high-performance liquid chromatography (HPLC). A Waters modular liquid chromatography system (Waters, Milford, MA, USA) equipped with two M510 pumps, a 717 plus autosampler, and an M996 photodiode array detector (PDA) was used. HPLC was run by the data system of the Millenium 2010 Chromatography Manager.

An Atlantis C18 5 μm, 25 cm × 4.6 mm i.d. (Waters) column was used. Elution was performed using a gradient between 175 mM phosphoric acid in water and 175 mM acetic acid in acetonitrile as the mobile phase. Flow rate was 0.6 mL/min. Samples, before being analyzed, were filtered using a 0.45 μm pore size cellulose acetate membrane, and the filtrate was analyzed for HPLC. Diclofenac was monitored at 270 nm. The retention time of diclofenac was about 6.4 min.

### *2.5. Advanced Oxidation Treatments*

#### 2.5.1. Sodium Hypochlorite

Treatment with sodium hypochlorite was performed with 1000 mL of filtered diclofenac solution. Four different amounts of available chlorine, namely, 1.0, 2.0, 3.0, and 4.0 mL of a 5% (*w*/*v*) sodium hypochlorite solution (Panreac, Barcelona, Spain) were added to each sample. Experiments were carried out in sealed glass bottles. They were incubated for 2 h until the moment of their quantification in a dark isothermal chamber at 20 °C, with slow agitation. Lastly, samples were filtered and analyzed by HPLC.

#### 2.5.2. UV Light Treatments

UV radiation was applied using an 8 watt low-pressure UV lamp (UV-C 8 watt Strahler–Messner, Kalletal, Germany), with a flow range of 0.5 to 1.1 $L \cdot h^{-1}$ and a wavelength of 254 nm. The facility was designed to be used with 500 mL of a sample, irradiated for short periods of time of 2, 5, 10, and 15 min. According to these irradiation durations, the provided doses were 15, 37, 73, and 110 $mJ \cdot cm^{-2}$, respectively. Before each treatment, the lamp was cleaned with deionized water. After irradiation, samples were filtered with a 0.45 μm filter and analyzed by HPLC.

#### 2.5.3. UV Radiation Combined with Sodium Hypochlorite (UV/HOCl) and Hydrogen Peroxide (UV/$H_2O_2$)

UV/HOCl and UV/$H_2O_2$ assays were conducted as batch experiments using a concentration of 5 $mg \cdot L^{-1}$ sodium hypochlorite and 5 $mg \cdot L^{-1}$ of hydrogen peroxide, respectively, which were added into the initial sample and subjected to different UV exposure times (2, 5, 10, and 15 min) before filtration and analysis of diclofenac concentration was performed by HPLC.

### *2.6. Chemical Oxygen Demand*

Chemical oxygen demand (COD) was determined using the methods described in APHA [45]. For COD measurements, a Spectroquant Nova 30 spectrophotometer (Merck, Darmstadt, Germany) was used.

## 3. Results and Discussion

### *3.1. Characterization of Biomass Samples*

The surface characteristics of the different biomass samples used in diclofenac bioadsorption studies were investigated using scanning electron microscopy (SEM) and energy-dispersive X-ray spectroscopy (EDS), as shown in Figures 1–3. SEM is considered to be one of the most widely used tools in the scientific field for studying the surface characteristics of biomass used as biosorbent [46]. In the different micrographs that were made in the SEM, important differences in the surface characteristics of the different studied agrifood waste types were observed, presenting a majority composition of carbon and oxygen in all cases.

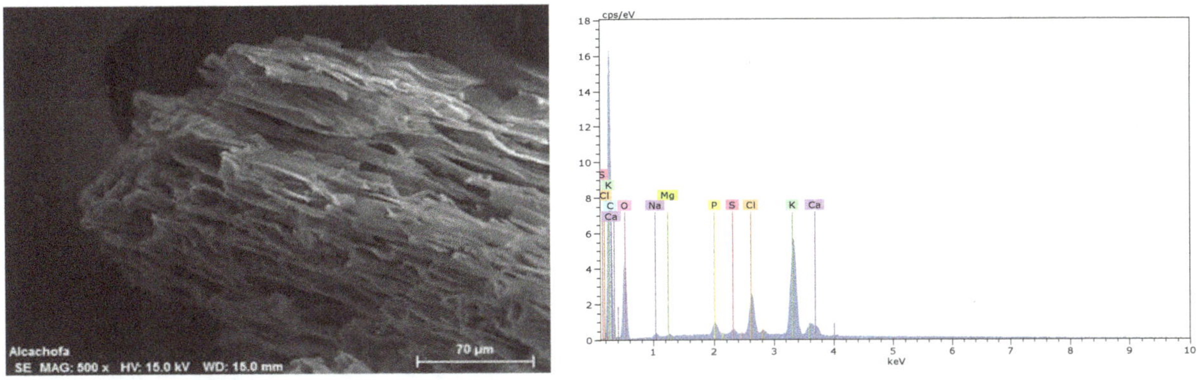


**Figure 1.** SEM image and EDS analysis of artichoke-agrowaste biomass.

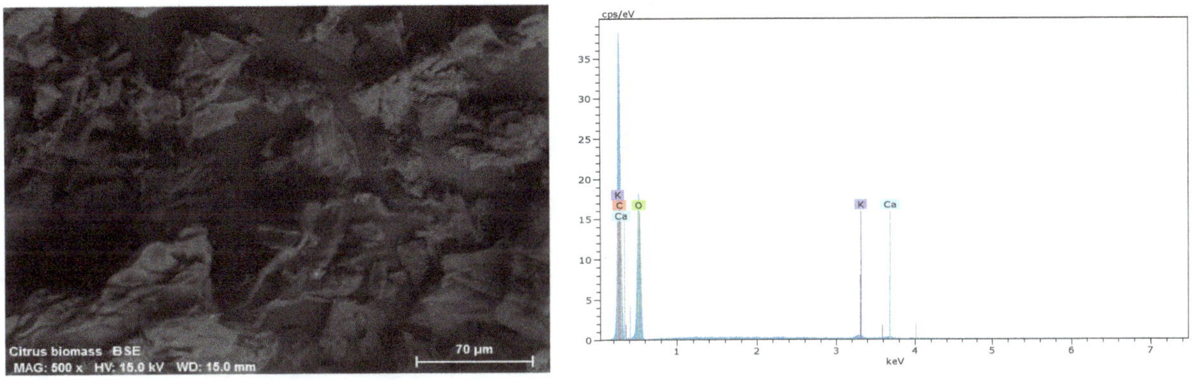


**Figure 2.** SEM image and EDS analysis of citrus-waste biomass.

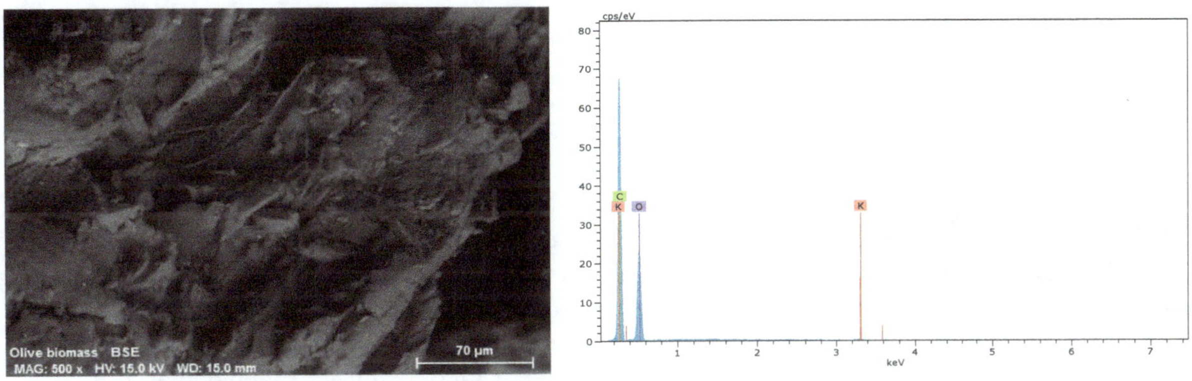


**Figure 3.** SEM image and EDS analysis of olive-mill residue biomass.

Figure 1 shows that artichoke biomass had filamentous structures made up of fiber cells that were longitudinally aligned. This fibrous structure provides consistency, resistance, and a large surface area [47] that, together with the small channels, allows for the rapid movement of fluids through the fibers, increasing the possibility of retaining diclofenac molecules. SEM analysis also showed that it was a rough surface, with the presence of holes of irregular shape and size [47] that could improve the retention of diclofenac.

Citrus-waste biomass (Figure 2) presented an irregular surface, but with a much more compact appearance and with fewer cavities than in the previous case, although with cracks that provided a large surface for the retention of diclofenac. Olive-mill residues (Figure 3) also presented an irregular surface with a fairly compact appearance, bit with few cavities. In all cases, the presence of cavities of irregular shape and dimensions would enhance the fixation of contaminants to be retained.

FTIR analysis allowed for inferring the structure of the sorbents on the basis of the locations and shapes of the bands in the spectra (Figure 4). The three sorbents showed a broad band at 3600–3100 $cm^{-1}$

that was assigned to –OH stretching vibrations of hydroxyl and carboxyl groups. A –CH aliphatic stretching vibration band was also observed at 2900–2850 $cm^{-1}$ in all sorbents. Around 1630 $cm^{-1}$, a band attributed to C=O stretching vibrations and typical skeletal vibrations of aromatic rings was presented. In the region of 1030–1000 $cm^{-1}$, a broad band, an indication of C–O and C=C vibrational stretching, was revealed. These are, in fact, typical FTIR spectra, corresponding to nonmodified plant biomass samples, where the lignocellulosic constituents are particularly outstanding [32,47–49].

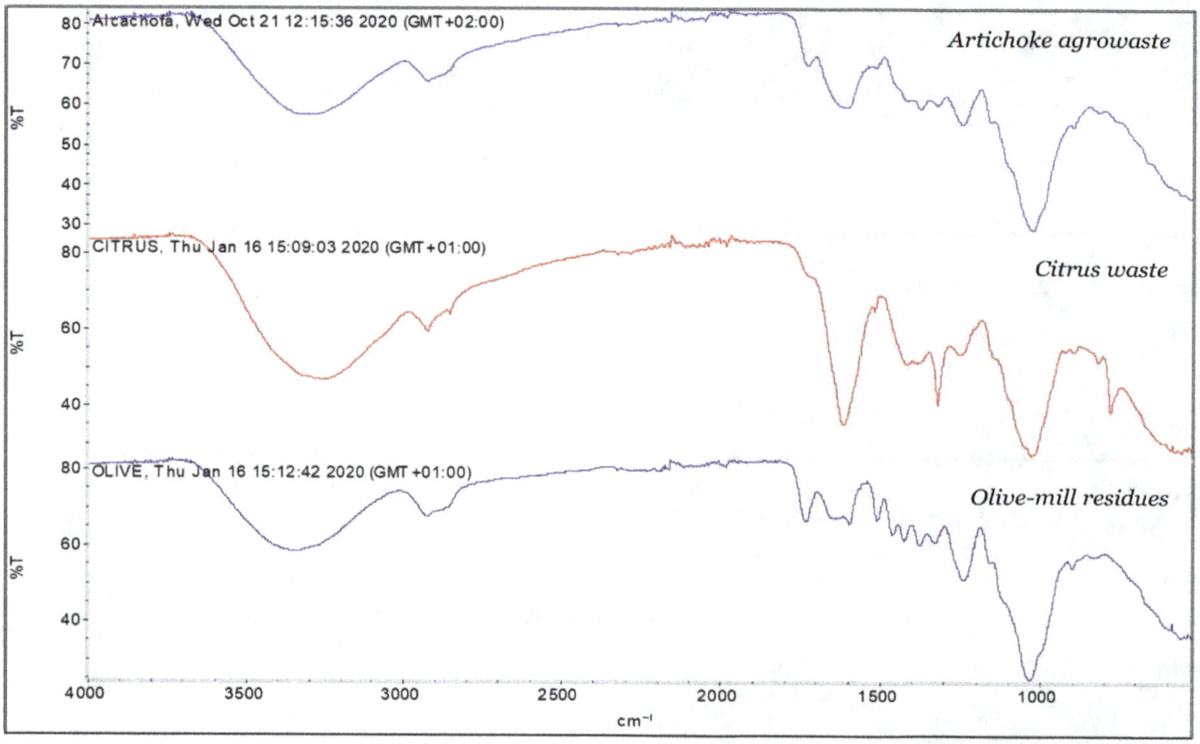


**Figure 4.** FTIR spectra between 4000 and 400 $cm^{-1}$ for sorbent samples.

### *3.2. Diclofenac Biosorption*

Table 1 shows the characterization of the working water prepared from diclofenac stock solution and filtered effluent from a treatment plant. The initial concentration ($C_o$) of diclofenac for the different experiments was 37.6 $mg{\cdot}L^{-1}$ and an initial COD value of 282 mg $O_2{\cdot}L^{-1}$.

**Table 1.** Initial composition of working solution. COD, chemical oxygen demand.

| Parameter | Value |
|---|---|
| pH | 6.6 ± 0.2 * |
| Conductivity ($\mu S{\cdot}cm^{-1}$) | 12.5 ± 2.5 |
| TSS (mg/L) | 98.2 ± 8.6 |
| Turbidity (NTU) | 4.7 ± 0.3 |
| $NO_3^-$ ($mg{\cdot}L^{-1}$) | 16.8 ± 2.7 |
| $PO_4^{3-}$ ($mg{\cdot}L^{-1}$) | 5.6 ± 1.1 |
| BOD (mg $O_2{\cdot}L^{-1}$) | 113.3 ± 10.2 |
| COD (mg $O_2{\cdot}L^{-1}$) | 282.0 ± 15.0 |
| Diclofenac initial ($mg{\cdot}L^{-1}$) | 37.6 |

* mean ± SD of 3 determinations.

The sorption capacities and percentages of diclofenac removal in the experiments carried out for the different used sorbents and doses (Table 2) revealed that citrus-waste biomass was the most effective. Citrus residues, with a zero point of charge of 4.2, reduced the pH of the solution during the sorption process to 5.7 by the end of the experiment. These pH reductions were less pronounced in the olive-mill-residue ($pH_{ZPC}$ = 5.4) and in the artichoke-agrowaste ($pH_{ZPC}$ = 4.8) sorption tests, of which the final pH values were 6.4 and 6.1, respectively. In all cases, the modifications in pH were very slight, hardly affecting the solubility of the adsorbate and sorption-capacity values. The biomass of artichoke agrowaste showed the least capacity to retain diclofenac. In the case of citrus waste, for a dose of 2 g/L, all the diclofenac is retained, obtaining a sorption capacity ($q_e$) of 9 mg·g$^{-1}$. Olive-mill residues had intermediate and very similar retention values regardless of the dose used in the experiments.

**Table 2.** Elimination of diclofenac for different biomass samples.

| Biomass | Sorbent Dosage (g·L$^{-1}$) | Removal % | $q_e$ (mg·g$^{-1}$) |
|---|---|---|---|
| Artichoke agrowaste | 1 | 18.7 ± 2.4 * | 3.9 ± 0.5 |
| | 2 | 19.9 ± 3.1 | 4.2 ± 0.4 |
| Olive-mill residues | 1 | 46.4 ± 4.4 | 4.2 ± 0.6 |
| | 2 | 46.3 ± 5.8 | 4.2 ± 0.8 |
| Citrus-waste | 1 | 87.1 ± 5.7 | 7.8 ± 1.1 |
| | 2 | 99.7 ± 0.5 | 9.0 ± 0.3 |

* mean ± SD of 3 determinations.

The results of this study revealed that the reuse of raw lignocellulosic biomass from agrifood waste for the sorption of diclofenac may be an efficient and common method of wastewater treatment. The sorption capacities of diclofenac obtained in this study were lower than those reported for microalgae biomass (20–28 mg·g$^{-1}$) [29] and grape bagasse (24 mg·g$^{-1}$) [49], but comparable for that shown with activated carbon from olive stones (11 mg·g$^{-1}$) [50]. In the present investigation, the agrifood waste used as sorbent was not previously modified or subjected to any thermal treatment. The obtained results in the present investigation reveal that this is a new line of research that is worth further exploring, and it may be of interest to test some cost-effective activation mechanism that enhances the bioadsorptive capacity of waste to specific contaminants. In this context, future studies will be made in application in real systems, where fixed-bed citrus-waste adsorbers can be implemented for the treatment of wastewater with diclofenac.

### 3.3. *Advanced Oxidation Treatments*

As summarized in Table 3, when the working solution ($C_o$ = 37.6 mg·L$^{-1}$) was exposed to different exposure times of UV radiation ranging from 2 to 15 min, significant reductions in the initial diclofenac concentration were achieved, ranging from 9.78% removal for 2 min of exposure and 88.49% removal for 15 min of exposure, without being able to achieve total removal. A significant reduction in COD values was also achieved, in a similar way to that in other reported investigations [51].

**Table 3.** Treatment with ultraviolet (UV) radiation.

| Contact Time (min) | Diclofenac (mg·L$^{-1}$) | Removed Diclofenac % | COD Reduction % |
|---|---|---|---|
| 2 | 33.9 ± 2.8 * | 9.8 ± 2.4 | 44.3 ± 3.6 |
| 5 | 16.6 ± 2.1 | 55.9 ± 3.9 | 62.1 ± 3.5 |
| 10 | 6.1 ± 1.1 | 83.8 ± 4.6 | 62.4 ± 4.1 |
| 15 | 4.3 ± 0.9 | 88.5 ± 3.8 | 66.3 ± 3.9 |

* mean ± SD of 3 determinations.

In the treatment with sodium hypochlorite and a contact time of 2 h (Table 4), for a dose of 4 mg·L$^{-1}$, all diclofenac present in the starting sample was eliminated. As the dose of hypochlorite was increased, greater elimination was achieved. The reduction in COD was also quite significant.

**Table 4.** Treatment with sodium hypochlorite (contact time 2 h).

| Hypochlorite Concentration (mg·L$^{-1}$) | Diclofenac (mg·L$^{-1}$) | Removed Diclofenac % | COD Reduction % |
|---|---|---|---|
| 1 | 14.4 ± 3.3 * | 61.7 ± 4.7 | 24.1 ± 2.3 |
| 2 | 4.4 ± 1.2 | 88.4 ± 5.1 | 38.3 ± 2.8 |
| 3 | 3.3 ± 0.9 | 91.3 ± 4.4 | 44.0 ± 2.6 |
| 4 | n.d. † | 100 | 44.7 ± 1.8 |

* mean ± SD of 3 determinations; † not detected.

In the treatment that combined UV radiation with the addition of sodium hypochlorite (Table 5), practically all the diclofenac (93.59%) was eliminated for a radiation time of 10 min. When exposure time was increased (15 min), it was possible to eliminate it completely. In addition, the unreacted chlorine in this process provided residual protection for water reuse purposes [52]. Comparable reductions in COD were achieved as those with previous treatments. For the treatment in which UV radiation was combined with hydrogen peroxide (Table 6), the initial diclofenac was eliminated for all treatments. For an exposure time of 2 min of UV radiation, 99.20% of the initial diclofenac was degraded. The achieved elimination of COD was quite comparable in all performed treatments.

**Table 5.** Ultraviolet radiation treatment combined with sodium hypochlorite (5 mg·L$^{-1}$).

| Contact Time (min) | Diclofenac (mg·L$^{-1}$) | Removed Diclofenac % | COD Reduction % |
|---|---|---|---|
| 2 | 29.1 ± 2.2 * | 22.6 ± 2.8 | 14.2 ± 1.9 |
| 5 | 13.5 ± 1.8 | 64.2 ± 3.1 | 22.0 ± 2.4 |
| 10 | 2.4 ± 0.5 | 93.6 ± 4.8 | 45.0 ± 3.8 |
| 15 | n.d. † | 100 | 47.9 ± 3.4 |

* mean ± SD of 3 determinations; † not detected.

**Table 6.** UV radiation treatment combined with $H_2O_2$ (5 mg·L$^{-1}$).

| Contact Time (min) | Diclofenac (mg·L$^{-1}$) | Removed Diclofenac % | COD Reduction % |
|---|---|---|---|
| 2 | 0.3 ± 0.1 * | 99.2 ± 4.4 | 23.4 ± 1.7 |
| 5 | n.d. † | 100 | 25.9 ± 2.3 |
| 10 | n.d. | 100 | 36.5 ± 2.0 |
| 15 | n.d. | 100 | 40.1 ± 3.1 |

* mean ± SD of 3 determinations; † not detected.

It is scientifically accepted that advanced oxidation processes are highly effective novel methods that accelerate the oxidation and the degradation of a wide range of organic and inorganic pollutants that are resistant to conventional treatment methods [53]. In our experiments, diclofenac was 88% by UV-C degraded irradiation; when the wastewater was treated for a time as short as 15 min, this result verified that this compound absorbs UV irradiation well at 254 nm and can be easily photolyzed [54]. When UV-C irradiation treatment was combined with 5 mg·L$^{-1}$ of sodium hypochlorite, the degradation of diclofenac was increased to 100%. Nevertheless, UV/$H_2O_2$ was the most efficient oxidation process, combining the immediate UV effect and the action of the HO$^{\cdot}$ radicals produced from the homolytic disruption of $H_2O_2$. The levels of diclofenac degradation under the adopted experimental conditions were higher than the values reported in the literature for oxidation processes, such as ozonation (32%) and UV/$H_2O_2$ (39%), after 39 min treatment [55]. The use of hypochlorite to degrade diclofenac in aqueous effluents was also reported with successful results [56]. Recent research in a Fenton-like system with FeCeOx–$H_2O_2$ achieved 84% degradation of diclofenac within 40 min [57]. All these results are in accordance with those obtained in our experimental design and confirm the feasibility of the tested methods. A significant reduction in chemical oxygen demand (COD) (40–48%) was also achieved with these oxidation treatments, although there was no direct relationship between the percentage of diclofenac degradation and the decrease in COD.

## 4. Conclusions

This paper offers a comparative study between the application of biosorption and advanced oxidation techniques to remove diclofenac from wastewater at high concentrations.

Artichoke agrowaste, olive-mill residues, and citrus waste were used as low-cost biosorbents to remove diclofenac from aqueous effluents. Citrus-waste biomass, at a sorbent dosage of 2 $g \cdot L^{-1}$, showed the highest sorption capacity (9 $mg \cdot g^{-1}$). This value was achieved with raw biomass without being subjected to any chemical or thermal activation process. The obtained results highlight the sustainable reutilization of agrifood waste for producing cost-effective sorbents, providing additional income for the agroindustrial sector. Waste valorization can efficiently help in reducing environmental stress by decreasing unwarranted pollution and promoting economy circularization.

UV-driven advanced oxidation treatments were effective in the degradation of diclofenac in wastewater. The UV/$H_2O_2$ (5 $mg \cdot L^{-1}$) oxidation process was faster than UV/HOCl (5 $mg \cdot L^{-1}$) to reach the same degradation rate. Chemical oxygen demand (COD) was considerably reduced (40–48%) with these oxidation treatments.

The obtained results in this study confirm biosorption as a low-cost and environmentally sustainable technique, and the high efficiency demonstrated by advanced oxidation techniques, may lead to applying specifically designed techniques that combine biosorption and advanced oxidation to eliminate specific emerging contaminants in aqueous effluents.

**Author Contributions:** Conceptualization, J.M.A. and J.A.F.-L.; methodology, J.M.A. and M.J.R.; formal analysis, J.M.A. and J.A.F.-L.; investigation, M.J.R.; writing—original-draft preparation, J.M.A.; writing—review and editing, J.A.F.-L. All authors have read and agreed to the published version of the manuscript.

**Acknowledgments:** This research is part of the Environmental Engineering R&D group. The excellent technical assistance of the Technical Research Support Service of the UPCT is greatly acknowledged.

## References

1. Dalecka, B.; Oskarsson, C.; Juhna, T.; Rajarao, G.K. Isolation of fungal strains from municipal wastewater for the removal of pharmaceutical substances. *Water* **2020**, *12*, 524. [CrossRef]
2. Geisen, V.; Mol, H.; Klumpp, E.; Umlauf, G.; Nadal, M.; Van der Ploeg, M.; van de Zee, S.E.A.T.M.; Ritsema, C.J. Emerging pollutants in the environment: A challenge for water resource management. *Int. Soil Water Conserv. Res.* **2015**, *3*, 57–65. [CrossRef]
3. Sauvé, S.; Desrosiers, M. A review of what is an emerging contaminant. *Chem. Cent. J.* **2014**, *8*, 15. [CrossRef] [PubMed]
4. López-Doval, J.C.; Montagner, C.C.; de Alburquerque, A.F.; Moschini-Carlos, V.; Umbuzeiro, G.; Pompeo, M. Nutrients, emerging pollutants and pesticides in a tropical urban reservoir: Spatial distributions and risk assessment. *Sci. Total Environ.* **2017**, *575*, 1307–1324. [CrossRef] [PubMed]
5. Margenat, A.; Matamoros, V.; Díez, S.; Cañameras, N.; Comas, J.; Bayona, J.M. Ocurrence of chemical contaminants in peri-urban agricultural irrigation waters and assesment of their phytotoxicity and crop productivity. *Sci. Total Environ.* **2017**, *599*, 1140–1148. [CrossRef] [PubMed]
6. Tran, N.; Drogui, P.; Brar, S.K. Sonochemical techniques to degrade pharmaceutical organic pollutants. *Environ. Chem. Lett.* **2015**, *13*, 251–268. [CrossRef]

7. Carmalin, A.; Lima, E. Removal of emerging contaminants from the environment by adsorption. *Ecotoxicol. Environ.* **2018**, *150*, 1–17.
8. Munthe, J.; Brorstrom-Lunden, E.; Rahmberg, M.; Posthuma, L.; Alterburger, R.; Brack, W.; Bunke, D.; Engelen, G.; Gawlik, B.; Gils, J.; et al. An expanded conceptual framework for solution-focused management of chemical pollution in European waters. *Environ. Sci. Eur.* **2017**, *29*, 1072. [CrossRef]
9. Richardson, S.D.; Kimura, S.Y. Emerging environmental contaminants: Challenges facing our next generation and potential engineering solutions. *Environ. Technol. Innov.* **2017**, *8*, 40–56. [CrossRef]
10. Rodriguez-Narvaez, O.M.; Peralta-Hernandez, J.M.; Goonetilleke, A.; Bandala, E.R. Treatment technologies for emerging contaminants in water: A review. *Chem. Eng. J.* **2017**, *323*, 361–380. [CrossRef]
11. Petrovic, M.; Solé, M.; López, M.; Barceló, D. Endocrine disruptors in sewage treatment plants, receiving river waters, and sediments: Integration of chemical analysis and biological effects of feral carp. *Environ. Toxic. Chem.* **2002**, *21*, 2146–2156. [CrossRef]
12. Esplugas, S.; Bila, D.M.; Krause, L.G.T.; Dezotti, M. Ozonation and advanced oxidation technologies to remove endocrine disrupting chemicals (EDCs) and pharmaceuticals and personal care products (PPCPs) in water effluents. *J. Hazard. Mater.* **2007**, *149*, 631–642. [CrossRef] [PubMed]
13. Richardson, S.D. Environmental mass spectrometry: Emerging contaminants and current issues. *Anal. Chem.* **2012**, *84*, 747–778. [CrossRef] [PubMed]
14. Richardson, S.D.; Kimura, S.Y. Water analysis: Emerging contaminants and current issues. *Anal. Chem.* **2016**, *88*, 546–582. [CrossRef]
15. Blanco, M.; Pérez-Albadalejo, E.; Piña, B.; Kuspilic, G.; Milun, V.; Lille-Langoy, R. Assessing the environmental quality of sediments from Split coastal area (Croatia) with a battery of cell-based bioassays. *Sci. Total Environ.* **2018**, *624*, 1640–1648.
16. Boehler, S.; Strecker, R.; Heinrich, P.; Prochazka, E.; Northcott, G.L.; Ataria, J.M. Assessment of urban stream sediment pollutants entering estuaries using chemical analysis and multiple bioassays to characterize biological activities. *Sci. Total Environ.* **2017**, *593*, 498–507. [CrossRef]
17. Abbas, M.; Adil, M.; Ehtisham-ul-Haque, S.; Munir, B.; Yameen, M.; Ghaffar, A.; Abbas Shar, G.; Asif Tahir, M.; Iqbal, M. *Vibrio fischeri* bioluminescence inhibition assay for ecotoxicity assessment: A review. *Sci. Total Environ.* **2018**, *626*, 1295–1309.
18. Schereiber, B.; Fischer, J.; Schiwy, S.; Hollert, H.; Schulz, R. Towards more ecological relevance in sediment toxicity testing with fish: Evaluation of multiple bioassays with embryos of the benthic weatherfish (*Misgurnus fossilis*). *Sci. Total Environ.* **2018**, *619*, 391–400. [CrossRef]
19. Sang, Z.; Jiang, Y.; Tsoi, Y.-K.; Leung, K.S. Evaluating the environmental impact of artificial sweeteners: A study of their distributions, photodegradation ant toxicities. *Water Res.* **2014**, *52*, 260–274. [CrossRef]
20. Funke, J.; Prasse, C.; Eversloh, C.L.; Ternes, T.A. Oxypurinol-A novel marker for wastewater contamination of the aquatic environment. *Water Res.* **2015**, *74*, 257–265.
21. Lonappan, L.; Brar, S.K.; Das, R.K.; Verma, M.; Rao, Y. Diclofenac and its transformation products: Environmental occurrence and toxicity. *Environ. Int.* **2016**, *96*, 127–138. [CrossRef] [PubMed]
22. Grandclément, C.; Piram, A.; Petit, M.E.; Seyssiecq, I.; Laffont-Schwob, I.; Vanot, G.; Tiliacos, N.; Roche, N.; Doumenq, P. Biological removal and fate assessment of diclofenac using *Bacillus subtilis* and *Brevibacillus laterosporus* strains and ecotoxicological effects of diclofenac and 4′-hydroxy-diclofenac. *J. Chem.* **2020**, *2020*, 9789420. [CrossRef]
23. Vieno, N.; Silanpää, M. Fate of diclofenac in municipal wastewater treatment plants—A review. *Environ. Int.* **2014**, *69*, 28–39. [CrossRef] [PubMed]
24. Langford, K.H.; Reid, M.; Thomas, K.V. Multi-residue screening of prioritised human pharmaceuticals, illicit drugs and bactericides in sediments and sludge. *J. Environ. Monit.* **2011**, *13*, 2284–2291. [CrossRef] [PubMed]
25. Kunkel, U.; Radke, M. Fate of pharmaceuticals in rivers: Deriving a benchmark dataset at favorable attenuation conditions. *Water Res.* **2012**, *46*, 5551–5565. [CrossRef]
26. Gil, A.; Santamaría, L.; Korili, S.A. Removal of caffeine and diclofenac from aqueous solution by adsorption on multiwalled carbon nanotubes. *Coll. Interf. Sci. Commun.* **2018**, *22*, 25–28. [CrossRef]
27. Mabrouki, H.; Akretche, D.E. Diclofenac potassium removal from water by adsorption on natural and pillared clay. *Des. Water Treatm.* **2016**, *57*, 6033–6043. [CrossRef]

28. Thi Minh Tam, N.; Liu, Y.; Bashir, H.; Yin, Z.; He, Y.; Zhou, X. Efficient removal of diclofenac from aqueous solution by potassium ferrate-activated porous graphitic biochar: Ambient condition influences and adsorption mechanism. *Int. J. Environ. Res. Public Health* **2020**, *17*, 291. [CrossRef]
29. Coimbra, R.N.; Escapa, C.; Vázquez, N.C.; Noriega-Hevia, G.; Otero, M. Utilization of non-living microalgae biomass from two different strains for the adsorptive removal of diclofenac from water. *Water* **2018**, *10*, 1401. [CrossRef]
30. Li, Y.; Taggart, M.A.; McKenzie, C.; Zhang, Z.; Lu, Y.; Pap, S.; Gibb, S. Utilizing low-cost natural waste for the removal of pharmaceuticals from water: Mechanisms, isotherms and kinetics at low concentrations. *J. Clean. Prod.* **2019**, *227*, 88–97. [CrossRef]
31. Yusuf, M.; Elfghi, F.M.; Zaidi, S.A.; Abdullah, E.C.; Khan, M.A. Applications of graphene and its derivatives as an adsorbent for heavy metal and dye removal: A systematic and comprehensive overview. *RSC Adv.* **2015**, *5*, 50392–50420. [CrossRef]
32. Rosique, M.; Angosto, J.M.; Guibal, E.; Roca, M.J.; Fernández-López, J.A. Factorial design methodological approach for enhanced cadmium ions bioremoval by *Opuntia* biomass. *CLEAN Soil Air Water* **2016**, *44*, 959–966. [CrossRef]
33. Ben-Othman, S.; Jõudu, I.; Bhat, R. Bioactives from agri-food wastes: Present insights and future challenges. *Molecules* **2020**, *25*, 510. [CrossRef] [PubMed]
34. Nowicki, P.; Kazmierczak-Razna, J.; Pietrzak, R. Physico-chemical and adsorption properties of carbonaceous sorbents prepared by activation of tropical fruit skins with potassium carbonate. *Mater. Des.* **2016**, *90*, 579–585. [CrossRef]
35. Castrica, M.; Rebucci, R.; Giromini, C.; Tretola, M.; Cattaneo, D.; Baldi, A. Total phenolic content and antioxidant capacity of agri-food waste and by-products. *Ital. J. Anim. Sci.* **2019**, *18*, 336–341. [CrossRef]
36. Manna, S.; Toy, D.; Adhikari, B.; Thomas, S.; Das, P. Biomass for water defluoridation and current understanding on biosorption mechanisms: A review. *Environ. Prog. Sustain. Energy* **2018**, *37*, 1560–1572. [CrossRef]
37. Ummartyotin, S.; Pechyen, C. Strategies for development and implementation of bio-based materials as effective renewable resources of energy: A comprehensive review on adsorbent technology. *Renew. Sustain. Energy Rev.* **2016**, *62*, 654–664. [CrossRef]
38. Arief, V.O.; Trilestari, K.; Sunarso, J.; Indraswati, N.; Ismadji, S. Recent progress on biosorption of heavy metals from liquids using low cost biosorbents: Characterization, biosorption parameters and mechanism studies. *CLEAN Soil Air Water* **2008**, *36*, 937–962. [CrossRef]
39. Ternes, T.; Stüber, J.; Herrmann, N.; McDowell, D.; Ried, A.; Kampmann, M.; Teiser, B. Ozonation: A tool for removal of pharmaceuticals, contrast media and musk fragrances from wastewater? *Water Res.* **2003**, *37*, 1976–1982. [CrossRef]
40. Cerreta, G.; Roccamante, M.A.; Oller, I.; Malato, S.; Rizzo, L. Contaminants of emerging concern removal from real wastewater by UV/free chlorine process: A comparison with solar/free chlorine and UV/$H_2O_2$ at pilot scale. *Chemosphere* **2019**, *236*, 124354.
41. Lee, Y.; von Gunten, U. Oxidative transformation of micropollutants during municipal wastewaters treatment: Comparison of kinetic aspects of selective (chlorine, chlorine dioxide, ferrate VI and ozone) and non-selective oxidants (hydroxyl radicals). *Water Res.* **2010**, *39*, 555–566. [CrossRef] [PubMed]
42. Kanakaraju, D.; Glass, B.D.; Oelgemöller, M. Titanium dioxide photocatalysis for pharmaceutical wastewater treatment. *Environ. Chem. Lett.* **2014**, *12*, 27–47. [CrossRef]
43. Jiang, J.; Pang, S.Y.; Ma, J.; Liu, H. Oxidation of phenolic endocrine disrupting chemicals by potassium permanganate in synthetic and real waters. *Environ. Sci. Technol.* **2012**, *46*, 1774–1781. [CrossRef]
44. Fiol, N.; Villaescusa, I. Determination of sorbent point zero charge: Usefulness in sorption studies. *Environ. Chem. Lett.* **2009**, *7*, 79–84. [CrossRef]
45. Eaton, A.D.; American Public Health Association; American Water Works Association; Water Environment Federation. *Standard Methods for the Examination of Water and Wastewater*, 21st ed.; APHA-AWWA-WEF: Washington, DC, USA, 2005.
46. Amiri, H.; Karimi, K. Improvement of acetone, butanol, and ethanol production from woody biomass using organosolv pretreatment. *Bioproc. Biosyst. Eng.* **2015**, *38*, 1959–1972. [CrossRef] [PubMed]

47. Saavedra, M.I.; Doval Miñarro, M.; Angosto, J.M.; Fernández-López, J.A. Reuse potential of residues of artichoke (*Cynara scolymus* L.) from industrial canning processing as sorbent of heavy metals in multimetallic effluents. *Ind. Crops Prod.* **2019**, *141*, 111751. [CrossRef]
48. Mahato, N.; Sharma, K.; Sinha, M.; Baral, E.R.; Koteswararao, R.; Dhyani, A.; Cho, M.H.; Cho, S. Bio-sorbents, industrially important chemicals and novel materials from citrus processing waste as a sustainable and renewable bioresource: A review. *J. Adv. Res.* **2020**, *23*, 61–82. [CrossRef] [PubMed]
49. Antunes, M.; Esteves, V.I.; Guégan, R.; Crespo, J.S.; Fernandes, A.N.; Giovanela, M. Removal of diclofenac sodium from aqueous solution by Isabel grape bagasse. *Chem. Eng. J.* **2012**, *192*, 114–121. [CrossRef]
50. Larous, S.; Meniai, A.-H. Adsorption of diclofenac from aqueous solution using activated carbon prepared from olive stones. *Int. J. Hydrog. Energy* **2016**, *41*, 10380–10390. [CrossRef]
51. Plauta, M.; Tisler, T.; Toman, M.J.; Pintar, A. Efficiency of advanced oxidation processes in lowering bisphenol A toxicity and oestrogenic activity in aqueous samples. *Arch. Ind. Hyg. Toxicol.* **2014**, *65*, 77–87.
52. Zhao, Q.; Shang, C.; Zhang, X.; Ding, G.; Yang, X. Formation of halogenated organic byproducts during medium-pressure UV and chlorine coexposure of model compounds, NOM and bromide. *Water Res.* **2011**, *45*, 6545–6554. [CrossRef] [PubMed]
53. Stasinakis, A. Use of selected advanced oxidation processes (AOPs) for wastewater treatment—A mini review. *Glob. NEST J.* **2008**, *10*, 376–385.
54. De la Cruz, N.; Giménez, J.; Esplugas, S.; Grandjean, D.; de Alencastro, L.F.; Pulgarín, C. Degradation of 32 emergent contaminants by UV and neutral photo-fenton in domestic wastewater effluent previously treated by activated sludge. *Water Res.* **2012**, *46*, 1947–1957. [CrossRef] [PubMed]
55. Vogna, D.; Marotta, R.; Napolitano, A.; Andreozzi, R.; d'Ischia, M. Advanced oxidation of the pharmaceutical drug diclofenac with UV/$H_2O_2$ and ozone. *Water Res.* **2004**, *38*, 414–422. [CrossRef] [PubMed]
56. Luongo, G.; Guida, M.; Siciliano, A.; Libralato, G.; Saviano, L.; Amoresano, A.; Previtera, L.; Di Fabio, G.; Zarrelli, A. Oxidation of diclofenac in water by sodium hypochlorite: Identification of new degradation by-products and their ecotoxicological evaluation. *J. Pharmac. Biomed. Anal.* **2020**, 113762. [CrossRef] [PubMed]
57. Chong, S.; Zhang, G.; Zhang, N.; Liu, Y.; Huang, T.; Chang, H. Diclofenac degradation in water by $FeCeO_x$ catalyzed $H_2O_2$: Influencing factors, mechanism and pathways. *J. Hazard. Mat.* **2017**, *334*, 150–159. [CrossRef]

# 4

# Energy Recovery from Wastewater: A Study on Heating and Cooling of a Multipurpose Building with Sewage-Reclaimed Heat Energy

**Daniele Cecconet [1,*], Jakub Raček [2], Arianna Callegari [1] and Petr Hlavínek [2]**

1 Department of Civil Engineering and Architecture, University of Pavia, 27100 Pavia, Italy; arianna.callegari@unipv.it
2 AdMaS Research Centre, Faculty of Civil Engineering, Brno University of Technology, 61200 Brno, Czech Republic; racek.j@fce.vutbr.cz (J.R.); hlavinek.p@fce.vutbr.cz (P.H.)
* Correspondence: daniele.cecconet@unipv.it

**Abstract:** To achieve technically-feasible and socially-desirable sustainable management of urban areas, new paradigms have been developed to enhance the sustainability of water and its resources in modern cities. Wastewater is no longer seen as a wasted resource, but rather, as a mining ground from which to obtain valuable chemicals and energy; for example, heat energy, which is often neglected, can be recovered from wastewater for different purposes. In this work, we analyze the design and application of energy recovery from wastewater for heating and cooling a building in Brno (Czech Republic) by means of heat exchangers and pumps. The temperature and the flow rate of the wastewater flowing in a sewer located in the proximity of the building were monitored for a one-year period, and the energy requirement for the building was calculated as 957 MWh per year. Two options were evaluated: heating and cooling using a conventional system (connected to the local grid), and heat recovery from wastewater using heat exchangers and coupled heat pumps. The analysis of the scenarios suggested that the solution based on heat recovery from wastewater was more feasible, showing a 59% decrease in energy consumption compared to the conventional solution (respectively, 259,151 kWh and 620,475 kWh per year). The impact of heat recovery from wastewater on the kinetics of the wastewater resource recovery facility was evaluated, showing a negligible impact in both summer (increase of 0.045 °C) and winter conditions (decrease of 0.056 °C).

**Keywords:** heat recovery; urban wastewater; temperature; energy recovery; thermal energy; urban water cycle

## 1. Introduction

The nexus between water and energy is becoming ever more crucial in the development of innovative solutions to achieve technically-feasible and socially-desirable sustainable management for the growth and resilience of urban areas [1]. Novel strategies are being developed to enhance the sustainability of the water cycle and the recycling of its resources in modern cities [2]. Wastewater is no longer seen as a wasted resource, but rather, as a mining ground from which to obtain valuable chemicals [3], nutrients [4,5], and energy [6]. Therefore, urban wastewater systems should be viewed, planned, and built according to a new set of rules and performance expectations that deviate from current standard practices. These rules should be based on entirely new paradigms that, while maintaining current environmental protection practices, should address the now fundamental requirements of robustness, sustainability, and resource recovery through flexible and site-specific solutions, including decentralization and a wider array of possible applicable technologies [3]. Sewage can be considered a source of different types of energy: electrical energy from bioelectrochemical wastewater treatment processes [7,8], low-head hydroelectric energy [9], biogas from anaerobic digestion [10], renewable fuels from residual sludge processing [11,12], and heat energy [13]. The latter is particularly appealing in the framework of water–energy sustainable urban development; in particular, the heat energy recovery potential from wastewater could possibly be even higher than its recoverable chemical energy potential [14,15], and it does not require biological treatment to be converted into a directly usable form. Waste heat from wastewater is reportedly able to generate electrical energy via a thermoelectric generator [16], but the implementation of this is difficult within an urban context. A more accessible technology is the energy recovery from wastewater using heat exchangers installed directly in the sewage collection system [17]. A heat exchanger is installed in direct contact with the wastewater that serves as a heat source or sink, and is later connected to a heat pump and then to the heating and cooling system of a building situated in close proximity. Temperatures of civil wastewater may be more than 25–27 °C in domestic outflows [18], representing a significant thermal energy source [16]; at the inlet of water treatment plants or water resource recovery facilities (WRRFs), the temperature is far lower (15–25 °C or lower, depending on the climate) [16,19,20], and thus, its use for heat recovery may be less practical. Also, heat recovery at this stage should be carefully considered, as it might interfere with treatment processes: biological processes, including denitrification, are sensitive to wastewater temperature, as biological reaction rates depend on Arrhenius' law, and excessive heat extraction from sewage could harm their efficiency [21]. Recently, it has been proposed that heat recovery from wastewater be used to assist in conventional cooling systems in coping with heat waves [22].

This paper presents a case study concerning heat energy recovery from wastewater for heating and cooling a building in Brno (Czech Republic) by means of heat pumps and exchangers. The advantages of the planned solution over conventional heating or cooling systems are described. To the best of the authors' knowledge, this is the first example of planned heat energy recovery from sewage in this European country.

## 2. Case Study Premises

### 2.1. Target Building

The building selected for the case study is located in the Brno city center, Czech Republic, and is situated in the proximity of the main collector of the municipal sewer. The edifice consists of a multifunctional, 13-storey building, including four underground floors, with an area of 111.5 m by 50 m. The underground, ground, and first floors are reserved for parking, the second and third floors are dedicated to a shopping gallery, while the rest of the building is occupied by administration offices, with a total rented area of 20,000 m$^2$.

The annual heating energy consumption of the building was calculated to be just over 500 MWh, while the annual cooling consumption was just under 457 MWh. Therefore, the overall annual energy demand of the building for heat conditioning is around approximately 957 MWh. Peak power requirements were determined as 800 and 1000 kW, respectively, for heating and cooling. An analysis of the concurrent heating and cooling power demand is summarized in Figure 1.

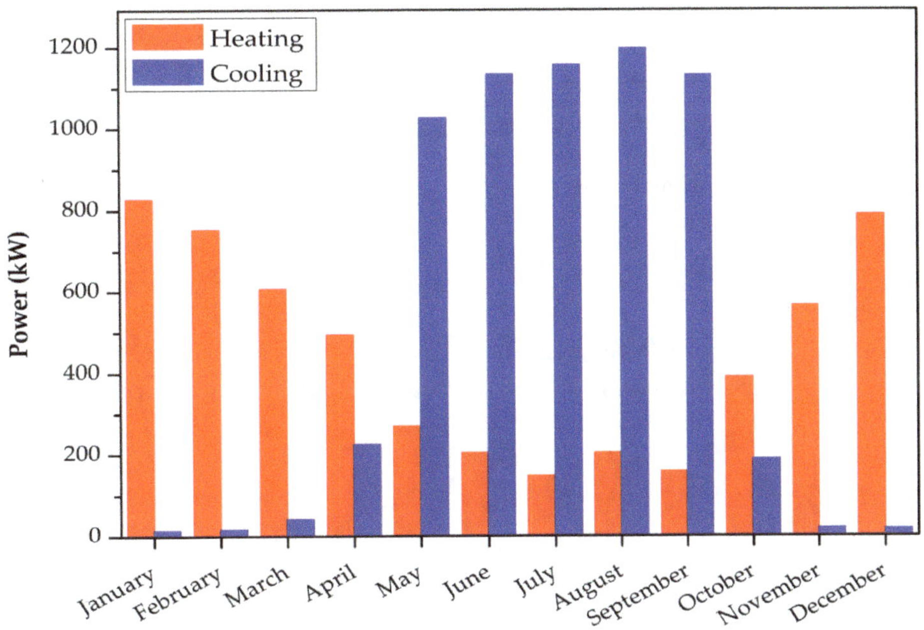


**Figure 1.** Monthly power demand for cooling and heating in the multipurpose building.

### 2.2. *The Sewer System*

The main sewer is located below the street in the proximity of the multipurpose building; the pipe size is 5.1 (width) by 3 (height) m, with a bottom gradient of 3.25‰. The wastewater flow rate and temperature were monitored for one year (from 16 June 2015 to 15 June 2016) before the beginning of the actual design process to preliminarily assess the feasibility and the extent of potential heat recovery from wastewater. The average dry weather flow was 291 L $s^{-1}$, with daily variability shown in Figure 2; high variations in hourly data may be ascribed to the extended time range of measurements, including winter and summer holidays. Several universities are located in Brno, and the city hosts around 70,000 students, with an additional 90,000 daily commuters. These facts are reflected in the aforementioned high hourly variability and low flow rate during university holidays (July and August). Monthly variability is shown in Table 1.

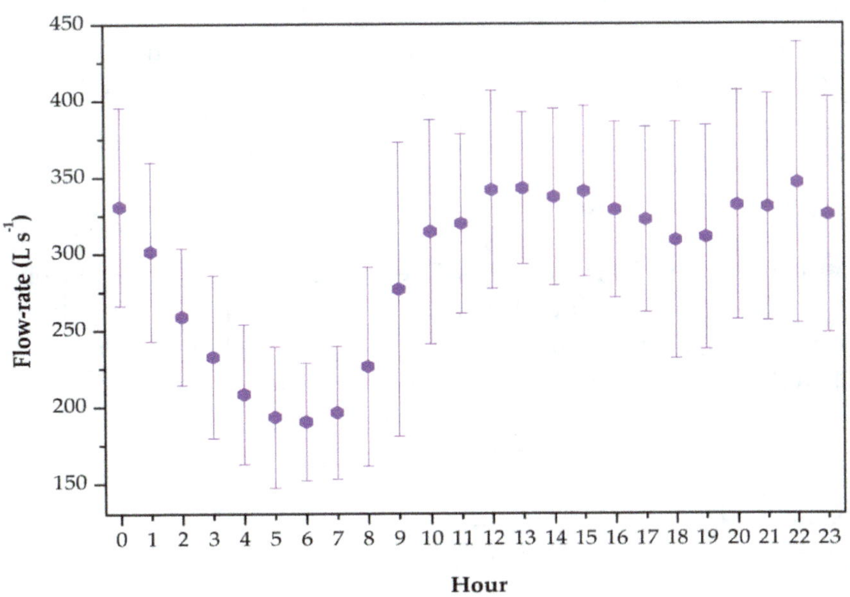


**Figure 2.** Mean hourly flow rate in the sewer during monitoring in the study year. Bars represent the standard deviations of the data during the observation period.

**Table 1.** Maximum, minimum, and average values for the sewer's flow rate during monitoring. SD: standard deviation.

| Q | Month | | | | | | | | | | | Year |
|---|---|---|---|---|---|---|---|---|---|---|---|---|
| (L s$^{-1}$) | 1 | 2 | 3 | 4 | 5 | 6 | 7 | 8 | 9 | 10 | 11 | 12 | - |
| Max | 427.9 | 454.8 | 338.0 | 389.7 | 344.7 | 425.6 | 299.8 | 313.3 | 389.7 | 470.6 | 405.4 | 427.9 | 470.6 |
| Min | 167.2 | 200.9 | 171.7 | 160.4 | 182.9 | 135.7 | 158.2 | 147.0 | 144.7 | 164.9 | 131.2 | 167.2 | 131.2 |
| Mean | 302.8 | 354.1 | 272.6 | 286.1 | 268.5 | 287.7 | 233.4 | 240.0 | 272.8 | 391.7 | 304.3 | 302.8 | 291.9 |
| SD | 72.5 | 76.8 | 57.3 | 66.1 | 52.3 | 84.1 | 40.1 | 53.2 | 67.4 | 95.1 | 91.6 | 72.5 | 83.4 |

The average temperature recorded during the monitoring period was 15.2 °C, with a minimum of 11.8 °C in February and a maximum of 18.7 °C in September, following the pattern shown in Figure 3.

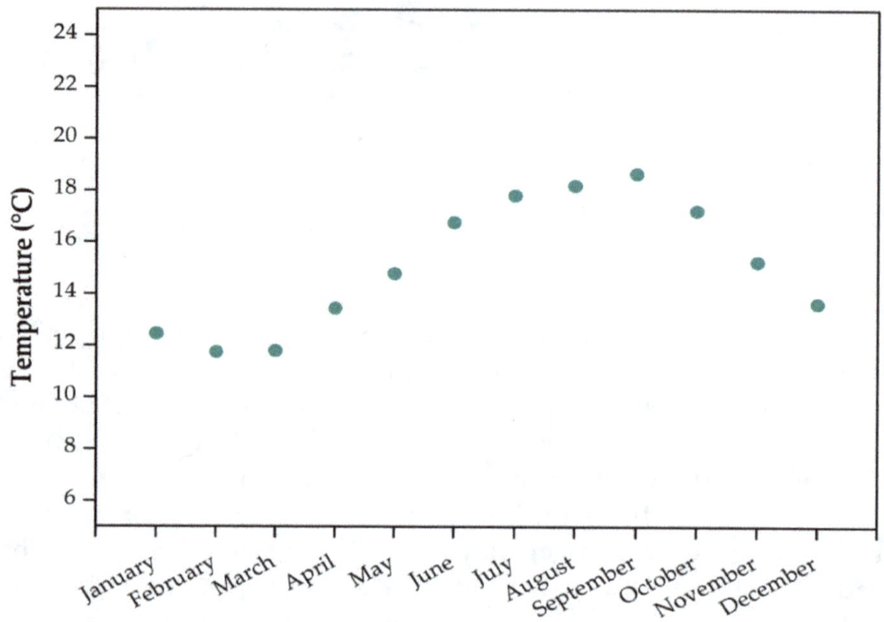


**Figure 3.** Monthly temperature of the wastewater in the sewer, recorded in the proximity of the building.

The temperature trend observed in the sewer was in accordance with the data reported by Cipolla and Maglionico [17], Wanner et al. [23], and Kretschmer et al. [19], respectively, in Bologna (Italy), Zurich (Switzerland), and in a 24,000-population town in Austria.

### 2.3. Heat Recovery: Influence on the Wastewater Temperature

Based on the aforementioned data, it is possible to use Equation (1) (modified from the original equation proposed by Kretschmer et al. [19]) to graphically depict the relationships between the wastewater flow rate diverted at the recovery site for heat exchange ($Q_{RS}$, L s$^{-1}$), available heat potential ($P_{RS}$, kW), and temperature difference or decrease due to heat recovery ($\Delta T_{RS}$, K), as shown in Figure 4:

$$P_{RS} = Q_{RS} \cdot c \cdot \Delta T_{RS} \cdot \rho \quad (1)$$

where c is the specific heat capacity of wastewater (4.18 kJ kg$^{-1}$ °C$^{-1}$ [24]), and ρ represents the wastewater density (assumed to be 1000 kg m$^{-3}$).

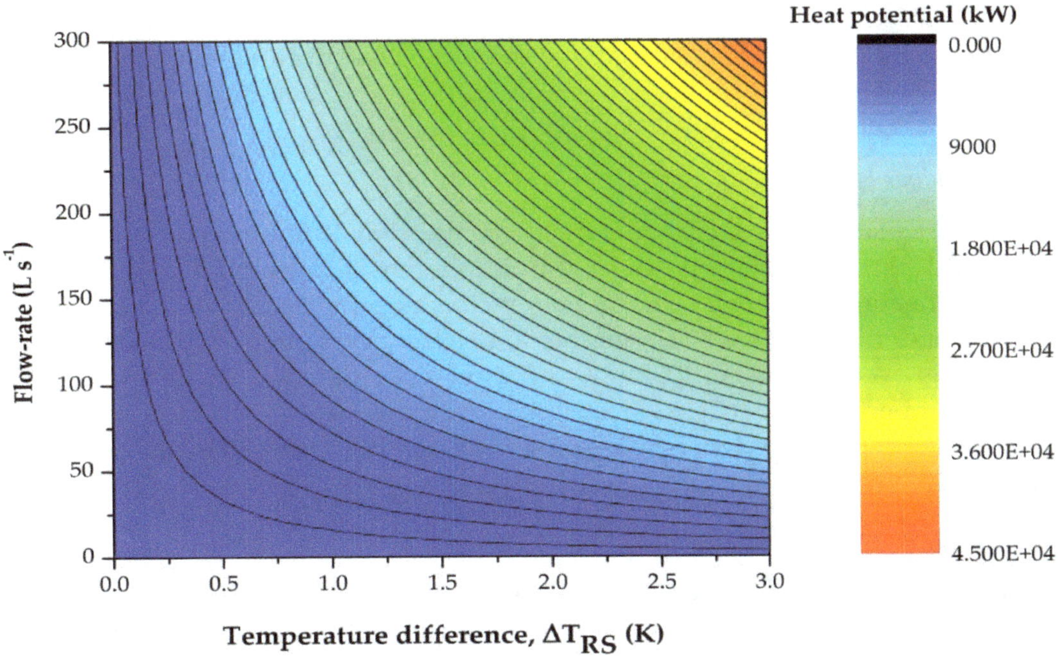


**Figure 4.** Relationship between the heat exchanger flow rate, temperature, and power recovered.

The change of temperature in the sewer immediately after the reintroduction of the heat-recovered wastewater ($\Delta T_{SEWER}$, °C) can be calculated using Equation (2) [19]:

$$\Delta T_{SEWER} = (Q_{RS} \cdot \Delta T_{RS})/Q_{SEWER}, \quad (2)$$

where $Q_{SEWER}$ is the flow rate in the sewer in the proximity of the building (L $s^{-1}$). Similarly, the change in the temperature of the wastewater at the inflow of the WRRF ($\Delta T_{WRRF}$) due to the heat recovery process can be computed using Equation (3):

$$\Delta T_{WRRF} = (Q_{RS} \cdot \Delta T_{RS})/Q_{WRRF}, \quad (3)$$

where $Q_{WRRF}$ is the flow rate at the inflow of the WRRF (L $s^{-1}$).

Considering an 800-kW heat power demand, coupled with a flow rate of 90 L $s^{-1}$, the decrease in the wastewater temperature would be 2.12 °C, calculated using Equation (1). Taking into consideration the winter months, when heat demand is higher (December, January, and February, as shown in Figure 1), the temperature drop calculated using Equation (2) would be in the range of 0.42–1.45 °C, in the case of the maximum and minimum flow rates recorded in that period.

In the case of a 1000-kW power demand for cooling, the temperature drop calculated using Equation (1) would be 2.65 °C at a flow of 90 L $s^{-1}$. Similar to the heating condition, the change in temperature after heat exchange may be computed using Equation (2). Considering the summer period flow rate, a temperature increase in the range of 0.52–1.76 °C could be expected.

Based on the heat exchanged from wastewater, it is possible to calculate the temperature drop or increase at the inflow of the local wastewater treatment plant, applying Equation (3). The flow rate at the Brno WRRF is reported as 4.222 $m^3$ $s^{-1}$ [25]; therefore, the flow to the heat exchanger is 2.13% of the WWRF inflow, while during maximum sewer flow (470 L $s^{-1}$), the percentage increases to 11.1%. The temperature decrease due to heat exchange for heating is equal to 0.045 °C, while the increase due to cooling is 0.056 °C, both calculated at the inflow of the WWRF. Based on that, it is possible to state that the application of heat recovery from wastewater for the building conditioning would not affect the kinetics of the processes of WWRF. A different situation, which should be specifically evaluated, would consist of a diffuse application of heat exchange in a multitude of buildings within the city, a condition that may lead to significant changes in the influent wastewater temperature to the plant.

## 3. Indoor Heating and Cooling Options

### *3.1. Conventional Indoor Heating and Cooling*

According to calculations based on a preliminary conventional indoor conditioning system scheme, the building would require an 824-kW thermal unit to be installed, including a 3% power margin on peak power for heating. The total heating energy consumption is considered equal to the annual building heat demand of 500,293 kWh. Under the same hypotheses, the cooling system would consist of three units with an overall power of 1030 kW, including the same margin on the peak. With a cooling system coefficient of performance (COP) of 3.9 (declared by the manufacturer), this would imply an energy consumption of 120,183 kWh per year. The sum of energy consumed by the heating and cooling system would thus imply an overall energy consumption of 620,475 kWh per year (option 1).

### *3.2. Heat Recovery from Wastewater*

The second option considered is heat recovery from wastewater. In this case, a commercial HUBER™ ThermWin system (HUBER SE, Berching, Germany [26]) was installed, as shown in Figure 5. The system is composed of a connection with the bottom of the sewer pipe, a shaft equipped with a screen, an aboveground heat exchanger, and a heat pump. Different from other applications in which the heat pump is located directly in the sewer conduits [18,27], this solution foresees a shaft installed near the sewer bottom, connected purely by gravity. The shaft serves as a sump for the pump feeding the heat exchanger, and houses a coarse screen to avoid unwanted material carried by the wastewater from reaching the heat exchanger. Collected solids would then be sent back to the mains by a vertical screw equipped with brushes. A dedicated hydrodynamic study was carried out on the sewer–shaft intersection to evaluate the likelihood of any operational issues [28]. The hydrodynamics of the setup were computed using the FLOW-3D model software (Flow-Science Inc., Santa Fe, NM, USA) in order to ensure the availability of a flow rate of 90 L $s^{-1}$, which is necessary for the system's operation under any flow-rate conditions, as shown in Figure 6.

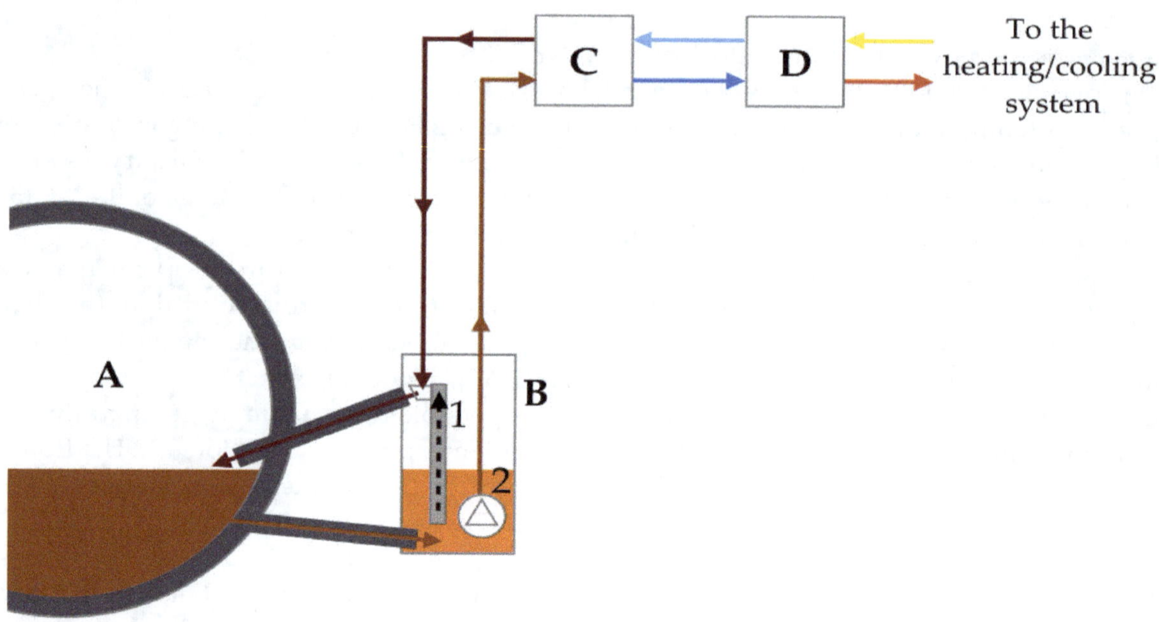


**Figure 5.** Scheme of the heat recovery system. A: Sewer collector; B: pump with pretreatment (2) and vertical screen (1); C: heat exchanger; D: heat pump. The sewer and pump shaft are underground, while the heat exchanger and heat pump are located aboveground.

**Figure 6.** Scheme of the intersection between the inflow and outflow pipes and the main sewer, at a 323 L $s^{-1}$ flow rate. Reproduced with permission from [28].

The system consists of two heat pumps connected to three heat exchangers. The installed power of heat pumps and exchangers is rated at 700 kW; in addition, a supplementary 180 kW of power is required to operate the system. A supplementary conventional heating system is provided, with an installed power of 800 kW. The system is designed to work in combined mode: 10% of the demand is provided by conventional heating and 90% by wastewater-recovered heat. This would result in an annual energy consumption for the conventional heating fraction of 50,029 kWh (10% of the yearly consumption), and 112,566 kWh for the heat pumps–exchangers assembly, based on a manufacturer-declared 3.9 COP. The annual energy requirement for the heating demand under option 2 is therefore estimated to be 162,595 kWh.

Considering cooling requirements, a similar pattern is replicated: 90% of the energy demand is covered by wastewater-recovered energy and 10% by conventional cooling. An additional 400 kW unit is planned to assist the 700-kW-rated wastewater energy recovery unit. The former unit works at specifications rated 4.2 COP, while the heat pump–exchanger system is rated 6.6 COP. The total energy required for heating or cooling is then determined as the sum of energy for the booster conventional cooling system (10,874 kWh) and for the heat pump–exchanger system (62,276 kWh), with a total of 73,150 kWh. Additional energy requirements derive from the operation of blowers, heat exchangers, pumps, and electric drive shears, and are calculated as 23,406 kWh $y^{-1}$.

Thus, the overall annual direct energy consumption for option 2, including cooling, heating, and related energy needs, is estimated to be 259,151 kWh, in addition to the fraction recovered from wastewater. This sums to more than 58% lower energy requirements than in an equivalent conventional system.

## 4. Comparison and Discussion

Based on the results shown in Sections 3.1 and 3.2, the energy requirements in the case of the use of heat pumps and exchangers for wastewater heat recovery are almost 59% lower than those required for a conventional solution. As shown in Figure 7, the option of energy recovery from wastewater could be considered advantageous for both heating and cooling, even though the advantage is more prominent in the former case.

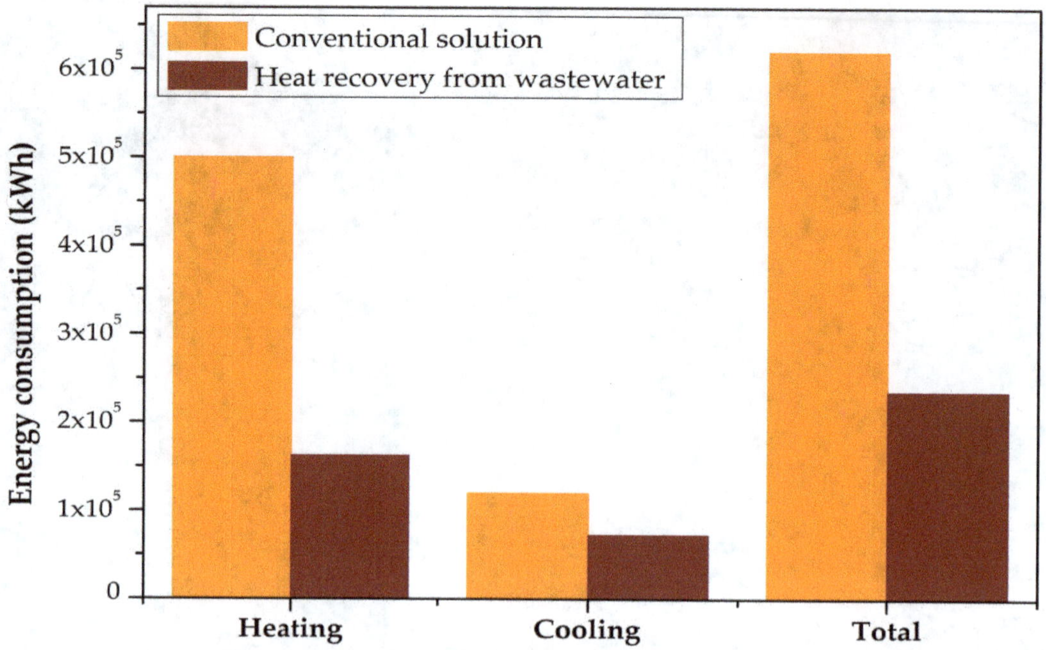


**Figure 7.** Energy consumption for heating, cooling, and overall energy consumption in the two options considered.

The results suggest the feasibility of energy recovery from wastewater to reduce the energy demands of the study building. However, from technical and operational points of view, it should be highlighted that such a new and locally-untested application may be more complex than a conventional heating system, and can only be feasible when the main sewer is located in close proximity to the building. In addition, it would be complicated to retrofit an existing building with such a solution, suggesting that this approach should be implemented in new urbanizations.

Additional issues should be considered in relation to similar applications. One is related to the "ownership" of the wastewater-embedded heat. As a "waste," wastewater is currently considered a liability, with costs associated with its treatment and disposal. Users normally pay a treatment fee, usually associated with their water consumption. If this pilot application becomes successful, wastewater would gain a sort of "redemption" value, and appropriate schemes would have to be developed to apportion this resource and collect revenue from its exploitation. The water utility should be able to charge an appropriate heat recovery fee to users, as they would be enjoying savings from lower heating and conditioning bills, to maintain sewerage system sustainability and fair conditions to all users. If too many users attempt to exploit a similar scheme in an uncontrolled, unregulated fashion, an excessive wastewater temperature reduction could occur, with a negative impact on WRRF performance, and thus, on all users and on the sewerage system itself. Therefore, a multidisciplinary approach should be followed with all the stakeholders involved when planning extensive heat recovery from wastewater, [29,30].

Even though it may seem like a free resource, embedded heat exploitation is, in fact, the cause of indirect costs, even to water utility companies. The kinetics of the biological sewage treatment processes follow Arrhenius' law, i.e., they are dependent on environmental (water) temperature, such that lower temperatures imply slower biodegradation rates [31]:

$$k = k_{20} \cdot \theta^{(T-20)}, \tag{4}$$

where k is the temperature-dependent reaction rate coefficient ($d^{-1}$), $k_{20}$ is the reaction rate at the temperature of 20 °C (1.104 $d^{-1}$, as proposed by Sheridan et al. [32]), θ is the heat coefficient (1.047 for aerobic wastewater treatment [33], adimensional), and T is the temperature of the wastewater (°C). Temperature influences a large number of biochemical processes in wastewater treatment, including oxygen transfer [34], nitrification [35], aerobic and anaerobic organic matter removal [36], and biomass decay [37].

Based on Equation (4), considering increases or drops in wastewater temperature at the WRRF inflow, and comparing the kinetics of biological processes to those in original conditions, it is possible to estimate the effects of heat recovery; in the situation considered, the difference would be minimal in both winter and summer, i.e., equal to a decrease of $2.28 \cdot 10^{-3}$ $d^{-1}$ and an increase of $2.84 \cdot 10^{-3}$ $d^{-1}$, respectively. The results of the simulation suggest that the influence of the heat recovery in a large single building would have negligible effects on the biochemical processes of the WRRF. Multiple heat recovery applications would have a different outcome. If recovery is too high, wastewater temperature drop may be significant, and may have adverse consequences on the efficiency of treatment processes, reducing carbon and especially nitrogen removal rates, and inducing the need to upgrade the possible treatment units. All these situations require a preliminary facility assessment to determine the limits to which wastewater heat energy recovery can be carried out without impairment [19]; recently, modeling has been proposed as a solution to evaluate the amount of heat that can be recovered from wastewater without harming the biological processes in WRRFs [38–40].

## 5. Conclusions

The planning of a wastewater-supplied heating and cooling system for a multipurpose building in Brno (Czech Republic) was carried out. The tested system required a far lower energy consumption (235,745 kWh), i.e., almost 59% less than a conventional solution. The impact analysis of wastewater heat recovery on the downstream WRRF biological processes proved such an impact to be negligible, i.e., an influent wastewater temperature decrease of 0.045 °C and an increase of 0.056 °C, respectively, in winter and summer. Therefore, the envisioned solution would enhance the overall sustainability of the building without impairing the wastewater treatment performance; however, performance and efficiency issues may develop, should the general application of this solution be attempted throughout the city.

In view of the possibility of a more generalized adoption of these systems, preliminary WRRF performance assessments should be carried out. In addition, sewer operators should consider introducing pricing schemes for the exploitation of wastewater-embedded heat energy, both to safeguard wastewater treatment facility performance and as a principle of fair treatment of all stakeholders.

**Author Contributions:** Conceptualization, J.R. and P.H.; software, J.R.; investigation, J.R. and P.H.; writing—original draft preparation, D.C.; writing—review and editing, D.C. and A.C.; supervision, A.C. and P.H.; funding acquisition, P.H. and A.C. All the authors contributed substantially to the work reported. All authors have read and agreed to the published version of the manuscript.

## References

1. Capodaglio, A.G.; Olsson, G. Energy issues in sustainable urban wastewater management: Use, demand reduction and recovery in the Urban Water Cycle. *Sustainability* **2019**, in press.
2. Capodaglio, A.G.; Ghilardi, P.; Boguniewicz-Zablocka, J. New paradigms in urban water management for conservation and sustainability. *Water Pract. Technol.* **2016**, *11*, 176–186. [CrossRef]
3. Capodaglio, A.G.; Callegari, A.; Cecconet, D.; Molognoni, D. Sustainability of decentralized wastewater treatment technologies. *Water Pract. Technol.* **2017**, *12*, 463–477. [CrossRef]
4. Daneshgar, S.; Callegari, A.; Capodaglio, A.G.; Vaccari, D. The Potential Phosphorus Crisis: Resource Conservation and Possible Escape Technologies: A Review. *Resources* **2018**, *7*, 37. [CrossRef]
5. Daneshgar, S.; Buttafava, A.; Callegari, A.; Capodaglio, A.G. Economic and energetic assessment of different phosphorus recovery options from aerobic sludge. *J. Clean. Prod.* **2019**, *223*, 729–738. [CrossRef]

6. Cecconet, D.; Molognoni, D.; Callegari, A.; Capodaglio, A.G. Agro-food industry wastewater treatment with microbial fuel cells: Energetic recovery issues. *Int. J. Hydrogen Energy* **2018**, *43*, 500–511. [CrossRef]
7. Cecconet, D.; Bolognesi, S.; Molognoni, D.; Callegari, A.; Capodaglio, A.G. Influence of reactor's hydrodynamics on the performance of microbial fuel cells. *J. Water Process Eng.* **2018**, *26*, 281–288. [CrossRef]
8. Capodaglio, A.G.; Molognoni, D.; Dallago, E.; Liberale, A.; Cella, R.; Longoni, P.; Pantaleoni, L. Microbial fuel cells for direct electrical energy recovery from urban wastewaters. *Sci. World J.* **2013**, *2013*, 634738. [CrossRef]
9. Zhou, D.; Deng, Z. (Daniel) Ultra-low-head hydroelectric technology: A review. *Renew. Sustain. Energy Rev.* **2017**, *78*, 23–30. [CrossRef]
10. Frijns, J.; Hofman, J.; Nederlof, M. The potential of (waste)water as energy carrier. *Energy Convers. Manag.* **2013**, *65*, 357–363. [CrossRef]
11. Callegari, A.; Hlavinek, P.; Capodaglio, A.G. Production of energy (biodiesel) and recovery of materials (biochar) from pyrolysis of urban waste sludge. *Rev. Ambiente Agua* **2018**, *13*, 1. [CrossRef]
12. Callegari, A.; Bolognesi, S.; Cecconet, D.; Capodaglio, A.G. Production technologies, current role, and future prospects of biofuels feedstocks: A state-of-the-art review. *Crit. Rev. Environ. Sci. Technol.* **2020**, *50*, 384–436. [CrossRef]
13. Meggers, F.; Leibundgut, H. The potential of wastewater heat and exergy: Decentralized high-temperature recovery with a heat pump. *Energy Build.* **2011**, *43*, 879–886. [CrossRef]
14. Hao, X.; Li, J.; van Loosdrecht, M.C.M.; Jiang, H.; Liu, R. Energy recovery from wastewater: Heat over organics. *Water Res.* **2019**, *161*, 74–77. [CrossRef]
15. Hao, X.; Wang, X.; Liu, R.; Li, S.; van Loosdrecht, M.C.M.; Jiang, H. Environmental impacts of resource recovery from wastewater treatment plants. *Water Res.* **2019**, *160*, 268–277. [CrossRef]
16. Zou, S.; Kanimba, E.; Diller, T.E.; Tian, Z.; He, Z. Modeling assisted evaluation of direct electricity generation from waste heat of wastewater via a thermoelectric generator. *Sci. Total Environ.* **2018**, *635*, 1215–1224. [CrossRef]
17. Cipolla, S.S.; Maglionico, M. Heat recovery from urban wastewater: Analysis of the variability of flow rate and temperature. *Energy Build.* **2014**, *69*, 122–130. [CrossRef]
18. Hepbasli, A.; Biyik, E.; Ekren, O.; Gunerhan, H.; Araz, M. A key review of wastewater source heat pump (WWSHP) systems. *Energy Convers. Manag.* **2014**, *88*, 700–722. [CrossRef]
19. Kretschmer, F.; Simperler, L.; Ertl, T. Analysing wastewater temperature development in a sewer system as a basis for the evaluation of wastewater heat recovery potentials. *Energy Build.* **2016**, *128*, 639–648. [CrossRef]
20. Mattsson, J.; Hedström, A.; Westerlund, L.; Dahl, J.; Ashley, R.M.; Viklander, M. Impacts on Rural Wastewater Systems in Subarctic Regions due to Changes in Inputs from Households. *J. Cold Reg. Eng.* **2018**, *32*, 04017019. [CrossRef]
21. Callegari, A.; Boguniewicz-Zablocka, J.; Capodaglio, A.G. Energy recovery and efficiency improvement for an activated sludge, agro-food WWTP upgrade. *Water Pract. Technol.* **2018**, *13*, 909–921. [CrossRef]
22. Guo, X.; Hendel, M. Urban water networks as an alternative source for district heating and emergency heat-wave cooling. *Energy* **2018**, *145*, 79–87. [CrossRef]
23. Wanner, O.; Panagiotidis, V.; Clavadetscher, P.; Siegrist, H. Effect of heat recovery from raw wastewater on nitrification and nitrogen removal in activated sludge plants. *Water Res.* **2005**, *39*, 4725–4734. [CrossRef] [PubMed]
24. Funamizu, N.A.; Iida, M.; Sakakura, Y.; Takakuwa, T. Reuse of heat energy in wastewater: Implementation examples in Japan. *Water Sci. Technol.* **2001**, *43*, 277–285. [CrossRef] [PubMed]
25. Adamcová, D.; Vaverková, M.; Brouskova, E. The toxicity of two types of sewage sludge from wastewater treatment plant for plants in Czech Republic. *J. Ecol. Eng.* **2016**, *17*, 33–37. [CrossRef]
26. Huber, Energy from Wastewater, HUBER Heat Exchanger RoWin. Available online: https://www.huber.de/products/energy-from-wastewater.html (accessed on 27 November 2019).
27. Culha, O.; Gunerhan, H.; Biyik, E.; Ekren, O.; Hepbasli, A. Heat exchanger applications in wastewater source heat pumps for buildings: A key review. *Energy Build.* **2015**, *104*, 215–232. [CrossRef]
28. Raček, J.; Úterský, M.; Ševčík, J.; Dufek, Z.; Hlavínek, P. Energy recovery from wastewater for heating and cooling of multifunctional building in Brno: Modeling the connection. In Proceedings of the International Multidisciplinary Scientific GeoConference Surveying Geology and Mining Ecology Management, SGEM, Albena, Bulgaria, 29 June–5 July 2017; Volume 17, pp. 519–526.

29. Kretschmer, F.; Neugebauer, G.; Stoeglehner, G.; Ertl, T. Participation as a key aspect for establishing wastewater as a source of renewable energy. *Energies* **2018**, *11*, 3232. [CrossRef]
30. Kretschmer, F.; Ertl, T. Chances and Barriers of Wastewater Heat Recovery from a Multidisciplinary Perspective. In *Frontiers in Water-Energy-Nexus Nature-Based Solutions, Advanced Technologies and Best Practices for Environmental Sustainability*; Springer: Cham, Switzerland, 2020; pp. 297–299.
31. Grady, C.P.L.; Daigger, G.T.; Love, N.G.; Filipe, C.D.M. *Biological Wastewater Treatment*, 3rd ed.; IWA Publishing & CRC Press: London, UK, 2011; ISBN 9780849396793.
32. Sheridan, C.; Petersen, J.; Rohwer, J. On modifying the arrhenius equation to compensate for temperature changes for reactions within biological systems. *Water SA* **2012**, *38*, 149–151. [CrossRef]
33. Stenstrom, M.K.; Gilbert, R.G. Effects of alpha, beta and theta factor upon the design, specification and operation of aeration systems. *Water Res.* **1981**, *15*, 643–654. [CrossRef]
34. Mines, R.O.; Callier, M.C.; Drabek, B.J.; Butler, A.J. Comparison of oxygen transfer parameters and oxygen demands in bioreactors operated at low and high dissolved oxygen levels. *J. Environ. Sci. Health Part A* **2017**, *52*, 341–349. [CrossRef]
35. Marais, G.V.R.; Ekama, G.A. The activated sludge process part I—Steady state behaviour. *Water SA* **1976**, *2*, 164–200.
36. Novak, J.T. Temperature-substrate in biological treatment. *Journal (Water Pollut. Control Fed.)* **1974**, *46*, 1984–1994.
37. van Handeel, A.C.; van der Lubbe, J.G.M. *Handbook of Biological Wastewater Treatment*, 2nd ed.; IWA Publishing: London, UK, 2012; ISBN 9781780400808.
38. Abdel-Aal, M.; Schellart, A.; Kroll, S.; Mohamed, M.; Tait, S. Modelling the potential for multi-location in-sewer heat recovery at a city scale under different seasonal scenarios. *Water Res.* **2018**, *145*, 618–630. [CrossRef] [PubMed]
39. Elías-Maxil, J.A.; Van Der Hoek, J.P.; Hofman, J.; Rietveld, L. Energy in the urban water cycle: Actions to reduce the total expenditure of fossil fuels with emphasis on heat reclamation from urban water. *Renew. Sustain. Energy Rev.* **2014**, *30*, 808–820. [CrossRef]
40. Pelda, J.; Holler, S. Methodology to evaluate and map the potential of waste heat from sewage water by using internationally available open data. *Energy Procedia* **2018**, *149*, 555–564. [CrossRef]

5

# Removal of Crotamiton from Reverse Osmosis Concentrate by a $TiO_2$/Zeolite Composite Sheet

**Qun Xiang [1], Shuji Fukahori [2], Naoyuki Yamashita [3], Hiroaki Tanaka [3] and Taku Fujiwara [4,*]**

1 The United Graduate School of Agricultural Sciences, Ehime University, 3-5-7 Tarumi, Matsuyama, Ehime 790-8566, Japan; s16dre09@s.kochi-u.ac.jp

2 Paper Industry Innovation Center of Ehime University, 127 Mendoricho Otsu, Shikokuchuo, Ehime 799-0113, Japan; fukahori.shuji.mj@ehime-u.ac.jp

3 Research Center for Environmental Quality Management, Kyoto University, 1-2 Yumihama, Otsu, Shiga 520-0811, Japan; yamashita@biwa.eqc.kyoto-u.ac.jp (N.Y.); htanaka@biwa.eqc.kyoto-u.ac.jp (H.T.)

4 Research and Education Faculty, National Sciences Cluster, Agriculture Unit, Kochi University, 200 Monobe Otsu, Nankoku, Kochi 783-8502, Japan

* Correspondence: fujiwarat@kochi-u.ac.jp.

**Abstract:** Reverse osmosis (RO) concentrate from wastewater reuse facilities contains concentrated emerging pollutants, such as pharmaceuticals. In this research, a paper-like composite sheet consisting of titanium dioxide ($TiO_2$) and zeolite was synthesized, and removal of the antipruritic agent crotamiton from RO concentrate was studied using the $TiO_2$/zeolite composite sheet. The RO concentrate was obtained from a pilot-scale municipal secondary effluent reclamation plant. Effective immobilization of the two powders in the sheet made it easy to handle and to separate the photocatalyst and adsorbent from purified water. The $TiO_2$/zeolite composite sheet showed excellent performance for crotamiton adsorption without obvious inhibition by other components in the RO concentrate. With ultraviolet irradiation, crotamiton was simultaneously removed through adsorption and photocatalysis. The photocatalytic decomposition of crotamiton in the RO concentrate was significantly inhibited by the water matrix at high initial crotamiton concentrations, whereas rapid decomposition was achieved at low initial crotamiton concentrations. The major degradation intermediates were also adsorbed by the composite sheet. This result provides a promising method of mitigating secondary pollution caused by the harmful intermediates produced during advanced oxidation processes. The cyclic use of the HSZ-385/P25 composite sheet indicated the feasibility of continuously removing crotamiton from RO concentrate.

**Keywords:** paper-like composite sheet; zeolite; photocatalysis; reverse osmosis concentrate; pharmaceutical; inhibitory effect; intermediate

---

## 1. Introduction

Reverse osmosis (RO) is a well-established technology for water desalination, the production of potable water, and more recently, tertiary wastewater treatment [1,2]. With increasing global water demand, it is predicted that the global market value of RO system components will reach 8.1 billion USD by 2018 [3]. Along with the purification of wastewater, the RO process produces a concentrate containing high levels of rejected pollutants (about 15–20% of the influent volume) [4]. Some of the emerging pollutants, such as pharmaceuticals and personal care products, are very persistent in sewage effluent, resulting in raised awareness of the environmental risk of RO concentrates [1,5,6]. Genotoxicity evaluation using the *SOS/umu* test has provided direct evidence that RO concentrates have much higher toxicological risk than RO influents [7]. Therefore, suitable technology needs to be developed for treating RO concentrates before discharging them into receiving water or recycling for other purposes. This requirement is especially important for large-scale RO treatment systems [8].

In a number of recent studies, $TiO_2$ photocatalysis has been used to treat pharmaceuticals in wastewater [9]. The nonselective oxidation ability of hydroxyl radicals enables effective degradation of various organic pollutants. However, the photocatalysis of target compounds can be inhibited by coexisting materials, such as inorganic ions and organic matter, in the wastewater [10–12]. In addition, toxic intermediates may be produced during photocatalysis, and the effects of pharmaceutical degradation products in the environment are of concern. Furthermore, when $TiO_2$ or nano-$TiO_2$ powder in water is exposed to ultraviolet (UV) radiation, radicals that are harmful to aquatic organisms are produced [13]. Therefore, the effective recovery of catalyst powder after wastewater treatment should be taken into consideration.

Wastewater treatment frequently involves adsorption processes, and various types of adsorbents have been developed to remove different pollutants [14–19]. The high-silica Y-type zeolite HSZ-385, which is a hydrophobic zeolite, has been used to remove sulfonamide antibiotics from wastewater and selectively removes sulfonamides even in the presence of high concentrations of coexisting materials [20]. However, after adsorption, the contaminants are permanently transferred to the sorbent and not destroyed, which can lead to problems with saturation of the adsorbent.

Attempts have been made to synthesize $TiO_2$-adsorbent composites that perform both photocatalysis and adsorption to remove pharmaceuticals from wastewater [21–23]. This synergistic effect has been confirmed for $TiO_2$ and zeolite in a $TiO_2$/zeolite composite powder that was used to remove sulfonamide antibiotics [23]. Wu et al. condensed nano-$TiO_2$ on the surfaces of carbon spheres through hydrothermal treatment to generate core–shell structures, and found that visible light absorption was enhanced compared with pure $TiO_2$ because of the interface formed between the two materials [21]. The activated carbon fiber felt (ACFF) in the $TiO_2$/ACFF porous composites significantly enhances the photocatalytic property of toluene by hindering the recombination of electron-hole pairs, reducing the $TiO_2$ band gap energy, and accelerating toluene adsorption [24]. Using a papermaking technique, Fukahori et al., prepared a paper-like composite sheet from $TiO_2$ and zeolite powder [25]. Under UV irradiation, bisphenol A was effectively degraded through the synergistic effect of the $TiO_2$ photocatalyst and zeolite adsorbent in these sheets [25]. In addition, the degradation intermediates of bisphenol A, which may be harmful to the environment, were temporarily captured by zeolite in the composite sheet and eventually decomposed through photocatalysis [26]. However, these studies were conducted using ultrapure water as the solvent, and the inhibitory effects of other components of the wastewater matrix have not been investigated.

In this study, we synthesized a $TiO_2$/zeolite composite sheet to remove of crotamiton from RO concentrate, and to recover the catalyst and adsorbent after water treatment. Crotamiton is a scabicide and antipruritic agent that has frequently been detected in sewage effluent in Japan because of its stable nature and wide consumption [27–29]. The effect of coexisting matter from the wastewater matrix on inhibiting crotamiton degradation was evaluated. In addition, the behavior of crotamiton degradation intermediates during photocatalysis was investigated.

## 2. Materials and Methods

### 2.1. Materials

HSZ-385 (surface area 600 $m^2/g$, mean particle size 4 μm, $SiO_2/Al_2O_3$ ratio 100:1) was purchased from Tosoh Ltd. (Tokyo, Japan). $TiO_2$ powder (P-25, 50 $m^2/g$, anatase) was purchased from Degussa (Dusseldorf, Germany) and F-type zeolite powder (F9, $SiO_2/Al_2O_3$ ratio 2.1:1) was purchased from Wako Pure Chemical Industries, Ltd. (Tokyo, Japan). Crotamiton (purity > 97%) and isotope-labelled surrogate crotamiton-d7 (purity 94.5%) were purchased from Sigma-Aldrich (St Louis, MO, USA) and Hayashi Pure Chemical (Osaka, Japan), respectively. Crotamiton-d7 was dissolved in methanol (purity > 99.8%; Kanto Chemical Co., Inc., Tokyo, Japan) to prepare an internal standard solution, which was stored at −20 °C. All other chemicals used were of reagent grade.

Composite sheets consisting of $TiO_2$ and zeolite (HSZ-385 or F-9) were prepared using a papermaking technique. $TiO_2$, zeolite (3.125 g each), and polyethylene terephthalate fiber (6.25 g) were suspended in water (1 L); a cationic flocculant [poly-(amideamine) epichlorohydrin, 0.05% of total solid] and an anionic flocculant (anionic polyacrylamide, 0.084% of total solid) were sequentially added and the final suspension was stirred. Hand sheets with a grammage of 200 $g/m^2$ were prepared according to JIS P8222 [30]. The sheets were dried at 120 °C. The mass ratio of $TiO_2$ to zeolite in the composite sheet was 1:1. The $TiO_2$/zeolite composite sheet used in this study contained 4 $mg/cm^2$ of $TiO_2$ and zeolite. Characterization of the $TiO_2$/zeolite composite sheet was performed by scanning electron microscopy-energy dispersive X-ray spectroscopy (SEM-EDS: ProX; Phenom World) as shown in Figure 1. The SEM and EDS images revealed the uniform distribution of $TiO_2$ and zeolite powder in the composite sheet.

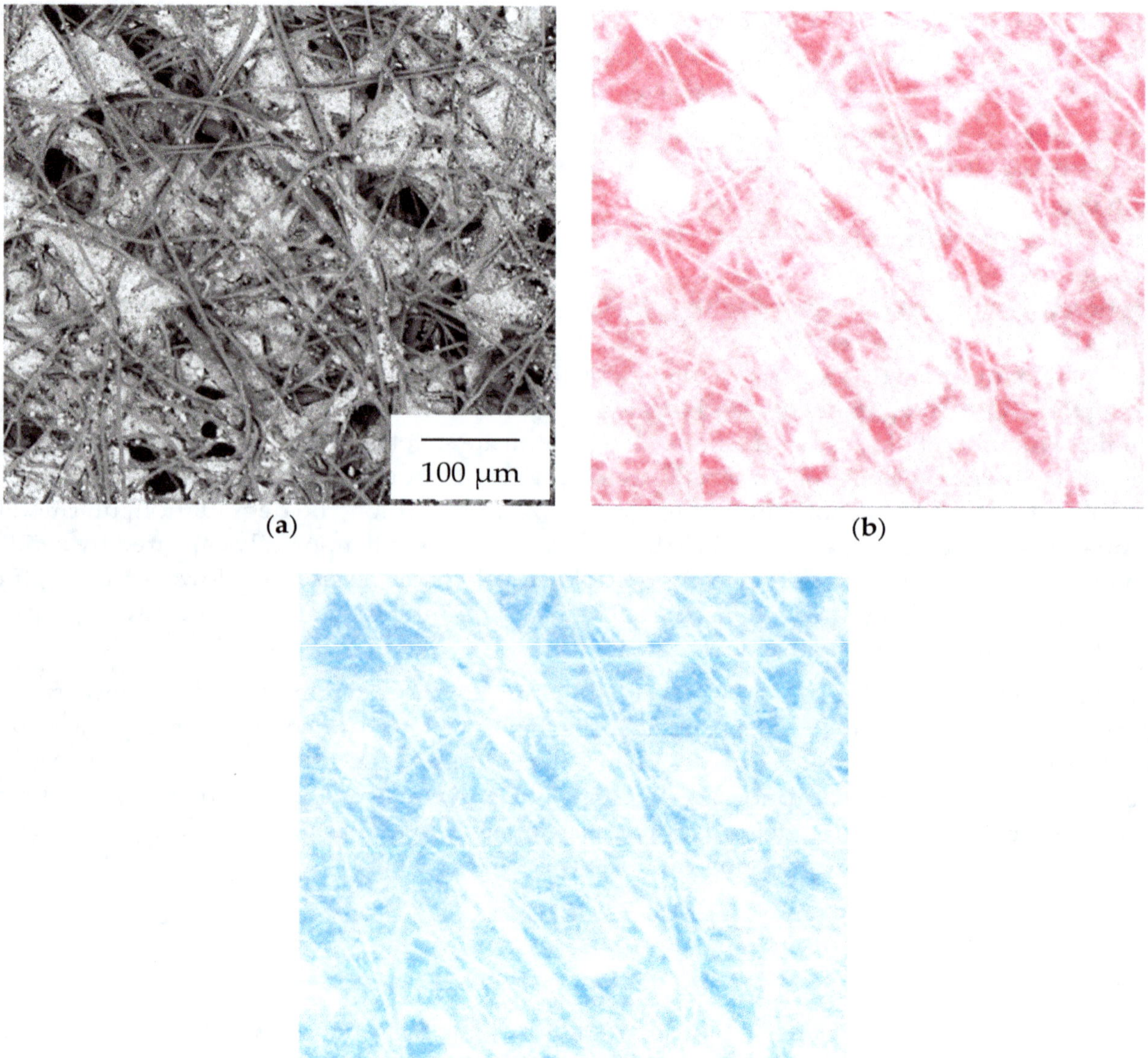


**Figure 1.** Scanning electron microscopy-energy dispersive X-ray spectroscopy (SEM-EDS) mapping images of $TiO_2$/zeolite composite sheet: SEM images (**a**); EDS mapping of Si (**b**) and Ti (**c**).

RO concentrate was collected from a nanofiltration/RO pilot-scale plant for municipal secondary effluent reclamation on 7 March 2017 in Japan, and stored at 4 °C. The RO concentrate was analyzed and the results are as shown in Table 1. Details of the quantitative analyses are given in the Supplementary materials. To clarify the mechanism for removing of crotamiton with the composite sheet, crotamiton solutions were prepared using either RO concentrate or ultrapure water (Millipore, Tokyo, Japan).

**Table 1.** Water quality analysis of the reverse osmosis concentrate.

| Parameter | Value | Ion | mg/L |
|---|---|---|---|
| pH | 7.8 | $Na^+$ | 223 |
|  |  | $NH_4^+$ | 25.7 |
| Conductivity (mS/m) | 170 | $K^+$ | 26.4 |
|  |  | $Mg^{2+}$ | 22.2 |
| TOC (mgC/L) [a] | 10.1 | $Ca^{2+}$ | 45.5 |
| $COD_{cr}$ (mg/L) [b] | 22 | $Cl^-$ | 316 |
|  |  | $NO_2^-$ | 15.6 |
| UV absorbance (λ = 365 nm) (1/cm) [c] | 0.049 | $NO_3^-$ | 46.8 |
| Alkalinity ($mgCaCO_3$/L) | 158 | $SO_4^{2-}$ | 87.5 |

[a] TOC, total organic carbon; [b] $COD_{cr}$, chemical oxygen demand; [c] UV, ultraviolet.

## 2.2. Quantitative Analyses

To determine the concentration of crotamiton in the RO concentrate solution, solid phase extraction (SPE) was carried out. The cartridges (Oasis HLB, 60 mg, 3 mL, Waters, Milford, MA, USA) were conditioned with 2 mL of methanol, followed by 2 mL of ultrapure water. Aqueous samples spiked with the internal standard solution were then loaded onto the cartridges. Next, the cartridges were washed with 2 mL of ultrapure water and dried with a GL-SPE vacuum manifold system (GL Science, Tokyo, Japan) for 30 min. The analyte was eluted first with 1 mL of 10% methanol and then with 4 mL of methanol. The average recovery rate of crotamiton was 97 ± 1.7% (mean ± standard deviation, $n = 3$).

The concentrations of crotamiton were determined with the internal standard addition method using liquid chromatography tandem mass spectrometry (LC/MS/MS, Acquity UPLC-Xevo TQ; Waters) after SPE. The intermediates were identified from the mass spectral patterns obtained by LC/MS/MS.

## 2.3. Methods

Adsorption experiments were carried out by submerging the $TiO_2$/zeolite composite sheets (2 × 5.5 $cm^2$) at a depth of 4 cm in 50 mL of the crotamiton solution (10 mg/L or 120 μg/L) at pH 7.0 ± 0.1 without ultraviolet irradiation (Figure S1). The mixture was stirred at a moderate speed at 25 °C. After a set treatment time, the treated solutions were passed through a DISMIC-13HP 0.2-μm membrane filter (Toyo Roshi Kaisha, Tokyo, Japan) to determine the crotamiton concentrations in the aqueous phase ($C_t$).

For the adsorption and photocatalytic degradation experiments, UV irradiation was applied perpendicular to the sheet surface (Figure S1) with a FL287-BL365 UV lamp (Raytronics, Tokyo, Japan), which had a maximum output wavelength of 365 nm. The UV intensity at the center of the reactor was controlled at 1000 μW/$cm^2$ using a UV-340C light meter (Custom, Tokyo, Japan). The other experimental conditions were the same as for the adsorption experiment. After a set irradiation time, the treated solutions were passed through 0.2-μm membrane filters, and the crotamiton concentrations ($C_t$) were then determined.

To measure the mass of crotamiton in the sheet, the composite sheet was soaked in methanol (purity > 99.8%). After ultrasonication for 60 min (38 kHz, 120 W; US-3KS; SND Co., Ltd., Nagano, Japan), the treated solutions were filtered through 0.2-μm membrane filters and analyzed by

LC/MS/MS ($C_t'$). The recovery rate for desorption was 107 ± 1%. The mass of crotamiton in the treated solution ($M_{in\ water}$) was calculated using Equation (1), in which $V$ is the solution volume.

$$M_{in\ water} = C_t \times V \quad (1)$$

Similarly, the mass of crotamiton in the composite sheet ($M_{in\ sheet}$) was calculated using Equation (2).

$$M_{in\ sheet} = C_t' \times V \quad (2)$$

The total mass of crotamiton remaining in the system ($M_{in\ system}$) was calculated using Equation (3).

$$M_{in\ sheet} = C_t' \times V \quad (3)$$

## 3. Results and Discussion

### 3.1. Adsorption of Crotamiton by the HSZ-385/P25 Composite Sheet

The HSZ-385/P25 composite sheet was applied to the adsorption of crotamiton in the RO concentrate. In preliminarily experiments, we confirmed that crotamiton was rapidly adsorbed by HSZ-385 zeolite powder (Figure S2). We also confirmed that crotamiton was not adsorbed by P25 [31]. The crotamiton concentrations were plotted against time (Figure 2). Similar performances of the sheet in RO concentrate and ultrapure water revealed that other components in the RO concentrate (Table 1) did not obviously affect the adsorption of crotamiton by the composite sheet within the 24-hr treatment period.

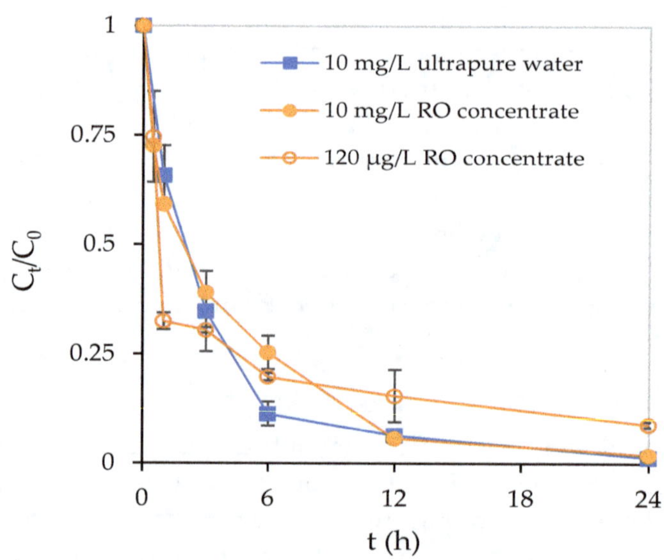


**Figure 2.** Adsorption of crotamiton using the HSZ-385/P25 composite sheet. Results are means ± standard deviations ($n$ = 2).

It has been reported that inorganic ions and organic materials affect the adsorption of pharmaceuticals [32]. Nevertheless, the adsorption of sulfonamide antibiotics from livestock urine and bisphenol A from landfill leachate by HSZ-385 was not affected by coexisting ions [14,20]. Meanwhile, even if the organic carbon content of porcine urine was two orders of magnitude higher than those of the sulfonamides, the sulfonamides were also effectively removed [20]. For bisphenol A, the removal efficiency decreased slightly when more than 50 mg/L humic acid was added [14]. Umar et al., reported that humic-like and fulvic acid-like matter in the RO concentrate were the major contributors to the color of the concentrate [33]. The RO concentrate used in this research appeared to be light brown. In the present study, when the HSZ-385/P25 composite sheet was used to adsorb the raw RO concentrate without crotamiton spiking, only approximately 10% of the total organic carbon (TOC) was removed. Therefore, in the RO concentrate, crotamiton would be removed by adsorption on the composite sheet prior to the raw organic matter. In addition, the removal results for different initial crotamiton concentrations in the RO concentrate were similar, which implies that crotamiton can be adsorbed by the HSZ-385/P25 composite sheet over a wide range of initial concentrations.

We previously investigated the mechanism involved in the adsorption of sulfonamides to HSZ-385. We found that HSZ-385 adsorbed neutral sulfonamides more effectively than non-neutral sulfonamides, and that hydrophobic interactions played important roles in the adsorption process [34]. Crotamiton is hydrophobic ($\log K_{OW} = 2.73$) and has no ionizable functional groups. Therefore, hydrophobic interactions may play an important role in the adsorption of crotamiton by the HSZ-385/P25 composite sheet.

### 3.2. Photocatalytic Degradation of Crotamiton by the F9/P25 Composite Sheet

To clarify the photocatalysis of crotamiton by the F9/P25 composite sheet, the composite sheet was synthesized to be similar to the HSZ-385 composite sheet. Both F9 zeolite and the F9/P25 composite sheet did not remove crotamiton by adsorption (Figure S2). The F9 zeolite is a hydrophilic zeolite, whereas the Y-type zeolite HSZ-385 is a hydrophobic zeolite. This is further evidence that crotamiton is removed by HSZ-385 mainly through hydrophobic interactions.

The photocatalytic degradation of crotamiton over time by the F9/P25 composite sheet is shown in Figure 3. Direct photolysis of crotamiton was not observed [31]. The removal efficiency of crotamiton from the RO concentrate was much lower than that from the ultrapure water. After 24 hr of UV irradiation, the majority of the crotamiton in the ultrapure water was degraded. In contrast, ca. 50% of the crotamiton was degraded in the RO concentrate. Linear relationships were found between $\ln(C_t/C_0)$ and UV irradiation time ($t$) (Figure 3). Therefore, the first-order kinetic model shown in Equation (4) was used to evaluate the photocatalysis of crotamiton. In that equation, $k_1$ is the pseudo-first-order rate constant. The $k_1$ value for crotamiton removal from the RO concentrate by the F9/P25 composite sheet was 0.048 $hr^{-1}$, and was only half of that in the ultrapure water (0.092 $hr^{-1}$). Obviously, the lower rate constant reflects the effect of other components in the RO concentrate on the photocatalytic degradation of crotamiton.

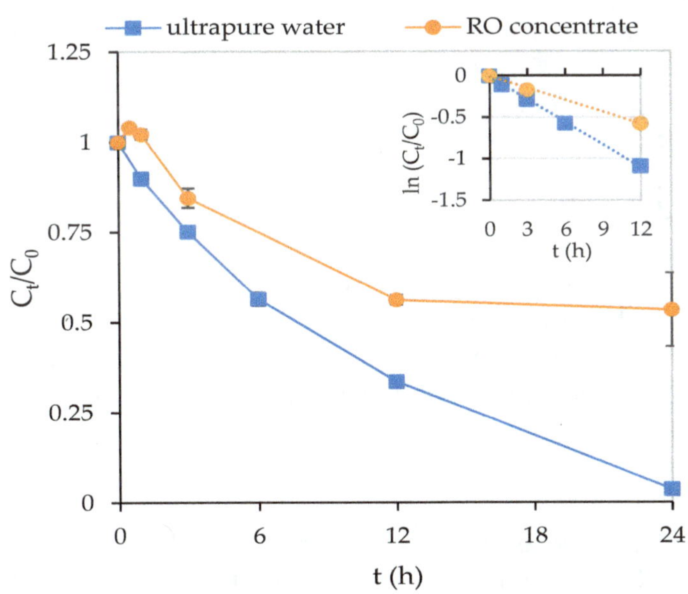


**Figure 3.** Photocatalytic degradation of crotamiton using the F9/P25 composite sheet with ultraviolet irradiation ($C_0 = 10$ mg/L). The inset shows the fitting results of the first-order kinetic model. Results are means ± standard deviations ($n = 2$).

$$\ln (C_t/C_0) = -k_1 t \tag{4}$$

UV absorbance is an important parameter affecting photocatalysis [10,35]. The maximum output wavelength of the UV lamp in our study was 365 nm, and the RO concentrate had an absorbance of 0.049 $cm^{-1}$ at 365 nm (Table 1). Based on the Beer-Lambert law and the distance from the reactor surface to the sheet, the light transmittance was 82.1%. Therefore, after passing through the RO concentrate in the batch reactor, the light intensity at the sheet surface decreased by 17.9%, which is called the screening effect [36]. This effect contributed to the decrease in the photocatalytic degradation efficiency. Furthermore, the RO concentrate was light brown, and the color of the composite sheet surface changed from white to light brown after treatment of the RO concentrate. This color change could be caused by adsorption of coexisting materials on the sheet surface, and this could negatively affect the performance of $TiO_2$ photocatalysis through occupation of the active sites on the surface of $TiO_2$ [37].

The $TiO_2$ photocatalysis could be inhibited via the scavenger effect by coexisting ions [10,38–40]. The ions $Cl^-$ and $HCO_3^-$ have been found to inhibit photocatalysis through the hydroxyl radical and valence band hole scavenging [10,40]. Rioja et al., reported a marked deactivation effect caused by added salts for two tested acidic drugs [41]. Furthermore, Tokumura et al., suggested that coexisting matter could mitigate the generation of hydroxyl radicals through direct reactions with holes in the valence band and electrons in the conduction band of the photocatalyst [37]. As reported by Song et al., $Cl^-$ can cause agglomeration of $TiO_2$ particles in a slurry by suppressing the stabilizing effect of electrostatic repulsion, reducing the effective contact surface between the photocatalyst and the pollutants [42]. In this research, the $TiO_2$/zeolite composite sheet was used instead of $TiO_2$ powder. Therefore, even if the RO concentrate contained 316 mg/L $Cl^-$ (Table 1), agglomeration of $TiO_2$ and its associated issues should be eliminated. However, the mechanism for this should be investigated in future research.

Organic matter in secondary effluent also competes with target pharmaceuticals during photocatalysis [43]. When the F9/P25 composite sheet was used to degrade raw RO concentrate without crotamiton spiking, approximately 15% of the initial TOC (10.1 mgC/L) was degraded after 24 h of UV irradiation. The TOC concentration in the RO concentrate was 10.1 mgC/L, and the TOC concentration for the 10 mg/L crotamiton solution in ultrapure water was 7.67 mgC/L theoretically. The coexisting organic matter may compete with crotamiton for consumption of the oxidizing agent during photocatalysis by the F9/P25 composite sheet.

Mineralization during photocatalytic degradation was evaluated by plotting $TOC/TOC_0$ against time at an initial crotamiton concentration of 10 mg/L (Figure 4). The TOC concentration provided by the residual crotamiton was also determined by performing stoichiometric calculations. During the photocatalytic degradation of crotamiton by the F9/P25 composite sheet in both ultrapure water and RO concentrate, the solution TOC did not obviously decrease with the degradation of crotamiton. This result implied that crotamiton was degraded step by step and that the intermediate compounds accumulated at the same time. Kuo et al., investigated the photocatalytic mineralization of methamphetamine in a UVA/$TiO_2$ system and found that TOC disappeared more slowly than methamphetamine because the methamphetamine intermediates took some time to be mineralized [44]. A more detailed discussion on the degradation intermediates is given in Section 3.4.

### *3.3. Adsorption and Photocatalytic Degradation of Crotamiton by the HSZ-385/P25 Composite Sheet*

In our previous research, we found that P25 was effective for photocatalytic degradation of crotamiton [31]. The HSZ-385/P25 composite sheet prepared in the present study combines adsorption and photocatalysis processes, which makes it possible to regenerate the adsorbent during treatment.

To clarify the adsorption and degradation performance of crotamiton by the HSZ-385/P25 composite sheet, the mass of crotamiton in the composite sheet ($M_{in\ sheet}$) was determined together with the mass of crotamiton in the aqueous phase ($M_{in\ water}$). The mass of crotamiton in the system

($M_{in\ system}$) was calculated as the sum of the residual mass of crotamiton in both the aqueous phase and in the sheet, which was the mass of undecomposed crotamiton remaining in the system.

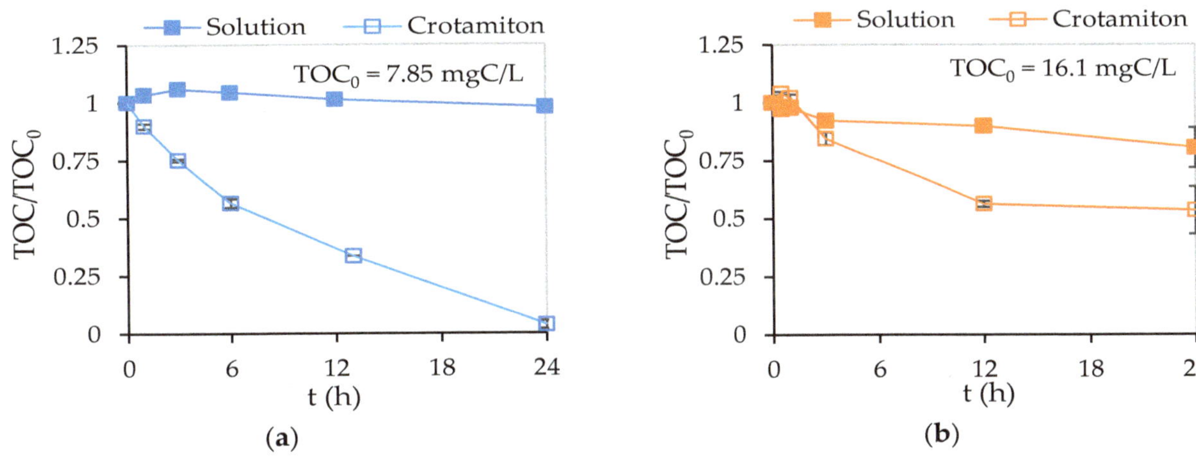


**Figure 4.** Removal of TOC in the solutions and the TOC derived from the residual crotamiton during the treatment by the F9/P25 composite sheet with ultraviolet irradiation in ultrapure water (**a**) and RO concentrate (**b**). Results are means ± standard deviations ($n$ = 2).

The masses of crotamiton in different phases were plotted against time (Figure 5) for solutions with initial crotamiton concentrations of 10 mg/L in ultrapure water (Figure 5a), 10 mg/L in RO concentrate (Figure 5b), and 120 µg/L in RO concentrate (Figure 5c). Although similar trends were observed for $M_{in\ water}$ in ultrapure water and RO concentrate (Figure 5a,b) with an initial crotamiton concentrations of 10 mg/L, the trends for $M_{in\ sheet}$ were very different. The highest value of $M_{in\ sheet}$ (ca. 0.12 mg) was observed after 3 hr treatment of crotamiton in the ultrapure water, and this then decreased with time (Figure 5a); at most, only 23% of the initial crotamiton was accumulated in the sheet, and all the crotamiton was eventually degraded by photocatalysis.

In removing of crotamiton from the RO concentrate, much more crotamiton (0.34 mg) was accumulated in the composite sheet after 6 h treatment (Figure 5b). After that, the mass of crotamiton in the sheet gradually decreased, and finally 0.25 mg remained in the sheet at 24 hr. The higher removal rate obtained with adsorption compared with photocatalysis led to the accumulation and long retention time of crotamiton in the composite sheet. In the treatment of both RO concentrate and ultrapure water, crotamiton could be effectively removed from the aqueous phase, thus purifying the water. The adsorption process was not greatly affected by the water matrix, but inhibition of photocatalysis resulted in low crotamiton degradation in the RO concentrate when using the HSZ-385/P25 composite sheet.

Removing of crotamiton from the RO concentrate with an initial crotamiton concentration of 120 µg/L was investigated (Figure 5c). The $M_{in\ sheet}$ values were maintained at a low level throughout the treatment, and the maximum accumulation of crotamiton in the sheet was only 6.7% of the initial crotamiton mass in the aqueous phase. A rapid decrease was observed in $M_{in\ system}$, showing that rapid decomposition of crotamiton occurred with the low initial crotamiton concentration. Inhibition of the degradation process with high initial crotamiton concentrations may be attributed to competition from intermediates produced by crotamiton degradation [37]. Jang et al., found that the target material (trichloroethylene) saturated the composite catalyst surface and reduced photon efficiency, leading to photocatalyst deactivation [45]. Kuo et al., showed that the degradation rates of codeine and methamphetamine increased with increasing initial concentration (100–250 µg/L) [44,46]. With the initial concentration at microgram per liter levels, the degradation rate may not be limited by the availability of catalytic sites but by contaminant concentration.

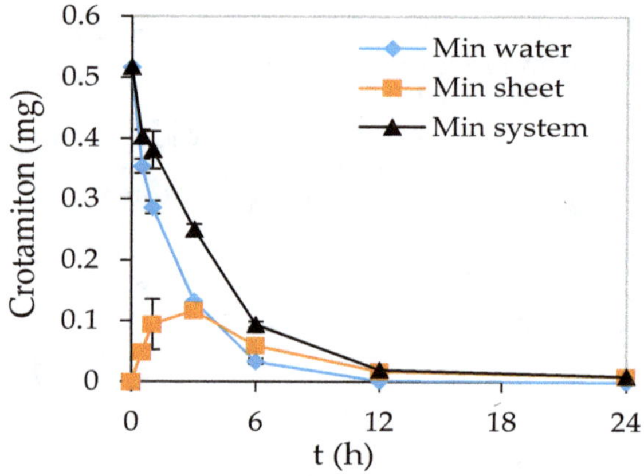

(**a**)

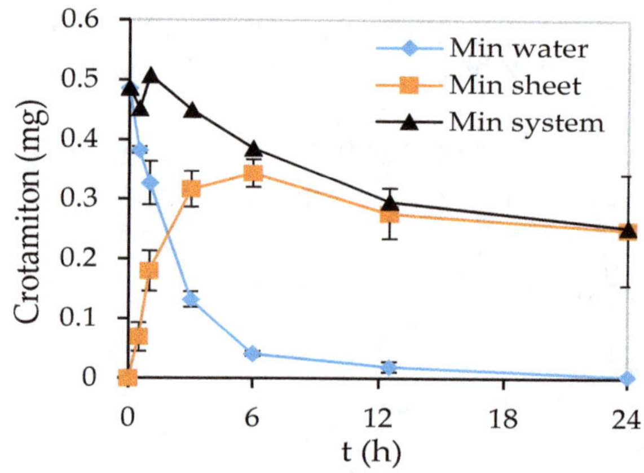

(**b**)

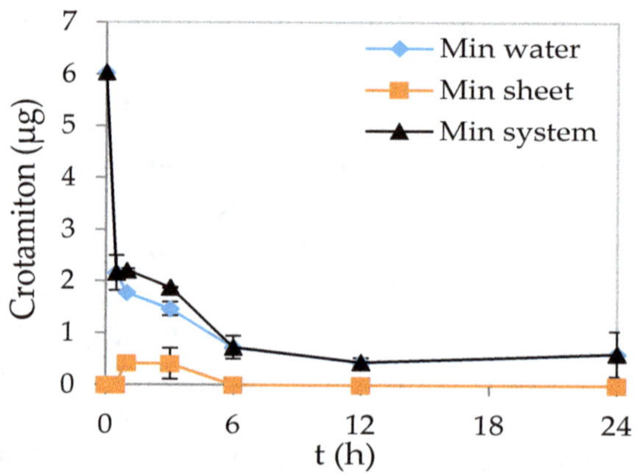

(**c**)

**Figure 5.** Removal of crotamiton using the HSZ-385/P25 composite sheet with ultraviolet irradiation for solutions of (**a**) 10 mg/L crotamiton in ultrapure water; (**b**) 10 mg/L crotamiton in RO concentrate; and (**c**) 120 μg/L crotamiton in RO concentrate. Results are means ± standard deviations ($n = 2$).

The $TOC/TOC_0$ ratios plotted against time in the HSZ-385/P25 composite sheet experiment are shown in Figure 6. With removal of crotamiton in the ultrapure water by the HSZ-385/P25 composite sheet, TOC was gradually removed and the removal efficiency reached up to 84% after 24 hr (Figure 6a), whereas the TOC removal efficiency was stable at ca. 51% after 6 hr of crotamiton treatment in the RO concentrate by the HSZ-385/P25 composite sheet (Figure 6b). The photocatalytic degradation of crotamiton was significantly inhibited by the other components in the RO concentrate. Furthermore, the low TOC removal by individual adsorption or degradation for the organic matter in the original RO concentrate is another important reason. After 24 hr treatment by the HSZ-385/P25 composite sheet, the majority of the crotamiton was removed, which was similar to that in the experiment using the F9/P25 composite sheet. A rather lower $TOC/TOC_0$ ratio was observed in the treatment using the HSZ-385/P25 composite sheet compared with the F9/P25 composite sheet. It has been assumed that accumulation of degradation intermediates in the experiment using the F9/P25 composite sheet resulted in the high residual TOC concentration in the aqueous phase (Figure 4a). The much lower $TOC/TOC_0$ in the treatment using the HSZ-385/P25 composite sheet indicated other TOC derived from degradation intermediates in the aqueous phase has been removed because of the function of the HSZ-385 in the composite sheet (Figure 6a). That is to say, not only crotamiton in the solution but also the degradation intermediates of crotamiton were removed by the HSZ-385 in the composite sheet.

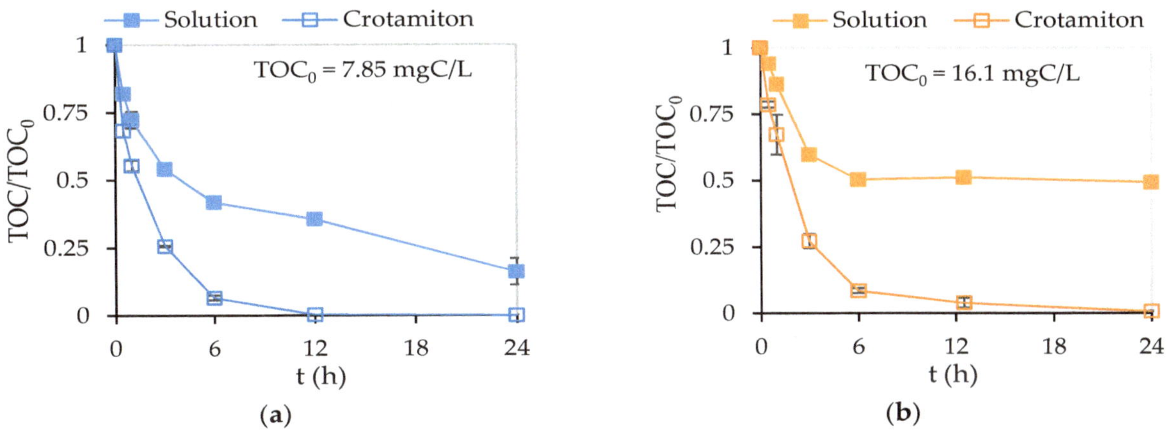


**Figure 6.** Removal of TOC in the solutions and the TOC derived from the residual crotamiton during the treatment by the HSZ-385/P25 composite sheet with ultraviolet irradiation in ultrapure water (**a**) and RO concentrate (**b**). Results are means ± standard deviations ($n = 2$).

### 3.4. Behavior of the Degradation Intermediates during Photocatalysis

The degradation intermediates were characterized following the methods described in our previous study [31]. We proposed that P25-catalyzed photodegradation of crotamiton could initially occur via hydroxylation of the aromatic ring, the double bond of the propenyl group, or the ethyl group. These reactions formed intermediates that we labeled as P189, P217, and P219. The peak areas of the intermediates were measured in the selected ion recording mode of LC/MS/MS.

With the degradation of crotamiton, the intermediates gradually accumulated and reached their highest levels in the treatment using the F9/P25 composite sheet (Figure 7). Intermediate P219, which was produced by hydroxylation of the aromatic ring of crotamiton, was the most noticeable degradation intermediate. The peak area of P219 was higher than those for P189 and P217. After 24 hr of treatment, most of the P189 and P217 had disappeared. In contrast, the peak area of P219 was still high after 24 hr of treatment with the F9/P25 composite sheet. This result corresponded well to the high $TOC/TOC_0$ level revealed in Figure 4a, validating the assumption of the accumulation of degradation intermediates during the treatment by the F9/P25 composite sheet.

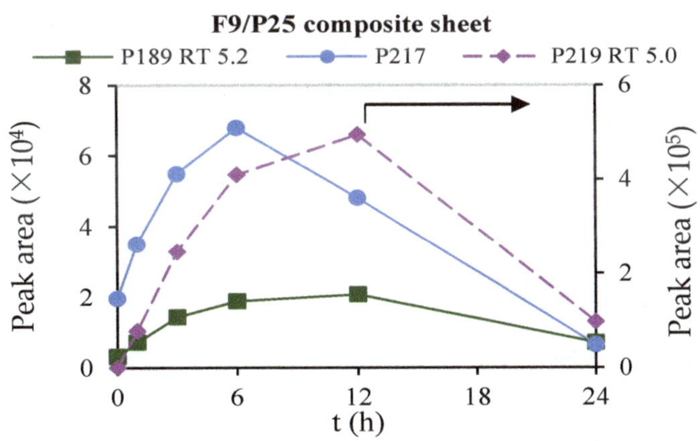


**Figure 7.** Changes in the peak areas of the major intermediates over time in ultrapure water when crotamiton was photocatalytically degraded using the F9/P25 composite sheet ($C_0$ = 10 mg/L). The squares are P189 (retention time 5.2 min) and the circles are P217. The diamonds with a dashed line are P219 (retention time 5.0 min, secondary $y$-axis).

Treatment with the HSZ-385/P25 composite sheet (Figure 8) was compared with that using the F9/P25 composite sheet. The peak areas for the three intermediates obtained with the HSZ-385/P25 composite sheet were clearly lower than those obtained with the F9/P25 composite sheet throughout the treatment, especially for intermediate P219. After 24 hr of treatment with the HSZ-385/P25 composite sheet, the majority of all three intermediates had disappeared. The lower peak areas of degradation intermediates as well as the lower $TOC/TOC_0$ ratios (Figure 6a) confirmed that the HSZ-385/P25 composite sheet captured the degradation intermediates.

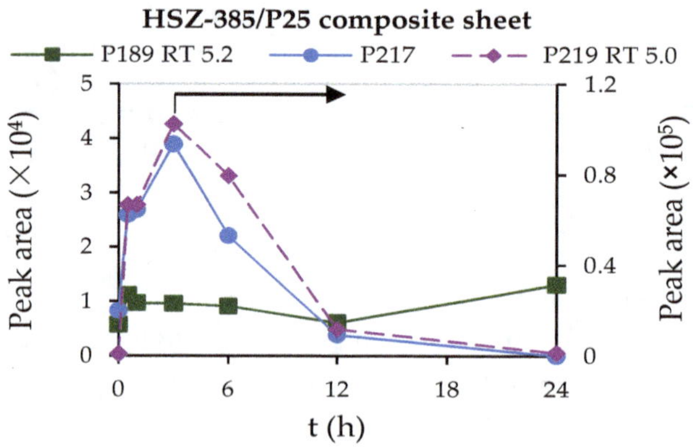


**Figure 8.** Changes in the peak areas of the major intermediates over time in ultrapure water when crotamiton was photocatalytically degraded using the HSZ-385/P25 composite sheet ($C_0$ = 10 mg/L). The squares are P189 (retention time 5.2 min) and the circles are P217. The diamonds with a dashed line are P219 (retention time 5.0 min, secondary *y*-axis).

Moreover, changes in the peak areas for the major intermediates in the HSZ-385/P25 composite sheet were evaluated through desorption treatment for the composite sheet after the adsorption and photocatalysis experiment. The methanol solution with the composite sheet after ultrasonic treatment contained crotamiton, as well as large amounts of degradation intermediates, retained in the composite sheet (Figure 9). The peak area for P219 was higher in the composite sheet during treatment than that in the aqueous phase when using the HSZ-385/P25 composite sheet shown in Figure 8. Even if the efficiency of desorption of the intermediates from the sheet could not be confirmed without the standard of every detected intermediate, the high peak areas for the intermediates provided direct evidence of the adsorption of intermediates on the composite sheet.

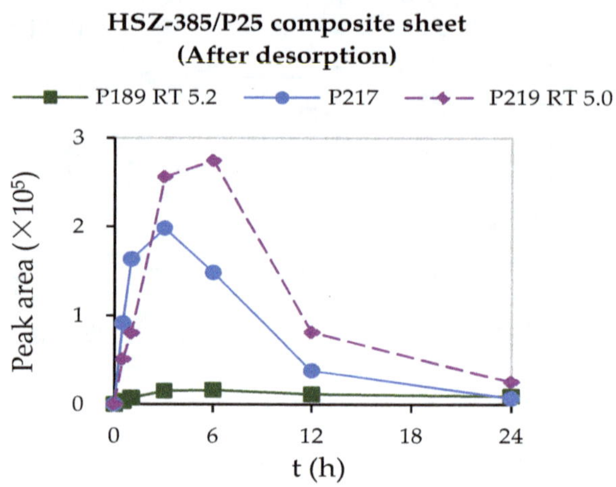


**Figure 9.** Changes in the areas of the peaks for the major intermediates over time in the HSZ-385/P25 composite sheet after desorption when crotamiton was photocatalytically degraded using the HSZ-385/P25 composite sheet ($C_0$ = 10 mg/L). The squares are P189 (retention time 5.2 min), the circles are P217, and the diamonds with a dashed line are P219 (retention time 5.0 min).

In conclusion, the HSZ-385/P25 composite sheet is effective not only for removing crotamiton, but also for capturing degradation intermediates produced by photocatalysis. This method mitigates the negative impact of harmful degradation intermediates produced by advanced oxidation processes.

### *3.5. Cyclic Use of the HSZ-385/P25 Composite Sheet*

To effectively apply the HSZ-385/P25 composite sheet to remove crotamiton from RO concentrate in practical treatment processes, it is essential that the composite sheet can remove pollutants even after several cycles of reuse. The efficiency of removing crotamiton from RO concentrate achieved by the HSZ-385/P25 composite sheet after 24 hr ultraviolet irradiation in three circles of reuse were all over 95% (Figure 10). The crotamiton amount in the composite sheet after three cycles of reuse was 0.40 mg, which was ca. 27% of the total amount of three cycles of treated crotamiton ($C_0$ = 10 mg/L, $V$ = 50 mL), demonstrating continuous crotamiton photocatalytic degradation. The TOC removal efficiency slightly decreased with an increase in the cycles of reuse. It can be concluded that the HSZ-385/P25 composite sheet is feasible for the cyclic removal of crotamiton in RO concentrate.

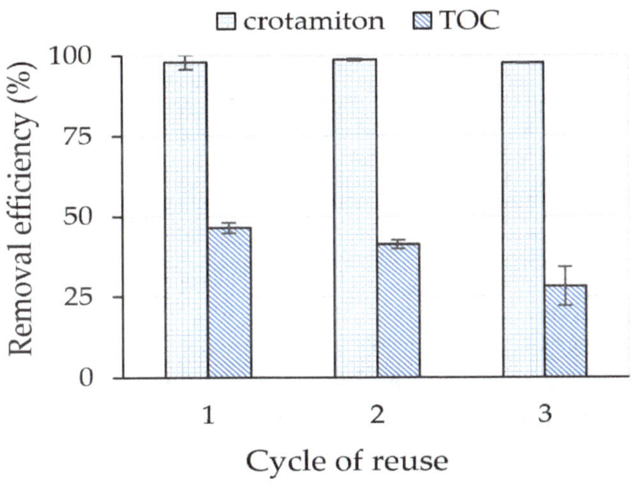


**Figure 10.** The removal efficiency of crotamiton and TOC in RO concentrate using the HSZ-385/P25 composite sheet after cyclic use. Results are means ± standard deviations ($n$ = 2).

Some other composite materials used as adsorbents and photocatalysts for treating organic pollutants are summarized in Table 2. No previous publications applied composite materials for the treatment of pollutants in RO concentrate. A few cycles of pollutant removal were carried out in studies using $TiO_2$–coconut shell powder composite [47], polyacrylic acid-grafted-carboxylic graphene/titanium nanotube composite [48], multi-walled carbon nanotubes/$Fe_3O_4$ composites [49], multi-walled carbon nanotube/$TiO_2$ composites [50] and nitrogen-doped-$TiO_2$/activated carbon composite [51] similar to that in this study. The stability of the photocatalyst and reusability of these materials were confirmed, thus making them promising cost-effective water purification materials.

Except for reusability, some of the composites were designed to promote degradation capability through improving visible light utilization, photon yield, and so on. The presence of $MoS_2$ in the $TiO_2$-$MoS_2$-reduced graphene oxide composite worked as a co-catalyst to reduce electron-hole pairs, and improved the photocatalytic performance of $TiO_2$ for BPA removal [52]. The nitrogen-doped-$TiO_2$/activated carbon composite was synthesized to regenerate spent powdered activated carbon using solar photocatalysis for cost-effective application in wastewater treatment [51]. The graphene/$TiO_2$/ZSM-5 composite material showed higher stability, stronger absorption of visible light, and lower band gap value [53].

**Table 2.** Composite materials synthesized for removing organic pollutants.

| Composite | Target Pollutant | Water Matrix | Reference |
| --- | --- | --- | --- |
| $TiO_2$–coconut shell powder composite | Carbamazepine, clofibric acid, and triclosan | Ultrapure water | [47] |
| Polyacrylic acid-grafted-carboxylic graphene/titanium nanotube composite | Enrofloxacin | Distilled water and simulated poultry farm effluent | [48] |
| Multi-walled carbon nanotubes/$Fe_3O_4$ composites | Bisphenol A | Doubly-distilled deionized water | [49] |
| Multi-walled carbon nanotubes/$TiO_2$ nanocomposite | Tetracycline | Pharmaceutical wastewater | [50] |
| Nitrogen-doped-$TiO_2$/activated carbon composite | Bisphenol-A, sulfamethazine, and clofibric acid | Ultrapure water | [51] |
| $TiO_2$-$MoS_2$-reduced graphene oxide composite | Bisphenol A | Not mentioned | [52] |
| Graphene/$TiO_2$/ZSM-5 composites | Oxytetracycline | Deionized water | [53] |

## 4. Conclusions

$TiO_2$/zeolite composite sheets were synthesized and used to remove crotamiton from RO concentrate. Crotamiton is effectively adsorbed by the HSZ-385/P25 composite sheet without obvious inhibition by other components of the RO concentrate. The photocatalytic decomposition of crotamiton in the RO concentrate is significantly inhibited by the water matrix at high initial concentrations of crotamiton, whereas rapid decomposition occurs at low initial concentrations. When the HSZ-385/P25 composite sheet is used with UV irradiation for the removal of crotamiton from RO concentrate, crotamiton is removed by adsorption and photocatalysis. The inhibition of photocatalytic degradation by other components resulted in crotamiton remaining in the composite sheet. The degradation intermediates are captured by the HSZ-385/P25 composite sheet, and this capture provides a way to mitigate the potential negative impact of intermediates from advanced oxidation processes. In addition, the HSZ-385/P25 can continually remove crotamiton from RO concentrate with repeated uses.

**Acknowledgments:** This work was supported by the Japan Society for the Promotion of Science Grants-in-Aid for Scientific Research Grant Number 16H02372. We appreciate the assistance of Suntae Lee with sampling of the RO concentrate. We thank Gabrielle David, from Edanz Group (www.edanzediting.com/ac) and Dennis Murphy of the United Graduate School of Agricultural Sciences, Ehime University, for editing a draft of this manuscript.

**Author Contributions:** Qun Xiang, Taku Fujiwara and Shuji Fukahori conceived and designed the experiments; Qun Xiang performed the experiments, analyzed the data and wrote the paper; Taku Fujiwara supervised the research; Shuji Fukahori contributed to the preparation and characterization of the composite sheet; Naoyuki Yamashita and Hiroaki Tanaka contributed to the sampling of the RO concentrate and provided advice on the experiments. All authors have read and approved the final manuscript.

## References

1. Joo, S.H.; Tansel, B. Novel technologies for reverse osmosis concentrate treatment: A review. *J. Environ. Manag.* **2015**, *150*, 322–335. [CrossRef] [PubMed]
2. Umar, M.; Roddick, F.; Fan, L. Recent advancements in the treatment of municipal wastewater reverse osmosis concentrate—An overview. *Crit. Rev. Environ. Sci. Technol.* **2015**, *45*, 193–248. [CrossRef]

3. Umar, M.; Roddick, F.; Fan, L. Comparison of coagulation efficiency of aluminium and ferric-based coagulants as pre-treatment for UVC/$H_2O_2$ treatment of wastewater RO concentrate. *Chem. Eng. J.* **2016**, *284*, 841–849. [CrossRef]
4. Umar, M.; Roddick, F.; Fan, L. Assessing the potential of a UV-based AOP for treating high-salinity municipal wastewater reverse osmosis concentrate. *Water Sci. Technol.* **2013**, *68*, 1994–1999. [CrossRef] [PubMed]
5. Westerhoff, P.; Moon, H.; Minakata, D.; Crittenden, J. Oxidation of organics in retentates from reverse osmosis wastewater reuse facilities. *Water Res.* **2009**, *43*, 3992–3998. [CrossRef] [PubMed]
6. Rodriguez-Mozaz, S.; Ricart, M.; Köck-Schulmeyer, M.; Guasch, H.; Bonnineau, C.; Proia, L.; de Alda, M.L.; Sabater, S.; Barceló, D. Pharmaceuticals and pesticides in reclaimed water: Efficiency assessment of a microfiltration-reverse osmosis (MF-RO) pilot plant. *J. Hazard. Mater.* **2015**, *282*, 165–173. [CrossRef] [PubMed]
7. Tang, F.; Hu, H.Y.; Wu, Q.Y.; Tang, X.; Sun, Y.X.; Shi, X.L.; Huang, J.J. Effects of chemical agent injections on genotoxicity of wastewater in a microfiltration-reverse osmosis membrane process for wastewater reuse. *J. Hazard. Mater.* **2013**, *260*, 231–237. [CrossRef] [PubMed]
8. Subramani, A.; Jacangelo, J.G. Treatment technologies for reverse osmosis concentrate volume minimization: A review. *Sep. Purif. Technol.* **2014**, *122*, 472–489. [CrossRef]
9. Yang, Y.; Ok, Y.S.; Kim, K.-H.; Kwon, E.E.; Tsang, Y.F. Occurrences and removal of pharmaceuticals and personal care products (PPCPs) in drinking water and water/sewage treatment plants: A review. *Sci. Total Environ.* **2017**, *596–597*, 303–320. [CrossRef] [PubMed]
10. Chong, M.N.; Jin, B.; Chow, C.W.K.; Saint, C. Recent developments in photocatalytic water treatment technology: A review. *Water Res.* **2010**, *44*, 2997–3027. [CrossRef] [PubMed]
11. Saha, S.; Wang, J.M.; Pal, A. Nano silver impregnation on commercial $TiO_2$ and a comparative photocatalytic account to degrade malachite green. *Sep. Purif. Technol.* **2012**, *89*, 147–159. [CrossRef]
12. Salaeh, S.; Perisic, D.J.; Biosic, M.; Kusic, H.; Babic, S.; Stangar, U.L.; Dionysiou, D.D.; Bozic, A.L. Diclofenac removal by simulated solar assisted photocatalysis using $TiO_2$-based zeolite catalyst; mechanisms, pathways and environmental aspects. *Chem. Eng. J.* **2016**, *304*, 289–302. [CrossRef]
13. Haynes, V.N.; Ward, J.E.; Russell, B.J.; Agrios, A.G. Photocatalytic effects of titanium dioxide nanoparticles on aquatic organisms—Current knowledge and suggestions for future research. *Aquat. Toxicol.* **2017**, *185*, 138–148. [CrossRef] [PubMed]
14. Chen, X.; Fujiwara, T.; Fukahori, S.; Ishigaki, T. Factors affecting the adsorptive removal of bisphenol A in landfill leachate by high silica Y-type zeolite. *Environ. Sci. Pollut. Res.* **2015**, *22*, 2788–2799. [CrossRef] [PubMed]
15. Yang, R.T. *Adsorbents: Fundamentals and Applications*; John Wiley & Sons, Inc.: Hoboken, NJ, USA, 2003; ISBN 0471297410.
16. Kyzas, G.Z.; Fu, J.; Lazaridis, N.K.; Bikiaris, D.N.; Matis, K.A. New approaches on the removal of pharmaceuticals from wastewaters with adsorbent materials. *J. Mol. Liq.* **2015**, *209*, 87–93. [CrossRef]
17. Anastopoulos, I.; Bhatnagar, A.; Hameed, B.H.; Ok, Y.S.; Omirou, M. A review on waste-derived adsorbents from sugar industry for pollutant removal in water and wastewater. *J. Mol. Liq.* **2017**, *240*, 179–188. [CrossRef]
18. Kyzas, G.Z.; Matis, K.A. Nanoadsorbents for pollutants removal: A review. *J. Mol. Liq.* **2015**, *203*, 159–168. [CrossRef]
19. Wei, X.; Wang, Y.; Hernández-Maldonado, A.J.; Chen, Z. Guidelines for rational design of high-performance absorbents: A case study of zeolite adsorbents for emerging pollutants in water. *Green Energy Environ.* **2017**, in press. [CrossRef]
20. Fukahori, S.; Fujiwara, T.; Funamizu, N.; Matsukawa, K.; Ito, R. Adsorptive removal of sulfonamide antibiotics in livestock urine using the high-silica zeolite HSZ-385. *Water Sci. Technol.* **2013**, *67*, 319–325. [CrossRef] [PubMed]
21. Wu, H.; Wu, X.-L.; Wang, Z.-M.; Aoki, H.; Kutsuna, S.; Jimura, K.; Hayashi, S. Anchoring titanium dioxide on carbon spheres for high-performance visible light photocatalysis. *Appl. Catal. B Environ.* **2017**, *207*, 255–266. [CrossRef]
22. Yap, P.-S.; Cheah, Y.-L.; Srinivasan, M.; Lim, T.-T. Bimodal N-doped P25-$TiO_2$/AC composite: Preparation, characterization, physical stability, and synergistic adsorptive-solar photocatalytic removal of sulfamethazine. *Appl. Catal. A Gen.* **2012**, *427–428*, 125–136. [CrossRef]

23. Fukahori, S.; Fujiwara, T. Modeling of sulfonamide antibiotic removal by $TiO_2$/high-silica zeolite HSZ-385 composite. *J. Hazard. Mater.* **2014**, *272*, 1–9. [CrossRef] [PubMed]
24. Li, M.; Lu, B.; Ke, Q.F.; Guo, Y.J.; Guo, Y.P. Synergetic effect between adsorption and photodegradation on nanostructured $TiO_2$/activated carbon fiber felt porous composites for toluene removal. *J. Hazard. Mater.* **2017**, *333*, 88–98. [CrossRef] [PubMed]
25. Fukahori, S.; Ichiura, H.; Kitaoka, T.; Tanaka, H. Photocatalytic decomposition of bisphenol A in water using composite $TiO_2$-zeolite sheets prepared by a papermaking technique. *Environ. Sci. Technol.* **2003**, *37*, 1048–1051. [CrossRef] [PubMed]
26. Fukahori, S.; Ichiura, H.; Kitaoka, T.; Tanaka, H. Capturing of bisphenol A photodecomposition intermediates by composite $TiO_2$–zeolite sheets. *Appl. Catal. B Environ.* **2003**, *46*, 453–462. [CrossRef]
27. Nakada, N.; Tanishima, T.; Shinohara, H.; Kiri, K.; Takada, H. Pharmaceutical chemicals and endocrine disrupters in municipal wastewater in Tokyo and their removal during activated sludge treatment. *Water Res.* **2006**, *40*, 3297–3303. [CrossRef] [PubMed]
28. Nakada, N.; Yasojima, M.; Okayasu, Y.; Komori, K.; Suzuki, Y. Mass balance analysis of triclosan, diethyltoluamide, crotamiton and carbamazepine in sewage treatment plants. *Water Sci. Technol.* **2010**, *61*, 1739–1747. [CrossRef] [PubMed]
29. Tamura, I.; Yasuda, Y.; Kagota, K.; Yoneda, S.; Nakada, N. Ecotoxicology and environmental safety contribution of pharmaceuticals and personal care products (PPCPs) to whole toxicity of water samples collected in effluent-dominated urban streams. *Ecotoxicol. Environ. Saf.* **2017**, *144*, 338–350. [CrossRef] [PubMed]
30. Japanese Industrial Standards Committee. Pulps-Preparation of Laboratory Sheets for Physical Testing-Conventional Sheet-Former Method, JIS P8222: 2015. Available online: http://www.jisc.go.jp/app/jis/general/GnrJISSearch.html (accessed on 26 July 2017).
31. Fukahori, S.; Fujiwara, T.; Ito, R.; Funamizu, N. Photocatalytic decomposition of crotamiton over aqueous $TiO_2$ suspensions: Determination of intermediates and the reaction pathway. *Chemosphere* **2012**, *89*, 213–220. [CrossRef] [PubMed]
32. Bui, T.X.; Choi, H. Influence of ionic strength, anions, cations, and natural organic matter on the adsorption of pharmaceuticals to silica. *Chemosphere* **2010**, *80*, 681–686. [CrossRef] [PubMed]
33. Umar, M.; Roddick, F.; Fan, L. Effect of coagulation on treatment of municipal wastewater reverse osmosis concentrate by UVC/$H_2O_2$. *J. Hazard. Mater.* **2014**, *266*, 10–18. [CrossRef] [PubMed]
34. Fukahori, S.; Fujiwara, T.; Ito, R.; Funamizu, N. pH-Dependent adsorption of sulfa drugs on high silica zeolite: Modeling and kinetic study. *Desalination* **2011**, *275*, 237–242. [CrossRef]
35. Egerton, T. UV-absorption—The primary process in photocatalysis and some practical consequences. *Molecules* **2014**, *19*, 18192–18214. [CrossRef] [PubMed]
36. Tsydenova, O.; Batoev, V.; Batoeva, A. Solar-enhanced advanced oxidation processes for water treatment: Simultaneous removal of pathogens and chemical pollutants. *Int. J. Environ. Res. Public Health* **2015**, *12*, 9542–9561. [CrossRef] [PubMed]
37. Tokumura, M.; Sugawara, A.; Raknuzzaman, M.; Habibullah-Al-Mamun, M.; Masunaga, S. Comprehensive study on effects of water matrices on removal of pharmaceuticals by three different kinds of advanced oxidation processes. *Chemosphere* **2016**, *159*, 317–325. [CrossRef] [PubMed]
38. Kudlek, E.; Dudziak, M.; Bohdziewicz, J. Influence of inorganic ions and organic substances on the degradation of pharmaceutical compound in water matrix. *Water* **2016**, *8*, 532. [CrossRef]
39. Zhou, T.; Lim, T.T.; Chin, S.S.; Fane, A.G. Treatment of organics in reverse osmosis concentrate from a municipal wastewater reclamation plant: Feasibility test of advanced oxidation processes with/without pretreatment. *Chem. Eng. J.* **2011**, *166*, 932–939. [CrossRef]
40. Burns, R.A.; Crittenden, J.C.; Hand, D.W.; Selzer, V.H.; Sutter, L.L.; Salman, S.R. Effect of inorganic ions in heterogeneous photocatalysis of TCE. *J. Environ. Eng.* **1999**, *125*, 77–85. [CrossRef]
41. Rioja, N.; Benguria, P.; Peñas, F.J.; Zorita, S. Competitive removal of pharmaceuticals from environmental waters by adsorption and photocatalytic degradation. *Environ. Sci. Pollut. Res.* **2014**, *21*, 11168–11177. [CrossRef] [PubMed]
42. Song, L.; Zhu, B.; Gray, S.; Duke, M.; Muthukumaran, S. Hybrid processes combining photocatalysis and ceramic membrane filtration for degradation of humic acids in saline water. *Membranes* **2016**, *6*. [CrossRef] [PubMed]

43. Ito, M.; Fukahori, S.; Fujiwara, T. Adsorptive removal and photocatalytic decomposition of sulfamethazine in secondary effluent using $TiO_2$-zeolite composites. *Environ. Sci. Pollut. Res.* **2014**, *21*, 834–842. [CrossRef] [PubMed]
44. Kuo, C.-S.; Lin, C.-F.; Hong, P.-K.A. Photocatalytic degradation of methamphetamine by UV/$TiO_2$-Kinetics, intermediates, and products. *Water Res.* **2015**, *74*, 1–9. [CrossRef] [PubMed]
45. Jang, D.; Ahn, C.; Choi, J.; Kim, J.; Kim, J.; Joo, J. Enhanced removal of trichloroethylene in water using nano-ZnO/polybutadiene rubber composites. *Catalysts* **2016**, *6*, 152. [CrossRef]
46. Kuo, C.S.; Lin, C.F.; Hong, P.K.A. Photocatalytic mineralization of codeine by UV-A/$TiO_2$-Kinetics, intermediates, and pathways. *J. Hazard. Mater.* **2016**, *301*, 137–144. [CrossRef] [PubMed]
47. Khraisheh, M.; Kim, J.; Campos, L.; Al-Muhtaseb, A.H.; Al-Hawari, A.; Al Ghouti, M.; Walker, G.M. Removal of pharmaceutical and personal care products (PPCPs) pollutants from water by novel $TiO_2$-coconut shell powder (TCNSP) composite. *J. Ind. Eng. Chem.* **2014**, *20*, 979–987. [CrossRef]
48. Anirudhan, T.S.; Shainy, F.; Christa, J. Synthesis and characterization of polyacrylic acid-grafted-carboxylic graphene/titanium nanotube composite for the effective removal of enrofloxacin from aqueous solutions: Adsorption and photocatalytic degradation studies. *J. Hazard. Mater.* **2017**, *324*, 117–130. [CrossRef] [PubMed]
49. Huang, Y.; Xu, W.; Hu, L.; Zeng, J.; He, C.; Tan, X.; He, Z. Combined adsorption and catalytic ozonation for removal of endocrine disrupting compounds over MWCNTs/$Fe_3O_4$ composites. *Catal. Today* **2017**, in press.
50. Ahmadi, M.; Motlagh, H.R.; Jaafarzadeh, N.; Mostoufi, A.; Saeedi, R.; Barzegar, G.; Jorfi, S. Enhanced photocatalytic degradation of tetracycline and real pharmaceutical wastewater using MWCNT/$TiO_2$ nano-composite. *J. Environ. Manag.* **2017**, *186*, 55–63. [CrossRef] [PubMed]
51. Yap, P.-S.; Lim, T.-T. Solar regeneration of powdered activated carbon impregnated with visible-light responsive photocatalyst: Factors affecting performances and predictive model. *Water Res.* **2012**, *46*, 3054–3064. [CrossRef] [PubMed]
52. Luo, L.; Li, J.; Dai, J.; Xia, L.; Barrow, C.J.; Wang, H.; Jegatheesan, J.; Yang, M. Bisphenol A removal on $TiO_2$ -$MoS_2$-reduced graphene oxide composite by adsorption and photocatalysis. *Process Saf. Environ. Prot.* **2017**, in press.
53. Hu, X.-Y.; Zhou, K.; Chen, B.-Y.; Chang, C.-T. Graphene/$TiO_2$/ZSM-5 composites synthesized by mixture design were used for photocatalytic degradation of oxytetracycline under visible light: Mechanism and biotoxicity. *Appl. Surf. Sci.* **2016**, *362*, 329–334. [CrossRef]

# 6

# Pyrolysis Process as a Sustainable Management Option of Poultry Manure: Characterization of the Derived Biochars and Assessment of their Nutrient Release Capacities

**Samar Hadroug [1,2], Salah Jellali [3,*], James J. Leahy [4], Marzena Kwapinska [4], Mejdi Jeguirim [5], Helmi Hamdi [6] and Witold Kwapinski [4]**

1 Wastewaters and Environment Laboratory, Water Research and Technologies Center, P.O. Box 273, Soliman 8020, Tunisia; samarhadroug@gmail.com
2 National Agricultural Institute of Tunisia, University of Carthage, Tunis 1082, Tunisia
3 Center for Environmental Studies and Research, Sultan Qaboos University, P.O. Box 31, Al-Khoud 123, Muscat, Oman
4 Department of Chemical Sciences, Bernal Institute, University of Limerick, V94 T9PX Limerick, Ireland; j.j.leahy@ul.ie (J.J.L.); marzena.kwapinska@ul.ie (M.K.); witold.kwapinski@ul.ie (W.K.)
5 CNRS, IS2M UMR 7361, University of Haute-Alsace, University of Strasbourg, F-68100 Mulhouse, France; mejdi.jeguirim@uha.fr
6 Center for Sustainable Development, College of Arts and Sciences, Qatar University, P.O. Box 2713 Doha, Qatar; hhamdi@qu.edu.qa
* Correspondence: s.jellali@squ.edu.om.

**Abstract:** Raw poultry manure (RPM) and its derived biochars at temperatures of 400 (B400) and 600 °C (B600) were physico-chemically characterized, and their ability to release nutrients was assessed under static conditions. The experimental results showed that RPM pyrolysis operation significantly affects its morphology, surface charges, and area, as well as its functional groups contents, which in turn influences its nutrient release ability. The batch experiments indicated that nutrient release from the RPM as well as biochars attains a pseudo-equilibrium state after a contact time of about 48 h. RPM pyrolysis increased phosphorus stability in residual biochars and, in contrast, transformed potassium to a more leachable form. For instance, at this contact time, P- and K-released amounts passed from 5.1 and 25.6 mg $g^{-1}$ for RPM to only 3.8 and more than 43.3 mg $g^{-1}$ for B400, respectively. On the other hand, six successive leaching batch experiments with a duration of 48 h each showed that P and K release from the produced biochars was a very slow process since negligible amounts continued to be released even after a total duration of 12 days. All these results suggest that RPM-derived biochars have specific physico-chemical characteristics allowing them to be used in agriculture as low-cost and slow-release fertilizers.

**Keywords:** raw poultry manure; pyrolysis; biochar; characterization; leaching; phosphorus; potassium

---

## 1. Introduction

Raw poultry manure (RPM) is a chicken waste that is produced from the normal operation of hatcheries, turkey, broiler, and egg laying [1]. Huge amounts of RPM are annually produced in the world. In the USA, the annual production of animal manure dry matter was estimated to be more than 19 million tons in 2018 [2]. In Tunisia, poultry production has seen a sharp increase during the last decade, generating about 400,000 tons/year of manure [3].

The composition of poultry manure is variable and depends mainly on the type of feed [4]. It generally contains high levels of organic carbon and nutrients, such as nitrogen, potassium, and phosphorus, which are indispensable macro-elements for soil fertility improvement and crop growth enhancement [5,6]. The sustainable management of this biowaste has been pointed out as an urgent priority due to its possible negative effects on human health and the environment [5,7]. In fact, the traditional methods for RPM management have various drawbacks [8]. Incineration has been the most widely used technique to reduce the huge volumes of RPM. However, the relatively high ash contents in RPM requires mitigation technologies [9]. Thermal gasification has also been investigated at the farm level, but gas quality is poor and requires extensive cleaning, which is expensive, particularly at the farm scale [10]. Likewise, the methane-fermentation process of malodorous RPM is generally limited by its excessive nitrogen contents [3]. Direct land application of fresh RPM or its composted form as organic amendments may result in: (i) The spread of pathogens [1], (ii) air pollution due to the emission of greenhouse gases and phytotoxic substances [11,12], and (iii) possible surface water resources eutrophication and groundwater pollution [3] due to nitrogen and phosphorus leaching [13]. One of the main potential methods used to reduce nutrient losses from RPM is thermal conversion by pyrolysis (in the absence of oxygen) into biogas and bio-oil that could have an important energetic added value [14–16] and a solid residue, named biochar. Biochars are highly porous carbonaceous materials that could be valorized for agricultural purposes as soil amendments [17,18], for climate change mitigation through carbon sequestration and a reduction of greenhouse gas emissions [19], and for environmental protection since they exhibit large capacities in removing various pollutants from both liquid and gas effluents [2,14,20,21]. All the above-cited effects are very dependent on the physico-chemical characteristics of the biochars that are mainly linked to the nature of the used feedstock and also to the pyrolysis experimental settings [22,23]. Therefore, in-depth characterization of produced biochars under given conditions has been pointed out as a key step in order to select the best method for their valorization [24]. On the other hand, according to our best of knowledge, no studies have investigated controlled RPM-derived biochars' capacities in releasing nutrients versus time by using the same biochars in successive leaching assays, which is close to what is happening in situ where biochar-amended soils are successively leached with rainfall and irrigation events. The majority of studies have attempted to monitor nutrient-release kinetics in a continuous way [22,24,25]. On the other hand, data on the relationships between the physico-chemical characteristics of RPM and RPM-derived biochars and their time-dependent capacity for phosphorus (P) and potassium (K) release in aqueous solutions are still missing.

Accordingly, the aims of the current research work were: (i) To study the thermal properties of RPM through thermogravimetric analyses, (ii) to determine the role of the selected pyrolysis temperatures on the physico-chemical properties of produced biochars, and (iii) to assess the importance of physico-chemical characteristics of RPM and its derived biochars on the kinetics and efficiency of phosphorus and potassium release under various experimental conditions.

## 2. Materials and Methods

### 2.1. *Preparation of Raw Poultry Manure*

The raw poultry manure was collected from a poultry farm in the city of Mornag (South of Tunis, Tunisia). Fresh RPM feedstock was first air-dried for about 10 days until a constant weight. Then, it was manually ground in order to have a relatively homogeneous particle size. Finally, RPM was mechanically sieved by using mechanical sieve shaker (Retch, Haan, Germany). Only fraction with a particle size less than 1 mm was collected and stored in bags for subsequent analysis.

### 2.2. *Thermogravimetric Analysis of RPM*

Thermogravimetric analysis (TGA) of RPM was performed using a thermal analyzer (LABSYS TG brand, Ireland) under the following thermo-analytical conditions: An RPM mass of 16.7 mg, a heating gradient of 10 °C $min^{-1}$ for temperatures varying from 30 to 810 °C, and a dynamic nitrogen atmosphere with a flow rate of 10 mL $min^{-1}$. These TGA measurements were carried out in triplicate and the mean values are reported in this study. The TG curves giving the variation of the residual RPM mass (%) versus time/temperature were derived (DTG curves) in order to get a better idea about the kinetic of RPM mass losses (mg $min^{-1}$) and to highlight even slight mass changes.

### *2.3. Biochar Production and Characterization*

The biochars used in this study were derived from pyrolysis of air-dried RPM at two different temperatures (400 and 600 °C). These biochars (named B400 and B600, consecutively) were produced using a laboratory pyrolysis furnace (Lenton brand oven) with dimensions of 67 cm length, 23 cm width, and 22 cm height. A pyrolysis metal container with a 13 cm length and 11 cm width equipped with four 0.5-cm holes in the lid was used, which permitted vapor elimination and maintained an air-absent environment in the container during the pyrolysis process. At the end of the pyrolysis operation, the container was taken out of the furnace and cooled down to room temperature. Solid residues (biochars) were removed from the container and placed in a plastic bag for use throughout this study.

Air-dried RPM, B400, and B600 were physically characterized through an assessment of: Particle size distribution by using a Malvern Mastersizer STD06 laser granulometry; surface area and pore volumes according to the Brunauer, Emmet, and Teller (BET) method, via the measurements obtained by $N_2$ adsorption using an Autosorb Quantachrome apparatus; and finally, their morphology and structure through a scanning electron microscope (SEM) apparatus (SU-70). These solid matrices were also chemically characterized by the determination of: (i) Crystalline phase through X-ray diffraction (XRD) analyses; (ii) pH of zero-point charge (pHZC) according to the methodology given by Azzaz et al. [26]; (iii) elemental composition (C, H, N, O, S) by using a "vario EL cube" elemental analyzer; (iv) moisture, volatile matter, ash, and fixed carbon contents according to the British Standard Institution (BSI) protocol; and (v) their main contained functional groups through a Fourier transform infrared spectroscopy apparatus (Agilent Tech Cary 630, Ireland). The spectral resolution of the FTIR was 1 $cm^{-1}$ measured between 400 and 4000 $cm^{-1}$.

### *2.4. Batch Nutrient Release Experiments*

The batch experiments consisted of determining the P and K release kinetics from RPM and its derived biochars, as well as the effect of some key parameters, such as solution pH and solid matrices dosages, on this release process. These experiments were conducted at 20 ± 2 °C in 120-mL capped flasks. During these assays, a defined amount of RPM or biochars was shaken in 50 mL of distilled water for a fixed time at 400 rpm using an IKA RT15 Power IKAMAG multi position magnetic stirrer for a contact time 48 h. Then, the resulting suspensions were centrifuged at 3000 rpm for 8 min using a ROTOFIX 32 apparatus and filtrated using a vacuum pumping system. Dissolved P and K concentrations in filtrates were analyzed by an inductively coupled plasma spectrometry (ICP) device (Agilent Tech 5100 ICP OES, Ireland). The released amount of P "$q_P$ (mg $g^{-1}$)" or K "$q_K$ (mg $g^{-1}$)" per gram of solid matrix were determined as follows:

$$q_P = \frac{C_P}{D}, \tag{1}$$

$$q_K = \frac{C_K}{D}, \tag{2}$$

where $C_P$, $C_K$, *and* $D$ are the measured released concentrations of P and K, and the used solid matrix dose, respectively.

All the following batch nutrient release assays were carried out in triplicate and each analysis data reported in this study is an average of at least three independent parallel sample solutions.

### 2.4.1. Effect of Contact Time

The kinetics of P and K release from RPM and the two derived biochars (B400 and B600) was monitored at 1, 5, 10, and 30 min and 4, 17, 24, and 48 h. During these assays, the solid matrices dosage, initial aqueous pH, and temperature were fixed to 10 g $L^{-1}$, 5.6 (natural without adjustment), and 20 ± 2 °C, respectively. The subsequent nutrient relse data were fitted to the most common models, namely pseudo-first order (PFO) and pseudo-second order (PSO) models. The corresponding equations and the related assumptions were given by Azzaz et al. [26]. In order to assess the applicability of these two models and their ability in fitting the current experimental data, average percentages errors (APEs) between the measured and calculated released amounts were estimated as follows [27]:

$$\mathrm{APE} = \frac{\sum_{i=1}^{N}\left|\frac{q_{exp,i}-q_{th,i}}{q_{exp,i}}\right|}{N}, \quad (3)$$

where $q_{exp,i}$ and $q_{th,i}$ are the experimental and theoretical released P or K quantities at a given time "$i$", respectively. N is the number of experimental data.

### 2.4.2. Effect of Initial Aqueous pH

The effect of initial aqueous pH on the rate of P and K release from RPM and the derived biochars was assessed for adjusted values of 3, 5.6 (natural pH), and 8. All these experiments were carried out at a constant dose of 10 g $L^{-1}$, a temperature of 20 ± 2 °C, and a contact time of 48 h. The aqueous pH adjustment was performed by using 0.1 M $HNO_3$ or NaOH solutions.

### 2.4.3. Effect of RPM and RPM-Derived Biochars Doses

The impact of RPM, B400, and B600 doses on P and K release was assessed for an initial aqueous pH of 5.6 (without adjustment), a temperature of 20 ± 2 °C and a contact time of 48 h. The tested doses were 2, 5, 10, and 15 g $L^{-1}$.

### 2.4.4. Successive Nutrient Release Experiments

In order to determine the slowness of the RPM and its two derived biochars' capacities in releasing P and K, six successive batch experiments were carried out for a total contact time duration of 12 days. For each leaching experiment, the same solid samples were put in contact with distilled water at a dose of 10 g $L^{-1}$, an initial pH of 5.6 (without adjustment), and a contact time of 48 h. At the end of each assay, these solid matrices were recovered through filtration by using a vacuum pumping system with 0.45-μm paper filters, and then dried overnight at 40 °C. These dried samples were reused for the next leaching experiment in the presence of new distilled water under the same experimental conditions cited above.

### *2.5. Statistical Analysis*

All the experimental data (except elemental analysis and granulometric distribution) for RPM and its derived biochars represent means of at least triplicate measurements. Data were analyzed using STATISTICA 8.0 (Statsoft, Tulsa, OK, USA). ANOVA with Duncan's multiple range test were applied for mean separation at $p < 0.05$.

## 3. Results and Discussion

### *3.1. Thermogravimetric Analysis of Raw Poultry Manure*

The TG analysis of RPM was carried out under the experimental conditions given in Section 2.2. Figure 1 reports the RPM's TG records and the corresponding first derivative curve (DTG).

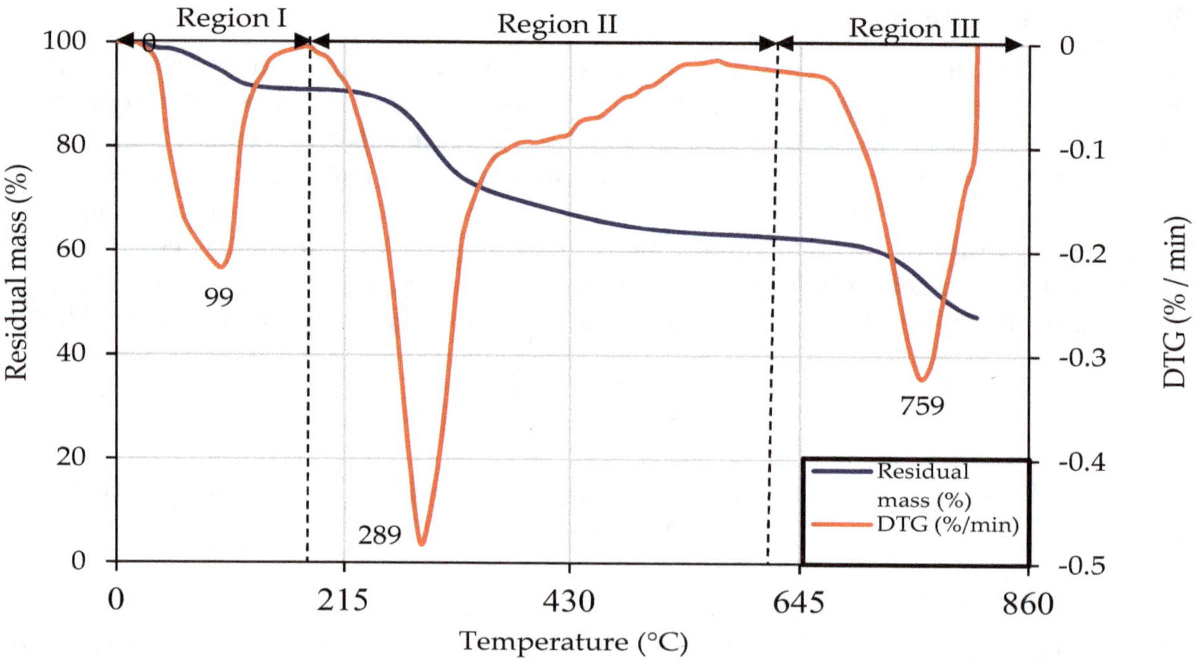


**Figure 1.** Thermograms (TGs) and its first derivative (DTG) of the used raw poultry manure (RPM dose = 16.7 mg; heating gradient = 10 °C min$^{-1}$).

Figure 1 shows the presence of three regions with different thermal behavior. Region I spans a temperature interval of 49 to 179 °C, with a progressive mass loss reaching a maximum rate of 9% at the end of this interval. In this region, the maximum loss rate (0.22% min$^{-1}$) was observed at a temperature of 99 °C and corresponds to moisture losses from the RPM. The second region is comprised between 179 and 609 °C, with a corresponding mass loss of about 27%. A maximum peak rate (0.48% min $^{1}$) was observed at a temperature of 289 °C. This peak apparition is mainly due to the thermo-oxidative degradation of highly oxygenated organic matter contained in RPM, such as labile alkyl systems, carbohydrates, and fatty acids [28,29]. The last thermal region corresponds to temperatures ranging between 609 to 810 °C, with a corresponding mass loss of 15%. A peak degradation rate was observed at a temperature of 759 °C (0.32% min$^{-1}$) and could be attributed to the oxidation of mineral and biogenic salts, such as calcium and potassium carbonate [30]. A global similar trend was reported by Cimo et al. [30] when studying the thermal behavior of RPM collected from a Sicilian chicken farm in Italy. It is important to underline that P speciation in biochars is very dependent on the pyrolysis temperature. At low temperatures, the organic carbon compounds are aromatic and therefore do not overlap with the P vibrations. For biochars produced at higher temperatures, prevailing P forms are tricalcium phosphate, hydroxyapatite, calcium, and iron phosphate precipitates [31].

### *3.2. Characterization of Raw Poultry Manure and Biochars*

The main physico-chemical properties of RPM as well as its two derived biochars at temperatures of 400 and 600 °C are given in Table 1. Granulometric characterization showed that the RPM and the corresponding biochars have a relatively fine texture with a mean particle size of 0.48, 0.43, and 0.31 mm, respectively. The uniformity coefficient (UC), which is the ratio of d60/d10, showed that RPM and B400 are heterogeneous media because their UCs are higher than 2. However, B600 could be considered as homogenous since its UC is lower than 2. The fine texture of the relative RPM and derived biochars indicates that they could play an important role in physico-chemical reactions with pollutants contained in rainwater or irrigation water [32].

**Table 1.** Main physico-chemical characteristics of raw poultry manure (RPM), B400, and B600 (*: In Wt.% dry basis). For each parameter, means with the same lowercase letters are not statistically different at $p < 0.05$.

| Physico-Chemical Property | RPM | B400 | B600 |
|---|---|---|---|
| Grains Size Distribution | | | |
| $d_{10}$ (mm) | 0.21 | 0.16 | 0.19 |
| $d_{50}$ (mm) | 0.48 | 0.43 | 0.31 |
| $d_{60}$ (mm) | 0.57 | 0.44 | 0.33 |
| Uniformity coefficient (UC = $d_{60}/d_{10}$) | 2.71 | 2.75 | 1.74 |
| Surface area ($m^2$ $g^{-1}$) | | | |
| BET | 0.88 | 4.30 | 5.34 |
| Charges density | | | |
| pHZC | 9.09 ± 0.03 a | 10.87 ± 0.008 b | 11.47 ± 0.015 c |
| Ultimate Analysis (%)* | | | |
| C | 25.56 ± 1.39 b | 22.04 ± 1.75 ab | 21.22 2.30a |
| H | 3.27 ± 0.79 b | 0.87 ± 0.015 a | 0.54 ± 0.04 a |
| N | 2.19 ± 0.17 b | 0.95 ± 0.19 a | 0.64 ± 0.11 a |
| S | 0.69 ± 0.0.17 a | 0.68 ± 0.46 a | 0.66 ± 0.005 a |
| O | 69.35± 2.43 a | 75.45 ± 2.31 b | 76.90 ± 2.38 b |
| C/H | 6.87 ± 1.85 a | 25.33 ± 1.59 b | 35.53 ± 7.61 c |
| Mineral Composition (mg $g^{-1}$) | | | |
| Al | 15.97 | 18.30 | 23.22 |
| Fe | 2.24 | 3.46 | 3.68 |
| K | 38.12 | 71.95 | 66.23 |
| Mg | 4.62 | 6.64 | 7.11 |
| Na | 18.98 | 20.17 | 28.67 |
| P | 17.00 | 20.33 | 43.17 |
| Si | 265.81 | 277.85 | 390.95 |
| Ti | 0.37 | 0.39 | 0.51 |
| Proximate Analysis (%)* | | | |
| Moisture | 16.69 ± 0.04 b | 0.76 ± 0.09 a | 0.80 ± 0.04 a |
| Ash | 51.35 ± 0.38 a | 78.64 ± 0.62 b | 79.74 ± 0.18 c |
| Volatile matter | 36.34 ± 0.18 b | 4.08 ± 0.12 a | 3.95 ± 0.13 a |
| Fixed carbon | 11.99 ± 0.05 a | 17.72 ± 0.13 c | 16.18 ± 0.24 b |

The BET surface area of RPM was assessed as 0.88 $m^2$ $g^{-1}$. Pyrolysis operation increased this value to 4.30 and 5.34 $m^2$ $g^{-1}$ for B400 and B600, respectively (Table 1). These relatively low surface areas could be imputed to the existence of inorganic ash materials at high concentrations that might fill or block access to micropores [22,33]. Generally, higher specific surfaces areas are obtained for higher carbonization temperatures of biomasses [23,24]. It is worth mentioning that the BET surface areas of biochars significantly affect not only the subsurface contaminant mobility but also the microbial activities [22,29]. These authors reported that the BET specific surface areas of their RPM and derived biochar at 400 °C were 0.90 and 3.94 $m^2$ $g^{-1}$, respectively, which are fairly close to the values found in the current study.

SEM analyses of the RPM and its derived biochars produced at temperatures of 400 and 600 °C are shown in Figure S1. RPM particles present a rough structure with a relatively oval shape (Figure S1). Besides, the pyrolysis process did not induce significant morphological changes. The presence of relatively small-sized particles in the three used solid materials (Figure S1) should play an important role in the nature of the physicochemical reactions with dissolved substances in liquid effluents [34].

The XRD spectra of RMP, B400, and B600 are given in Figure S2. Raw poultry manure showed an elevated background between $2\theta = 21°$, likely attributable to the organic matter content [35,36]. The peaks at $2\theta = 28.9°$ and $2\theta = 29.6°$ indicate the presence of low crystallinity (quartz and calcite) that already exists with a low intensity compared to its two biochars produced at 400 and 600 °C, respectively [37]. In addition to quartz and calcite, the sylvite (KCl), which is a cubic mesh mineral species of the chloride family, is present for B400 and B600 at peaks equal to $2\theta = 40.5°$. The phosphate mineral whitlockite ($(Ca,Mg)_3(PO_4)_2$) was identified only in B600 at a peak equal to $2\theta = 47.3°$ since phosphorus reached a more stable form [24,35].

The measured pHZC values of RPM, B400, and B600 were 9.09, 10.87, and 11.47, respectively (Table 1). This significant increase is mainly attributed to the concentration of the non-carbonized inorganic elements that were already present in the original feedstock [38]. The same behavior was reported by Song and Guo [22], who found that RPM carbonization at temperatures of 400 and 600 °C increased pHZC from 9.5 to 10.1 and 11.5, respectively. This observed alkaline property implies that these materials could be used as pH regulators for acidic soils when used as amendments or for acidic wastewaters when used as adsorbents for toxic substances. Such valorization pathways will significantly reduce the use of chemical reagents and hence the related expenses.

On the other hand, RPM pyrolysis significantly decreased C, H, N, and S contents as temperature increased (Table 1). For instance, original C, H, N, and S contents in RPM decreased respectively by 5.7%, 3.2%, 1.6%, and 0.02% in B600. These findings are mainly due to their transfer to bio-oil and/or incondensable exhaust during the pyrolysis process [24]. Furthermore, the C/H ratio increased with the increase of the pyrolysis temperature (Table 1) as a result of dehydration and decarboxylation reactions [33]. A similar trend was reported by Cimo et al. [30] when studying the pyrolysis of RPM collected from a chicken farm. They found C/H ratios of 7, 21, and 42 for RPM, and derived biochars at temperatures of 450 and 600 °C, respectively. The larger the C/H ratio is, the higher the polycondensation degree of organic systems. This confirms that an increasing pyrolysis temperature favors the formation of polycondensation aromatic rings.

The analysis of inorganic elements showed that silicon (Si), potassium (K), sodium (Na), phosphorus (P), aluminum (Al), and magnesium (Mg) exist in RPM with relatively high contents. The concentrations of these elements increase when increasing the pyrolysis temperature (Table 1). For instance, initial Mg, P, and K contents in RPM increased by 43.7% and 53.9%, 19.5% and 153.9%, 88.7% and 73.7% for B400 and B600, respectively. The presence of these nutrients in the produced biochars would be very useful for plant development and growth after land application.

Volatile matter (VM) contents significantly decreased from 36.3% for RPM to only 4.1% and 4.0% for B400 and B600. These changes are mainly due to the conversion of the RPM's organic matter into gaseous products during the pyrolysis process [39]. In contrast, ash contents increased from 51.4% for RPM to 78.6% and 79.7% for B400 and B600, respectively. The high ash contents of the three materials might be attributed to the physically incorporated soils and intrinsic inorganic elements, such as phosphorus and potassium, that are present in the raw manure and derived biochars [1]. Furthermore, the increase of the pyrolysis temperature leads to the dissociation of minerals from biomass [40]. As for ash, the fixed carbon contents increased after RPM pyrolysis, resulting in a more stable phase [39,41]. Unexpectedly, the measured fixed carbon content at 600 °C was slightly higher than the one measured at 400 °C (5.6%), which could be due to a combination of the samples' heterogeneity and analyses' uncertainties. Similar findings were reported by Enders et al. [42]. Finally, due to water evaporation during the pyrolysis process, the measured RPM moisture decreased from 16.7% to about 0.8% for both B400 and B600. Similar trends were also reported by Akdeniz and Novais et al. [2,37].

The FTIR spectra of raw manure feedstock and both biochars are given in Figure S3. For raw poultry manure, the major band near 3853 $cm^{-1}$ (H-bonded OH) was attributed to the stretching vibration of hydrogen bonds of carboxylic acid and alcohols [30]. Peaks at 2992 $cm^{-1}$ and 2865 $cm^{-1}$ revealed asymmetric and symmetric (–CH) stretching bands of $CH_3$ and $CH_2$ groups, respectively and are associated with the aliphatic functional groups existing in lipids [43]. The peak at 1650 $cm^{-1}$ concerns the C=O stretching of $CO_2H$ and C=O of primary amides, which indicates the presence of protein. A $CH_2$ deformation peak registered at 1416 $cm^{-1}$ is characteristic of saturated fatty acids and cellulose [44]. The peak at 1233 $cm^{-1}$ of C-O is associated with the stretching of carbohydrates. Finally, the small peak appearing between 871 and 1024 $cm^{-1}$ in the raw feedstock was attributed to stretching -CH and represents the aromatic structures' out-of-plane bending vibrations [39].

The pyrolysis of RPM at 400 and 600 °C generated new transformational products. Compared to RPM, the spectra of produced biochars reveal the presence of higher aromatic and lower carbohydrates' structures. In this context, the intensity of the H-bonded OH peak significantly decreased with the increase of the pyrolysis temperature. This is likely related to the dehydration process during the pyrolysis operation and suggests that the carboxylic acid might be dissolved and hydrolyzed. Furthermore, the aliphatic C-H stretching of $CH_3$ and $CH_2$ groups observed respectively at 2992 and 2865 $cm^{-1}$ almost disappeared in the biochars' spectra, which could be imputed to the hydrolysis of aliphatic chain and dissolution. Similarly, the peak at 1650 $cm^{-1}$, which is associated with C=O stretching of primary amides, presents a clear intensity decrease with an increasing pyrolysis temperature. This behavior is probably due to the decomposition of the C=O bonding that generally enhances the formation of monoxide carbon and dioxide carbon [43], and causes a change in N functional groups by pyrolysis [45,46]. The observed peak at 1416 $cm^{-1}$, which is associated with $CH_2$ deformation for both biochars, presents an intensity increase with temperature [39]. Also, the bonding of C-O increased at 1233 $cm^{-1}$, which is associated with the stretching of –COH in alcohols and –COR in aliphatic ethers [44]. The peak between 871 and 1024 $cm^{-1}$ in the raw poultry manure is associated to aromatic –CH out-of-plane bending vibrations. This peak increased with the pyrolysis temperature, indicating a gradual conversion of aromatic structures during this process [44]. The presence of the various above-cited functional groups may play an important role in nutrient leaching from these biochars and especially pollutant adsorption from wastewaters [45,46]. For instance, P ion adsorption was confirmed by the apparition of a new P-O stretching vibration at 1050 $cm^{-1}$ [47].

### *3.3. Phosphorus and Potassium Release*

The release of P and K from RPM and biochars was assessed in the batch mode under the various experimental conditions detailed in Section 2.4.

#### 3.3.1. Effect of Contact Time—Kinetic Release Study

In order to assess the effect of contact time on the release efficiency of P and K from RPM, B400, and B600, batch experiments were performed for an initial pH of 5.6 (without adjustment), a constant dose of 10 g $L^{-1}$, and contact times varying between 1 min and 48 h (see Section 2.4.1). Outcomes indicated that P and K release from the three studied solid matrices was clearly a time-dependent process (Figure 2a,b and Figure 6a,b).

The P release kinetics were fast at the beginning of the experiment (until 4 h) as shown in Figure 2b. This is probably attributed to an ion exchange or desorption process and/or dissolution of some poorly crystalline or amorphous phosphates [25]. At this time, the released amount was estimated to 4.81 mg $g^{-1}$, which represents 94% of the overall released quantity. Then, the released amount continued to increase but at much slower rates until an equilibrium state was reached. This latter state, which is characterized by an approximate released amount, was observed after approximately 17 h. A similar trend was also observed for potassium release (Figure 3a,b). The observed stiff slope at the beginning of the experiments indicates a quick and instantaneous release related to the abundance of low-energy-retained P or K (P/K) elements on the surface of the solid matrices. At this step, the release rate was relatively fast because P/K was essentially desorbed from the easily accessible sites to water located at the exterior surfaces of RPM and the derived biochars. When the release of P/K from the exterior surfaces reached its limits, P/K could be further released from the more inaccessible sites located at the inner surfaces of the tested materials.

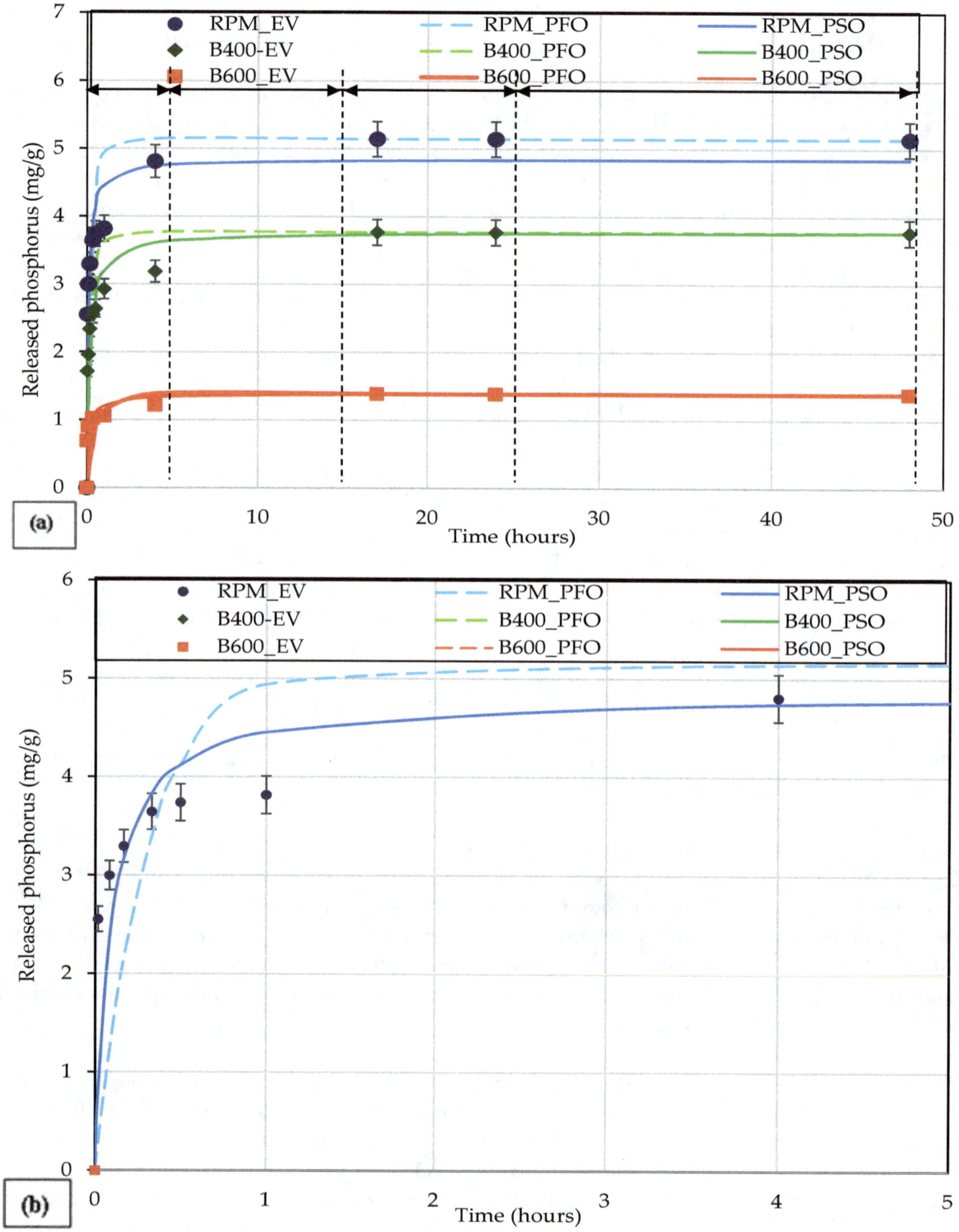


**Figure 2.** Comparison between the experimental quantities and theoretical ones of phosphorus released by raw poultry manure (RPM) and its derived biochars (EV: Experimental values; PFO: pseudo-first order; PSO: pseudo-second order) for 48 h (**a**) and zoom for 5 h (**b**).

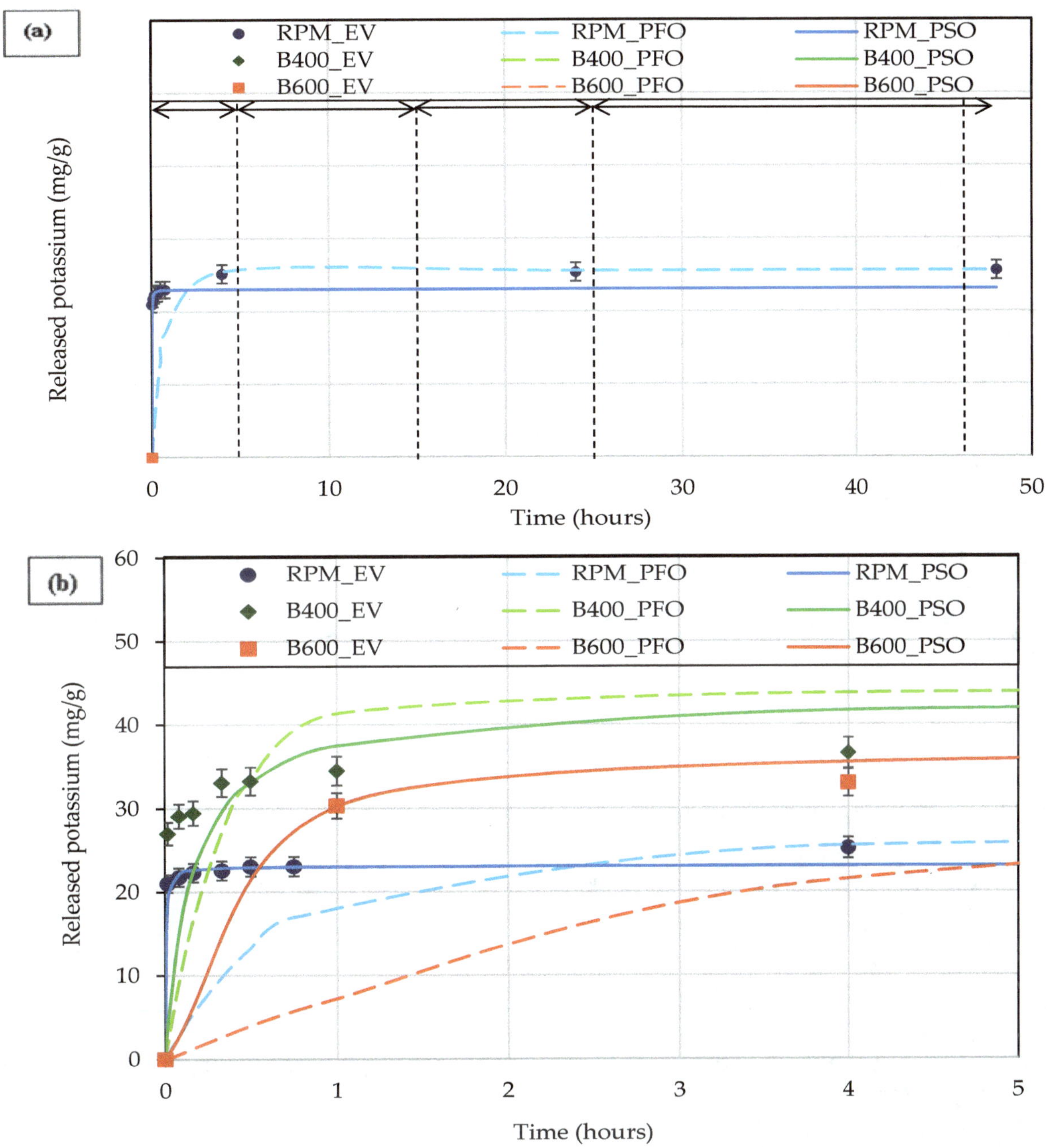


**Figure 3.** Comparison between the experimental and theoretical quantities of potassium released by raw poultry manure (RPM) and its derived biochars (EV: Experimental values; PFO: pseudo-first order; PSO: pseudo-second order) for 48 h (**a**) and zoom for 5 h (**b**).

Even if the pyrolysis process had concentrated P in the biochars compared to RPM (Table 1), at equilibrium, the released P amounts from RPM (5.14 mg $g^{-1}$) were much higher than those corresponding to B400 (3.77 mg $g^{-1}$) and B600 (1.39 mg $g^{-1}$). These amounts correspond to leachable P percentages of 30.2%, 18.5%, and 3.2% for RPM, B400, and B600, respectively. These rates are likely due to the fact that the pyrolysis process had converted a portion of RPM-containing phosphorus into water non-extractable forms, such as crystalline metal phosphates, which generally dissolve much more slowly and are less available [45,48]. In this context, Wang et al. [24] reported that RPM wastes collected from local hen houses (USA) released an average phosphorus amount of 6.48 mg $g^{-1}$. This quantity decreased to about 1.04 mg $g^{-1}$ for its derived biochar at a temperature of 600 °C. Similarly, Peak et al. [48] also reported that the released amount of phosphorus from an RPM sample (collected from the region of Avila, Spain) was about 2.57 times higher than the one measured for its derived biochar produced at 500 °C.

The required time to attain an equilibrium state for K was much longer compared to P since it took about 48 h of contact time (Figure 3a). Furthermore, it appears that the pyrolysis process converted the initial potassium in RPM into a more leachable and water-soluble form in biochars through dissociation mechanisms [40]. As such, released K amounts were 25.6, 43.3, and 37.7 mg $g^{-1}$ for RPM, B400, and B600, respectively. Accordingly, Song and Guo [22] showed that the released potassium amount from an RPM-derived biochar at a temperature of 600 °C (44.61 mg $g^{-1}$) was 1.9 higher than that observed for the raw biomass.

According to Tables 2 and 3, and Figures 2b and 3b, the confrontation of P/K release experimental data to those determined by the pseudo-first-order (PFO) and pseudo-second-order (PSO) models showed that the PFO model completely fails to reproduce the experimental values for the three tested solid materials. In fact, the calculated P/K determination coefficients as well as the related average percentage errors between the measured and the calculated values were not at all satisfactory. For instance, the "determination coefficients" and the APE of P were two by two 0.786 and 23.7%, 0.822 and 23.4%, and 0.734 and 29.6% for RPM, B400, and B600, respectively.

**Table 2.** Kinetic parameters of phosphorus release by RPM and its two derived biochars at 400 and 600 °C.

| Solid Support | Pseudo-First-Order Model (PFO) | | | Pseudo-Second-Order Model (PSO) | | |
|---|---|---|---|---|---|---|
| | $K_{1.PFO}$ ($h^{-1}$) | $R^2_{PFO}$ | $APE_{PFO}$ (%) | $K_{2.PSO}$ (g $mg^{-1}$ $h^{-1}$) | $R^2_{PSO}$ | $APE_{PSO}$ (%) |
| RPM | 3.238 | 0.786 | 23.714 | 2.318 | 0.867 | 14.467 |
| B400 | 3.242 | 0.822 | 23.431 | 1.435 | 0.879 | 17.766 |
| B600 | 1.688 | 0.734 | 29.566 | 4.602 | 0.836 | 19.860 |

**Table 3.** Kinetic parameters of potassium release by RPM and its two derived biochars at 400 and 600 °C.

| Solid Support | Pseudo-First-Order Model (PFO) | | | Pseudo-Second-Order Model (PSO) | | |
|---|---|---|---|---|---|---|
| | $K_{1.PFO}$ ($h^{-1}$) | $R^2_{PFO}$ | $APE_{PFO}$ (%) | $K_{2.PSO}$ (g $mg^{-1}$ $h^{-1}$) | $R^2_{PSO}$ | $APE_{PSO}$ (%) |
| RPM | 3.618 | 0.474 | 27.130 | 1.087 | 0.743 | 17.837 |
| B400 | 2.886 | 0.668 | 26.199 | 0.150 | 0.750 | 19.517 |
| B600 | 0.039 | 0.422 | 56.748 | 0.109 | 0.995 | 1.737 |

On the contrary, the PSO model succeeded to sufficiently reproduce the experimental data since the calculated determination coefficients (R2) for P or K were higher than that of the pseudo first order. Moreover, the calculated APE between the experimental and theoretical released P or K values were significantly lower than those calculated by the PFO model (Tables 2 and 3). For the three solid materials and the PSO model, the maximum values of APE for P and K were estimated at 19.9% (Table 2) and 19.5% (Table 3), respectively. However, the maximum values of this parameter were 29.6% and 56.7% for P and K, respectively. The highest non-accordance between the experimental and theoretical data corresponds to the initial phases of the experiments where the nutrient release slopes are maximal (Figures 2b and 3b). The fact that the experimental data were better fitted by the PSO suggests that nutrient leaching from RPM and its derived biochars might be controlled by a chemical process [26].

### 3.3.2. Effect of pH

The effect of initial aqueous pH on P and K release from the three materials was assessed under the experimental conditions described in Section 2.4.2. The results are illustrated in Figure 4a,b.

P-released amounts from RPM were approximately constant, with an average value of 5.51 mg $g^{-1}$. However, for B400 and B600, P-released amounts increased with the increase of pH values. For instance, the released amount from B600 increased by about 43.5% when the pH increased from 3 to 8. Globally, a similar variation trend was observed for K. The maximum released amount (41.33 mg $g^{-1}$) was observed for B400 at an initial aqueous leaching pH of 8. It is worth mentioning that high aqueous pH values could lead to the precipitation of the released phosphorus and potassium and therefore affect their availability to plants. For instance, depending on the aqueous pH value, phosphorus can exist as tricalcium phosphate, hydroxyapatite, calcium, and iron phosphates compounds [31]. However, it is very difficult for the precipitation of potassium as potassium hydroxide, which could be observed for very high pH values, to occur under the tested experimental conditions. It is important to underline that for high pH values, phosphorus could precipitate into calcium phosphate and magnesium phosphate, which makes P assimilation by crops more difficult [34,49]. These results confirmed that for a wide pH range, the biochars derived from RPM represent an important source of potassium that could be assimilated by various crops after soil application.

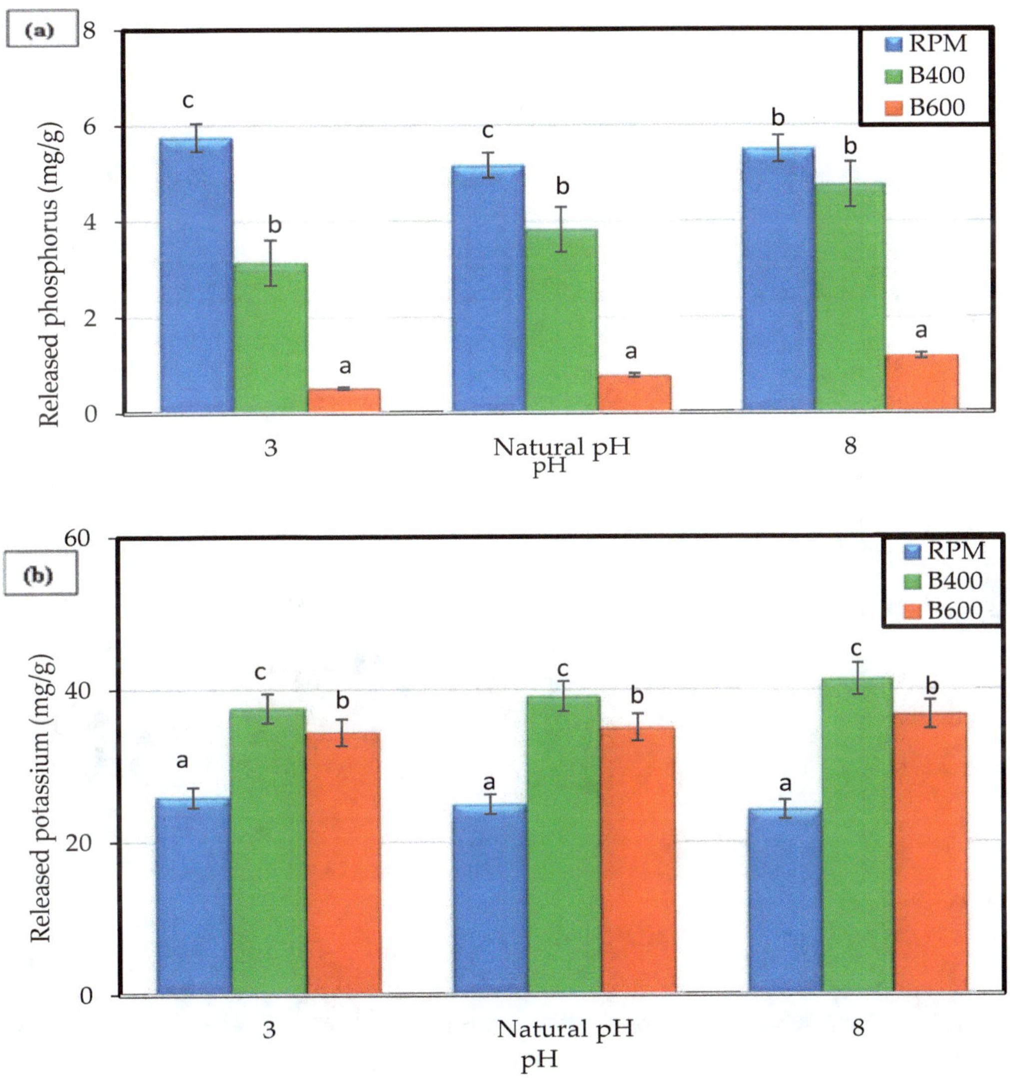


**Figure 4.** Effect of initial aqueous pH on the phosphorus (**a**) and potassium (**b**) release efficiency from raw poultry manure and its biochars derived at temperatures of 400 and 600 °C (dosage = 10 g $L^{-1}$, T = 20 ± 2 °C). At each pH value, means with the same lowercase letters are not statistically different at $p < 0.05$.

### 3.3.3. Effect of Dose

The effect of RPM, B400, and B600 doses on P and K release efficiency was determined according to the experimental conditions given in Section 2.4.3 and are illustrated in Figure 5a,b.

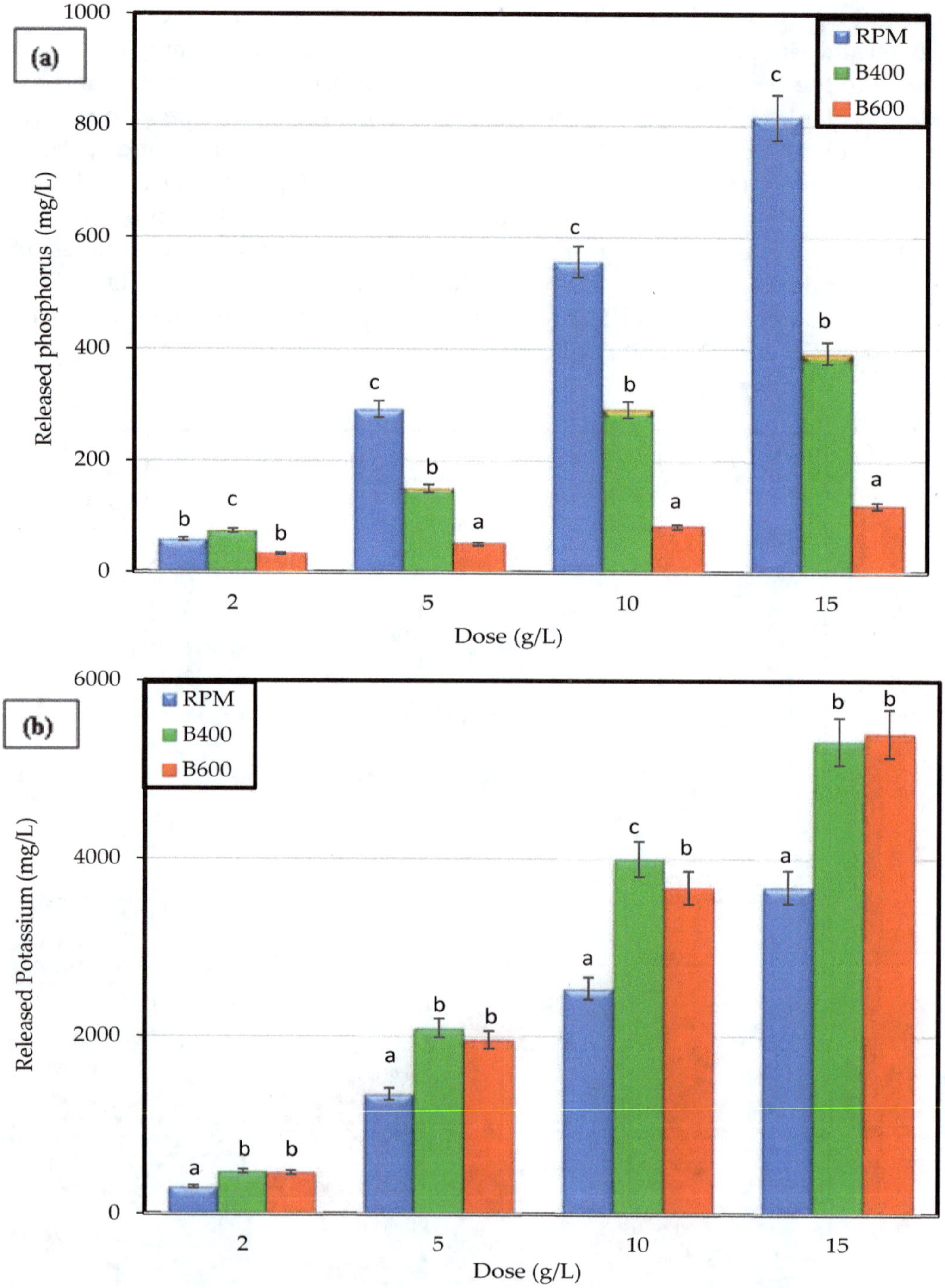


**Figure 5.** Effect of the used solid matrix doses on the release efficiency of phosphorus (**a**) and potassium (**b**) (pH = 6.2, T = 20 ± 2 °C). At each dose, means with the same lowercase letters are not statistically different at $p < 0.05$.

Released P or K concentrations increased significantly with the increase of the used dose. For example, P/K released concentrations increased from 58.58/304.83 to 815.47/3680.1 mg $L^{-1}$ for RPM, from 74.45/47.88 to 393.6/5321.8 mg $L^{-1}$ for B400, and from 33.18/464.7 to 119.13/5409.5 mg $L^{-1}$ for B600 when the material dose increased from 2 to 15 g $L^{-1}$, respectively. This is mainly due to the increase of the available surface area of the solid matrices that could interact with the aqueous solutions [40]. Similar behavior was observed by Wang et al. [24] when studying the release of phosphorus from an RPM-derived biochar at 600 °C. They found that increasing the biochar concentration from 5 to 10 and then to 25 g $L^{-1}$ resulted in an increase of the released P from 8.8 to 14.0 and 26.8 mg $L^{-1}$, respectively.

#### 3.3.4. Successive Leaching Experiments—Nutrient Slow Release

The cumulative release of P and K from RPM, B400, and B600 for six successive leaching experiments (duration of 48 h for each one) is shown in Figure 6a,b, respectively.

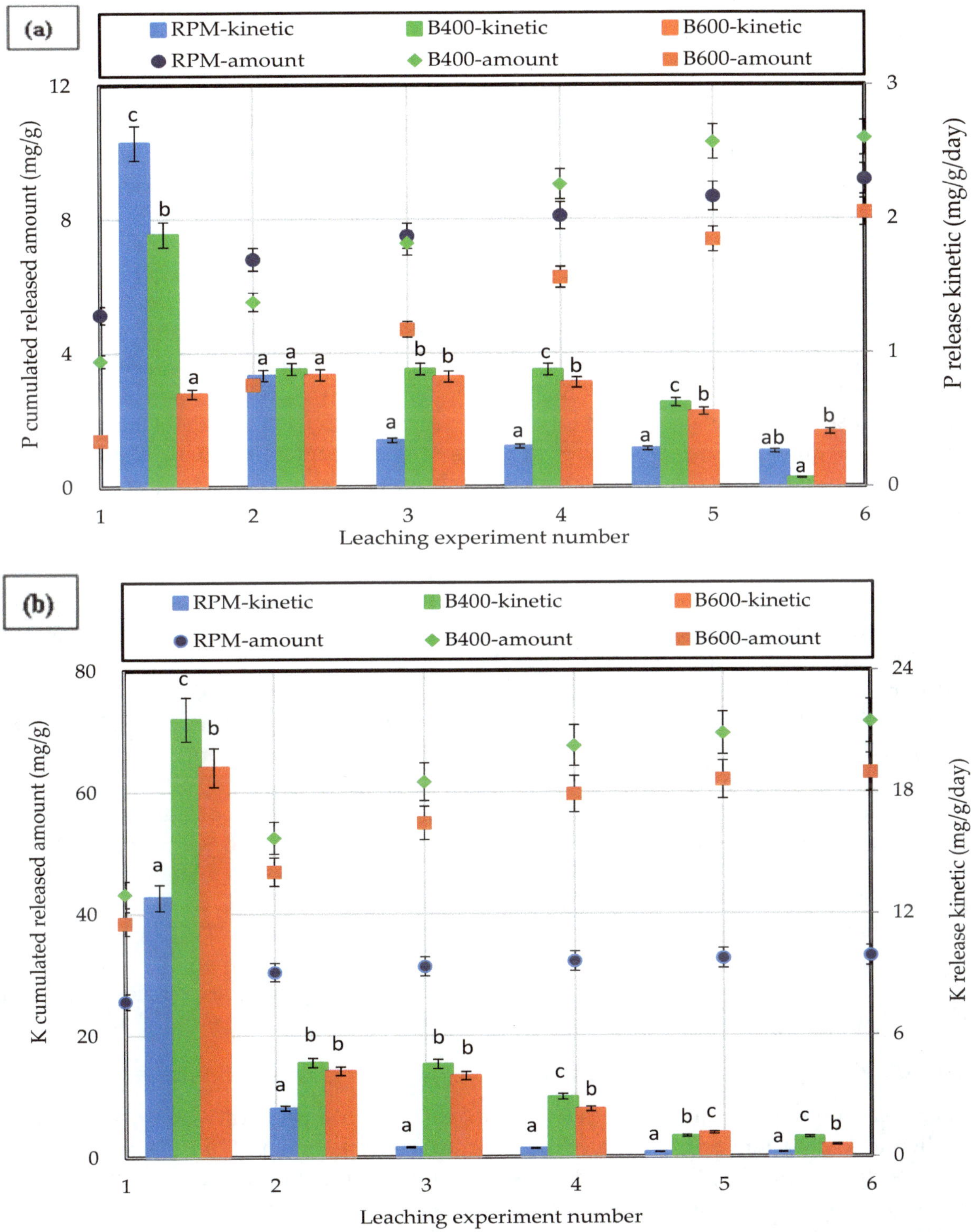


**Figure 6.** Effect of successive leaching experiments on the release of phosphorus (**a**) and potassium (**b**) from RPM and its two derived biochars: B400 and B600 (dosage = 10 g L-1; pH = 6.2; T = 20 ± 2 °C). At each leaching experiment number, means with the same lowercase letters are not statistically different at $p < 0.05$.

It can be clearly seen that for all the studied solid matrices, P release significantly increased with the operated number of leaching experiments. At the sixth assay, the maximal released amounts attained were approximately 9.2, 10.4, and 8.2 mg $g^{-1}$ for RPM, B400, and B600, respectively (Figure 6a), which are 78.7%, 176.3%, and 489.5% higher than those observed in the first leaching experiment. These released amounts correspond to 54.0%, 51.2%, and 19.0% of the initial contained P in RPM, B400, and B600, respectively (Table 1). These percentages are very important compared to the ones given in the scientific literature from continuous (not successive) leaching experiments. In fact, they were assessed as only 5.5% [24] and 6.2% [22] for RPM-derived biochars collected from broiler farms in Seaford (USA) at temperatures of 400 and 600 °C and contact times of 12 and 3 days, respectively. Besides, Liang et al. [25] estimated the P release percentage to be about 10% for a biochar derived from RPM collected from a Florida dairy farm at a temperature of 450 °C and a continuous contact time of 10 days.

Compared to the first leaching experiment, cumulated P amounts considerably increased at the sixth successive leaching assay by 78.7%, 176.3%, and 489.5% for RPM, B400, and B600, respectively (Figure 6a). Concerning the kinetic release, in contrast to RPM and the fourth leaching experiment, B400 and B600 continued to release P with relatively high average release kinetic rates of 0.878 and 0.812 mg $g^{-1}$ $day^{-1}$. Even after the six successive leaching experiment (12 days), B400 and B600 were still releasing P att rates of 0.062 and 0.407 mg $g^{-1}$ $day^{-1}$, respectively (Figure 6a).

Similar release trends were observed for potassium. For both B400 and B600 materials, released K amounts significantly increased with the progress of the leaching experiment, attaining values of 71.6 and 63.2 mg $g^{-1}$, respectively, at the sixth leaching assay, corresponding to overall increase percentages of 65.7% and 64.3%, respectively, compared to the first leaching experiment (Figure 6b). These amounts represent 99.5% and 95.4% of the contained K in B400 and B600, respectively, confirming its transformation to a more leachable form and its slow release. These percentages were important when compared to continuous leaching assays where they were evaluated as only 45% to 49% for RPM-derived biochars collected from broiler farms in Seaford (USA) at temperatures varying between 400 and 600 °C, and a continuous contact time of 3 days [22].

Moreover, compared to RPM, until four successive leaching experiments, relatively high kinetic release rates of 4.071 and 3.531 mg $g^{-1}$ $day^{-1}$ were recorded for B400 and B600, respectively (Figure 6b). Even in the last leaching experiment, B400 and B600 were still releasing K at 0.968 and 0.587 mg $g^{-1}$ $day^{-1}$, respectively (Figure 6b).

These outcomes indicate that the two RPM-derived biochars could be considered as slow-release fertilizers, where P and K could be available for several seasons in real conditions [50]. Unlike commercial fertilizers, where nutrient dissolution is very fast—even instantaneous for some products [51]—the use of RPM biochars as amendments for agricultural soils constitutes an important asset for both optimal plant growth and groundwater pollution reduction [52]. As crops also need nitrogen and other micronutrients, combining biochar application with chemical fertilizers or fresh composts has the benefits of simultaneously improving soil physical properties, nutrient retention, and plant yield [53].

## 4. Conclusions

This research work demonstrated that raw poultry manure pyrolysis significantly affects its physico-chemical characteristics, especially the chemical composition, pH of zero-point charge, and surface properties, including the specific surface area and functional groups. Altogether, these characteristics play an important role on the nutrient release efficiency. Nevertheless, the chemical state of these nutrients in these solid matrices, which is very dependent on the used pyrolysis temperature, might be the most important factor influencing this process. The pyrolysis process favors P conversion into a more stable phase through its probable complexation with magnesium and/or calcium and therefore reduces its release rate in the aqueous phase. In contrast, RPM pyrolysis converts potassium into a more leachable form. On the other hand, compared to RPM, RPM-derived biochars could be considered as an attractive soil conditioner with very slow nutrient release capacities. This promising property will result in a more efficient nutrient uptake by crops and limited leaching to both groundwater and surface runoff. Further investigation is currently being carried out to monitor the impact of biochar doses in agricultural soils under dynamic conditions on nutrient leaching ability and assimilation by selected plants.

**Supplementary Materials:** Figure S1: SEM images of surface surfaces morphologies of RPM (a), B400 (b) and B600 (c), Figure S2: XRD images of crystalline phases of RPM, B400 and B600 (OM: organic matter; Q: quartz; C: calcite; S: sylvite and W: whitlockite), Figure S3: FTIR Spectra of RPM and its two derived biochars at temperatures of 400 and 600 °C.

**Author Contributions:** Conceptualization, S.H., S.J., W.K, and M.K.; methodology, S.H., S.J., J.J.L. and W.K.; software, H.H.; validation, S.J., W.K. and M.J.; formal analysis, S.H., M.J.; investigation, S.H., S.J. and W.K.; resources, M.K., J.J.L. and H.H.; data curation, S.H., S.J., M.J., H.H. and W.K.; writing—original draft preparation, S.H.; writing—review and editing, S.J., M.J., H.H. and W.K.; visualization, S.H., S.J. and W.K.; supervision, S.J. and W.K.; project administration, S.J. and W.K.; funding acquisition, S.J. and W.K.

**Acknowledgments:** This research work was carried out in close collaboration between CERTE (Tunisia) and the University of Limerick (Ireland) within the framework of the ERASMUS+ International Credit Mobility Programme. The authors would like to gratefully thank the financial contribution of this Programme and the Tunisian Ministry of Higher Education and Scientific Research.

## References

1. Mante, O.D.; Agblevor, F.A. Influence of pine wood shavings on the pyrolysis of poultry litter. *Waste Manag.* **2010**, *30*, 2537–2547. [CrossRef] [PubMed]
2. Akdeniz, N. A systematic review of biochar use in animal waste composting. *Waste Manag.* **2019**, *88*, 291–300. [CrossRef] [PubMed]
3. Sellami, F.; Jarboui, R.; Hachicha, S.; Medhioub, K.; Ammar, E. Co-composting of oil exhausted olive-cake, poultry manure and industrial residues of agro-food activity for soil amendment. *Bioresour. Technol.* **2008**, *99*, 1177–1188. [CrossRef] [PubMed]
4. Turan, N.G. The effects of natural zeolite on salinity level of poultry litter compost. *Bioresour. Technol.* **2008**, *99*, 2097–2101. [CrossRef] [PubMed]
5. Rodic, V.; Peric, L.; Djukic-Stojcic, M.; Vukelic, N. The environmental impact of poultry production. *Biotechnol. Anim. Husb.* **2011**, *27*, 1673–1679. [CrossRef]
6. Lu, L.; Liao, X.D.; Luo, X.G. Nutritional strategies for reducing nitrogen, phosphorus and trace mineral excretions of livestock and poultry. *J. Integr. Agric.* **2017**, *16*, 2815–2833. [CrossRef]
7. Hu, Y.; Cheng, H.; Tao, S. Environmental and human health challenges of industrial livestock and poultry farming in China and their mitigation. *Environ. Int.* **2017**, *107*, 111–130. [CrossRef]
8. Cely, P.; Gascó, G.; Paz-Ferreiro, J.; Méndez, A. Agronomic properties of biochars from different manure wastes. *J. Anal. Appl. Pyrolysis* **2015**, *111*, 173–182. [CrossRef]
9. Lynch, D.; Henihan, A.M.; Kwapinski, W.; Zhang, L.; Leahy, J.J. Ash agglomeration and deposition during combustion of poultry litter in a bubbling fluidized-bed combustor. *Energy Fuels* **2013**, *27*, 4684–4694. [CrossRef]
10. Taupe, N.C.; Lynch, D.; Wnetrzak, R.; Kwapinska, M.; Kwapinski, W.; Leahy, J.J. Updraft gasification of poultry litter at farm-scale—A case study. *Waste Manag.* **2016**, *50*, 324–333. [CrossRef]
11. Kelleher, B.P.; O'Callaghan, M.N.; Leahy, M.J.; O'Dwyer, T.F.; Leahy, J.J. The use of fly ash from the combustion of poultry litter for the adsorption of chromium (III) from aqueous solution. *J. Chem. Technol. Biotechnol.* **2002**, *77*, 1212–1218. [CrossRef]
12. Agyarko-Mintah, E.; Cowie, A.; Singh, B.P.; Joseph, S.; Van Zwieten, L.; Cowie, A.; Harden, S.; Smillie, R. Biochar increases nitrogen retention and lowers greenhouse gas emissions when added to composting poultry litter. *Waste Manag.* **2017**, *61*, 138–149. [CrossRef] [PubMed]

13. Wongrod, S.; Simon, S.; Guibaud, G.; Lens, P.N.L.; Pechaud, Y.; Huguenot, D.; van Hullebusch, E.D. Lead sorption by biochar produced from digestates: Consequences of chemical modification and washing. *J. Environ. Manag.* **2018**, *219*, 277–284. [CrossRef]
14. Kwapinski, W.; Byrne, C.M.P.; Kryachko, E.; Wolfram, P.; Adley, C.; Leahy, J.J.; Novotny, E.H.; Hayes, M.H.B. Biochar from biomass and waste. *Waste Biomass Valoriz.* **2010**, *1*, 177–189. [CrossRef]
15. Mau, V.; Gross, A. Energy conversion and gas emissions from production and combustion of poultry-litter-derived hydrochar and biochar. *Appl. Energy* **2018**, *213*, 510–519. [CrossRef]
16. Hu, X.; Gholizadeh, M. Biomass Pyrolysis: A Review of the Process Development and Challenges from Initial Researches up to the Commercialisation Stage. *J. Energy Chem.* **2019**, 109–143. [CrossRef]
17. Ding, Y.; Liu, Y.; Liu, S.; Huang, X.; Li, Z.; Tan, X.; Zeng, G.; Zhou, L. Potential Benefits of Biochar in Agricultural Soils: A Review. *Pedosphere* **2017**, *27*, 645–661. [CrossRef]
18. Shashvatt, U.; Benoit, J.; Aris, H.; Blaney, L. $CO_2$-assisted phosphorus extraction from poultry litter and selective recovery of struvite and potassium struvite. *Water Res.* **2018**, *143*, 19–27. [CrossRef]
19. Ghanim, B.M.; Pandey, D.S.; Kwapinski, W.; Leahy, J.J. Hydrothermal carbonisation of poultry litter: Effects of treatment temperature and residence time on yields and chemical properties of hydrochars. *Bioresour. Technol.* **2016**, *216*, 373–380. [CrossRef]
20. Azzaz, A.A.; Jellali, S.; Akrout, H.; Assadi, A.A.; Bousselmi, L. Optimization of a cationic dye removal by a chemically modified agriculture by-product using response surface methodology: Biomasses characterization and adsorption properties. *Environ. Sci. Pollut. Res.* **2016**, *24*, 9831–9846. [CrossRef]
21. Abdallah, M.M.; Ahmad, M.N.; Walker, G.; Leahy, J.J.; Kwapinski, W. Batch and Continuous Systems for Zn, Cu, and Pb Metal Ions Adsorption on Spent Mushroom Compost Biochar. *Ind. Eng. Chem. Res.* **2019**, *58*, 7296–7307. [CrossRef]
22. Song, W.; Guo, M. Quality variations of poultry litter biochar generated at different pyrolysis temperatures. *J. Anal. Appl. Pyrolysis* **2012**, *94*, 138–145. [CrossRef]
23. Trazzi, P.A.; Leahy, J.J.; Hayes, M.H.B.; Kwapinski, W. Adsorption and desorption of phosphate on biochars. *J. Environ. Chem. Eng.* **2016**, *4*, 37–46. [CrossRef]
24. Wang, Y.; Lin, Y.; Chiu, P.C.; Imhoff, P.T.; Guo, M. Phosphorus release behaviors of poultry litter biochar as a soil amendment. *Sci. Total Environ.* **2015**, *512–513*, 454–463. [CrossRef] [PubMed]
25. Liang, Y.; Cao, X.; Zhao, L.; Xu, X.; Harris, W. Phosphorus Release from Dairy Manure, the Manure-Derived Biochar, and Their Amended Soil: Effects of Phosphorus Nature and Soil Property. *J. Environ. Qual.* **2014**, *43*, 1504–1509. [CrossRef]
26. Azzaz, A.A.; Jellali, S.; Assadi, A.A.; Bousselmi, L. Chemical treatment of orange tree sawdust for a cationic dye enhancement removal from aqueous solutions: Kinetic, equilibrium and thermodynamic studies. *Desalin. Water Treat.* **2016**, *57*, 22107–22119. [CrossRef]
27. Haddad, K.; Jellali, S.; Jeguirim, M.; Ben Hassen Trabelsi, A.; Limousy, L. Investigations on phosphorus recovery from aqueous solutions by biochars derived from magnesium-pretreated cypress sawdust. *J. Environ. Manag.* **2018**, *216*, 305–314. [CrossRef]
28. Xu, Y.; Chen, B. Investigation of thermodynamic parameters in the pyrolysis conversion of biomass and manure to biochars using thermogravimetric analysis. *Bioresour. Technol.* **2013**, *146*, 485–493. [CrossRef]
29. Zhao, L.; Cao, X.; Mašek, O.; Zimmerman, A. Heterogeneity of biochar properties as a function of feedstock sources and production temperatures. *J. Hazard. Mater.* **2013**, *256–257*, 1–9. [CrossRef]
30. Cimo, G.; Kucerik, J.; Berns, A.E.; Schaumann, G.E.; Alonzo, G.; Conte, P. Effect of Heating Time and Temperature on the Chemical Characteristics of Biochar from Poultry Manure. *J. Agric. Food Chem.* **2014**, *62*, 1912–1918. [CrossRef]
31. Bekiaris, G.; Peltre, C.; Jensen, L.S.; Bruun, S. Using FTIR-photoacoustic spectroscopy for phosphorus speciation analysis of biochars. *Spectrochim. Acta Part A Mol. Biomol. Spectrosc.* **2016**, *168*, 29–36. [CrossRef] [PubMed]
32. Kalderis, D.; Hawthorne, S.B.; Clifford, A.A.; Gidarakos, E. Interaction of soil, water and TNT during degradation of TNT on contaminated soil using subcritical water. *J. Hazard. Mater.* **2008**, *159*, 329–334. [CrossRef] [PubMed]
33. Jindo, K.; Mizumoto, H.; Sawada, Y.; Sanchez-Monedero, M.A.; Sonoki, T. Physical and chemical characterization of biochars derived from different agricultural residues. *Biogeosciences* **2014**, *11*, 6613–6621. [CrossRef]

34. Haddad, K.; Jellali, S.; Jaouadi, S.; Benltifa, M.; Mlayah, A.; Hamzaoui, A.H. Raw and treated marble wastes reuse as low cost materials for phosphorus removal from aqueous solutions: Efficiencies and mechanisms. *C. R. Chim.* **2015**, *18*, 75–87. [CrossRef]
35. Cao, X.; Harris, W. Properties of dairy-manure-derived biochar pertinent to its potential use in remediation. *Bioresour. Technol.* **2010**, *101*, 5222–5228. [CrossRef]
36. Wang, M.; Zhu, Y.; Cheng, L.; Andserson, B.; Zhao, X.; Wang, D.; Ding, A. ScienceDirect Review on utilization of biochar for metal-contaminated soil and sediment remediation. *J. Environ. Sci.* **2017**, *63*, 156–173. [CrossRef]
37. Novais, S.V.; Zenero, M.D.O.; Barreto, M.S.C.; Montes, C.R.; Cerri, C.E.P. Phosphorus removal from eutrophic water using modified biochar. *Sci. Total Environ.* **2018**, *633*, 825–835. [CrossRef]
38. Novak, J.M.; Busscher, W.J.; Laird, D.L.; Ahmedna, M.; Watts, D.W.; Niandou, M.A.S. Impact of Biochar Amendment on Fertility of a Southeastern Coastal Plain Soil. *Soil Sci.* **2009**, *174*, 105–112. [CrossRef]
39. Cantrell, K.B.; Hunt, P.G.; Uchimiya, M.; Novak, J.M.; Ro, K.S. Impact of pyrolysis temperature and manure source on physicochemical characteristics of biochar. *Bioresour. Technol.* **2012**, *107*, 419–428. [CrossRef]
40. Chen, X.; Lin, Q.; He, R.; Zhao, X.; Li, G. Hydrochar production from watermelon peel by hydrothermal carbonization. *Bioresour. Technol.* **2017**, *241*, 236–243. [CrossRef]
41. Domingues, R.R.; Trugilho, P.F.; Silva, C.A.; De Melo, I.C.N.A.; Melo, L.C.A.; Magriotis, Z.M.; Sánchez-Monedero, M.A. Properties of biochar derived from wood and high-nutrient biomasses with the aim of agronomic and environmental benefits. *PLoS ONE* **2017**, *12*, e0176884. [CrossRef] [PubMed]
42. Enders, A.; Hanley, K.; Whitman, T.; Joseph, S.; Lehmann, J. Characterization of biochars to evaluate recalcitrance and agronomic performance. *Bioresour. Technol.* **2012**, *114*, 644–653. [CrossRef] [PubMed]
43. Cao, W.; Cao, C.; Guo, L.; Jin, H.; Dargusch, M.; Bernhardt, D.; Yao, X. Hydrogen production from supercritical water gasification of chicken manure. *Int. J. Hydrogen Energy* **2016**, *41*, 22722–22731. [CrossRef]
44. Mau, V.; Quance, J.; Posmanik, R.; Gross, A. Phases' characteristics of poultry litter hydrothermal carbonization under a range of process parameters. *Bioresour. Technol.* **2016**, *219*, 632–642. [CrossRef] [PubMed]
45. Sun, J.; Hoon, S.; Jung, S.; Ryu, C.; Jeon, J.; Shin, M.; Park, Y. Production and utilization of biochar: A review. *J. Ind. Eng. Chem.* **2016**, *40*, 1–15.
46. Xiao, R.; Wang, J.J.; Gaston, L.A.; Zhou, B.; Park, J.H.; Li, R.; Dodla, S.K.; Zhang, Z. Biochar produced from mineral salt-impregnated chicken manure: Fertility properties and potential for carbon sequestration. *Waste Manag.* **2018**, *78*, 802–810. [CrossRef]
47. Halajnia, A.; Oustan, S.; Najafi, N.; Khataee, A.R.; Lakzian, A. Adsorption-desorption characteristics of nitrate, phosphate and sulfate on Mg-Al layered double hydroxide. *Appl. Clay Sci.* **2013**, *80–81*, 305–312. [CrossRef]
48. Peak, D.; Sims, J.T.; Sparks, D.L. Solid-state speciation of natural and alum-amended poultry litter using XANES spectroscopy. *Environ. Sci. Technol.* **2002**, *36*, 4253–4261. [CrossRef]
49. Barca, C.; Gérente, C.; Meyer, D.; Chazarenc, F.; Andrès, Y. Phosphate removal from synthetic and real wastewater using steel slags produced in Europe. *Water Res.* **2012**, *46*, 2376–2384. [CrossRef]
50. Mukherjee, A.; Zimmerman, A.R. Organic carbon and nutrient release from a range of laboratory-produced biochars and biochar-soil mixtures. *Geoderma* **2013**, *193*, 122–130. [CrossRef]
51. Liu, G.; Zotarelli, L.; Li, Y.; Dinkins, D.; Wang, Q. *Controlled-Release and Slow-Release Fertilizers as Nutrients Management Tools*; UF/IFAS Extension: Gainesville, FL, USA, 2017.
52. Sun, K.; Qiu, M.; Han, L.; Jin, J.; Wang, Z.; Pan, Z.; Xing, B. Speciation of phosphorus in plant- and manure-derived biochars and its dissolution under various aqueous conditions. *Sci. Total Environ.* **2018**, *634*, 1300–1307. [CrossRef] [PubMed]
53. Schulz, H.; Glaser, B. Effects of biochar compared to organic and inorganic fertilizers on soil quality and plant growth in a greenhouse experiment. *J. Plant Nutr. Soil Sci.* **2012**, *175*, 410–422. [CrossRef]

# 7

# Effect of Seasonal Temperature on the Performance and on the Microbial Community of a Novel AWFR for Decentralized Domestic Wastewater Pretreatment

**Juanhong Li and Xiwu Lu ***

School of Energy and Environment, Southeast University, Nanjing 210096, China; li_juanhong1@163.com
* Correspondence: xiwulu@seu.edu.cn.

Academic Editors: Faisal Ibney Hai, Kazuo Yamamoto and Jega Veeriah Jegatheesan

**Abstract:** Due to environmental burden and human health risks in developing countries, the treatment of decentralized domestic wastewater has been a matter of great concern in recent years. A novel pilot-scale three-stage anaerobic wool-felt filter reactor (AWFR) was designed to treat real decentralized domestic wastewater at seasonal temperature variations of 8 to 35 °C for 364 days. The results showed that the average chemical oxygen demand (COD) removal efficiencies of AWFR in summer and winter were 76 ± 7.2% and 52 ± 5.9% at one day and three days Hydraulic Retention Time (HRT), respectively. COD mass balance analysis demonstrated that even though COD removal was lower in winter, approximately 43.5% of influent COD was still converted to methane. High-throughput MiSeq sequencing analyses indicated that *Methanosaeta*, *Methanobacterium*, and *Methanolinea* were the predominant methanogens, whereas the genus *Bacillus* probably played important roles in fermentation processes throughout the whole operation period. The performance and microbial community composition study suggested the application potential of the AWFR system for the pretreatment of decentralized domestic wastewater.

**Keywords:** decentralized domestic wastewater; seasonal temperature; anaerobic wool-felt filter reactor; high-throughput MiSeq sequencing

## 1. Introduction

Water, land, and energy are important resources for the rapid growth and development of the global economy. However, in recent years, many developing countries face challenges and pressures of water and land pollution, and energy shortages. Particularly in rural areas of China, water pollution problems are increasingly aggravated due to the direct discharge of a large amount of untreated domestic wastewater [1,2]. Although centralized biological processes have been well developed to be used in urban municipal wastewater treatment plants, they are not suitable for rural areas owing to the dispersed population, poor wastewater collection, and weaker economy in these areas. In addition, a concept of wastewater-to-resource is receiving increased attention by researchers and engineers working on wastewater treatment technology development [3]. Therefore, it is highly desirable to select a sustainable, robust and cost-effective process for the treatment of decentralized domestic wastewater in developing countries [4].

Simultaneous energy recovery and sustainable wastewater treatment make the application of anaerobic biotechnology in decentralized domestic wastewater treatment interesting [5–7]. Upflow anaerobic sludge blanket reactors (UASB) [8,9], anaerobic baffled reactor (ABR) [10], and anaerobic membrane bioreactor (AnMBR) [7,11,12] are commonly used to treat domestic wastewater. Despite the effective chemical oxygen demand (COD) removal by these anaerobic biotechnologies, some challenges still exist, such as complex three-phase separator and long solids' retention time of UASB, membrane fouling and high energy costs of AnMBR, and high land footprint of ABR. Compared to these high-rate anaerobic reactors, the anaerobic filters (AFs) have drawn attention because of the following advantages [13–15]: (1) simple design configuration with low capital and operating costs, (2) excellent capability of high biomass retention with carriers, (3) stable operation, (4) greater tolerance to hydraulic loading rate and organic loading rate, and (5) low footprint. The potential of using AFs for treating wastewater have been well developed not only for industrial wastewater, but also for domestic wastewater [5,16–18]. However, most of these studies focused on the effects of carriers, hydraulic retention time (HRT), and organic loading rate on the performance of AFs, and there were a few studies that focused on the effects of seasonal temperature [5,18]. Previous studies have demonstrated that anaerobic biotechnology strongly depends on operation temperature, and AFs are commonly operated under mesophilic conditions [5,19]. Nevertheless, considering the energy consumption and capital expenditure, it is economically unviable to heat anaerobic systems in decentralized domestic wastewater treatment of rural areas. Therefore, the application of AFs at seasonal temperature is more useful due to less energy demand. In context to China, this study is inevitable as the seasonal temperature varied throughout the whole year. Additionally, carriers are an important component of AFs, which determine the biomass retention capacity and the performance of the system. Therefore, the choices of appropriate carriers play an important role in AF system performance, particularly in the rural area of developing countries. A variety of natural materials, including zeolite [20], ceramic [21], rock [5,22] and coconut shells [23] have been adopted as biofilm carriers. Even though these natural materials are low-cost, they are susceptible to clogging and require significant operation attention due to biofilm growth [22]. Wool felt is a class of natural porous materials that is a common waste product of paper making factories in China. Compared to solid materials, wool felt is a class of soft filler with a large specific surface area of about 950 $m^2/g$ that avoids clogging. Wool felt may be a viable option for AFs due to low cost and a large specific surface. Hence, wool felt has been applied before in a membrane reactor for the biosorption of heavy metals [24]. In response to those factors, a pilot-scale three-stage anaerobic wool-felt filter (AWFR) was designed to treat decentralized domestic wastewater.

It is well known that anaerobic biotechnology is a biological degradation process comprising mutual metabolic interaction among the bacterial and archaeal community. Microorganisms play important roles in the efficiency and stability of the anaerobic treatment [25]. However, most of the available studies have been operated as a "black box" without focusing on the role of the microbial communities [26]. To get comprehensive insights into the microbial community of the anaerobic process, molecular biology tools are used to determine the structures of the microorganisms in the system. To date, the molecular biology tools have been extensively developed. Nevertheless, most of the studies still applied the techniques of terminal restriction fragment length polymorphism (T-RFLP) [27], fluorescence in situ hybridization (FISH) [19,28], denaturing gradient gel electrophoresis (DGGE) [29–31] and quantitative polymerase chain reaction (q-PCR) [32]. Compared to the conventional ones, the high-throughput sequencing, especially Illumina MiSeq sequencing, is becoming one of the most popular molecular tools for microbial community analysis due to low-cost, fast turnaround time producing several gigabases of sequence and greater coverage [33]. Nevertheless, there is no study that focused on the microbial communities in the AWFR system.

The aim of this work was to evaluate the performance of a pilot-scale three-stage anaerobic wool-felt filter reactor (AWFR) in treating decentralized domestic wastewater seasonally. Additionally, this work also investigated the seasonal variations of microbial community in the AWFR system based on high-throughput MiSeq sequencing. The results provided a comprehensive understanding of the relationship between the microbial community and the performance of the AWFR system.

## 2. Materials and Methods

### 2.1. Experimental Setup and Operation

The experimental studies were conducted in a pilot-scale three-stage AWFR made of polyvinyl chlorine polymer. Each identical AWFR has a height of 2.5 m and an internal diameter of 170 mm as shown in Figure 1. In order to avoid clogging, each reactor with the effective volume of 50 L was packed with vertical wool felt carrier as biomass growth support media. The wool felt carriers had a high specific surface area (about 950 $m^2/g$) with a high porosity (>95%). The reactors were continuously fed with real decentralized domestic wastewater by using a peristaltic pump. Based on the seasonal temperature variations, the operation of the AWFR system was divided into five periods: start-up period (1–30 day, 15 ± 3.4 °C, $n$ = 30), spring period (31–90 day, 21 ± 3.0 °C, $n$ = 60), summer period (91–180 day, 31 ± 3.7 °C, $n$ = 90), autumn period (181–270 day, 25 ± 5.2 °C, $n$ = 90) and winter period (271–364 day, 10 ± 2.2 °C, $n$ = 94). The main operating conditions are summarized in Table 1.

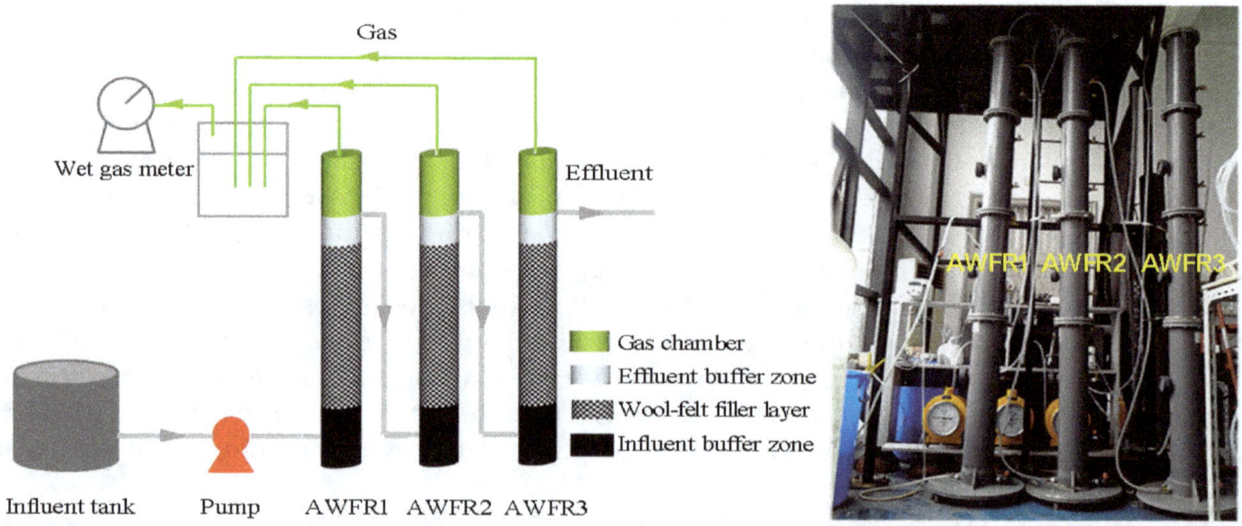


**Figure 1.** Schematic diagram and photo of the anaerobic wool-felt filter reactor (AWFR).

**Table 1.** The main operating conditions for the anaerobic wool-felt filter reactor (AWFR) system during the five operation stages.

| Stage | Phase (Days) | Duration (Days) | Hydraulic Retention Time (HRT) (Days) | Hydraulic Loading Rate (HLR) ($m^3/m^2$/day) | Organic Loading Rate (OLR) (mgCOD/L/day) [1] | Temperature (°C) [1] |
|---|---|---|---|---|---|---|
| start-up | 1–30 | 30 | 3 | 2.2 | 75 ± 13.6 ($n$ = 30) | 15 ± 3.4 ($n$ = 30) |
| spring | 31–90 | 60 | 2 | 4.4 | 143 ± 28.3 ($n$ = 60) | 21 ± 3.0 ($n$ = 60) |
| summer | 91–180 | 90 | 1 | 6.6 | 352 ± 61.8 ($n$ = 90) | 31 ± 3.7 ($n$ = 90) |
| autumn | 181–270 | 90 | 2 | 4.4 | 135 ± 32.9 ($n$ = 90) | 25 ± 5.2 ($n$ = 90) |
| winter | 271–364 | 94 | 3 | 2.2 | 82 ± 12.9 ($n$ = 94) | 10 ± 2.2 ($n$ = 94) |

[1] Values are given as mean ± standard deviation.

### 2.2. Decentralized Domestic Wastewater and Seed Sludge

The experimental setup was fed with real decentralized domestic wastewater, which was collected from the dormitories and restaurants of the Southeast University (Wuxi, China). The chemical characteristics of the decentralized domestic wastewater are presented in Table 2.

**Table 2.** The chemical characteristics of decentralized domestic wastewater (unit: mg/L, except for pH).

| Parameter | Chemical Oxygen Demand (COD) | Total Phosphorus (TP) | Total Nitrogen (TN) | pH |
|---|---|---|---|---|
| Value [1] | 284 ± 69.2 | 2.8 ± 0.8 | 32.2 ± 7.7 | 7.1 ± 0.2 |

[1] Values are given as mean ± standard deviation; number of measurements ($n$): $n$ = 364 for COD, TP, TN and pH.

The seeding anaerobic sludge used in this study was obtained from the anaerobic digester of a municipal wastewater treatment plant located in Wuxi, China. Approximately 10 L anaerobic sludge was inoculated into the wool felt carrier of each reactor before startup of the AWFR system. The initial total suspended solids (TSS) and volatile suspended solids (VSS) of the seeding anaerobic sludge were 28.2 g/L and 14.3 g/L, respectively.

### 2.3. *Analysis Methods*

#### 2.3.1. Chemical Analysis

Chemical oxygen demand (COD), total nitrogen (TN), total phosphorus (TP), total suspended solids (TSS) and volatile suspended solids (VSS) were measured according to the Standard Methods [34]. The concentrations of volatile fatty acids were analyzed using a gas chromatography GC 3900 (Tenghai, Shandong, China) equipped with an SE-30 capillary column (30 m × 0.32 mm × 0.25 μm) and a flame ionization detector. The operation temperatures of the injector port, column oven and detector were 200, 120 and 230 °C, respectively. Nitrogen gas was used as the carrier gas at a flow rate of 40 mL/min. The biogas production was measured by using an LML-1 wet gas meter (Changchun Automobile Filter Co., Ltd., Changchun, China). The biogas composition was determined using a gas chromatography GC 2001 (Tenghai, Shandong, China) equipped with a thermal conductivity detector and a 4 m × 3 mm inside diameter stainless-steel column packed with TDX-01 (80/100 mesh). The operation temperature of the injector port, column oven and detector were 150, 150 and 180 °C, respectively. Argon gas was used as the carrier gas at a flow rate of 25 mL/min. The pH and temperature were measured using a portable YSI-pH 100 meter (YSI Co., Yellow Springs, OH, USA).

#### 2.3.2. COD Mass Balance Calculation

The COD mass balance of the AWFR system was conducted during spring, summer, autumn and winter. Seasonal periods were characterized by the following operation HRTs: 2 days (spring period covering Day 31–90), 1 day (summer period covering Day 91–180), 2 days (autumn period covering Day 181–270), and 3 days (winter period covering Day 271–364).

The COD mass balance was calculated using the Equation (1):

$$COD_{in} = COD_{VFAs} + COD_{CH4(g)} + COD_{CH4(s)} + COD_{others}, \quad (1)$$

where: $COD_{in}$ (g/day) represents the average COD concentration of real decentralized domestic wastewater; $COD_{VFAs}$ (g/day) represents the average COD concentration of acetate and propionate in the effluent; $COD_{CH4(g)}$ (g/day) represents the average COD concentration of methane produced in biogas; $COD_{CH4(s)}$ (g/day ) represents the average COD concentration of methane dissolved in the effluent; $COD_{others}$ (g/day) represents the organic matter that has been utilized for biomass formation, the complex organic matter that is not biodegradable, COD consumed by sulphate reducing bacteria (SRB), and COD removed and converted to $CO_2$.

Dissolved methane was calculated using Equations (2) and (3) suggested by a previous study [35]:

$$CH_4(s) = 4 \times K_H \times P_{gas} \times flow(Q), \quad (2)$$

$$K_H = 0.384 \times t + 36.44, \quad (3)$$

where $K_H$ (mg/L/atm) is Henry's constant, $P_{gas}$ (atm) is the partial pressure of the gas above the liquid, flow(Q) (L/day) is feed flow and *t* (°C) is the temperature.

#### 2.3.3. Scanning Electron Microscopy (SEM)

Scanning electron microscopy (SEM) was used to investigate the development and structure of the anaerobic biofilm in the system. Representative samples (about 1 $cm^2$ size) were taken from the AWFR system on days 15, 90, 180, 270 and 340. The samples were firstly fixed with 3% (*v*/*v*) glutaraldehyde in 0.1 M phosphate buffer (pH 6.8) for 4 h at 4 °C. Then, these samples were washed with 0.1 M phosphate buffer for three times. Subsequently, these samples were dehydrated through graded ethanol (30%, 50%, 70%, 90%, and 100% *v*/*v*, 10 min for each concentration). After that, the samples were replaced by isoamyl acetate (twice, 10 min for each time), dried at a critical point and then coated with gold. Finally, these samples were examined via the scanning electron microscope (S-4800, Hitachi, Tokyo, Japan).

#### 2.3.4. Microbial Community Analysis by Illumina MiSeq Sequencing

To assess the complete microbial community structures, anaerobic biofilm samples were collected from the AWFR system on days 15, 90, 180, 270 and 340. The total DNA from the anaerobic biofilm samples was extracted using the OMEGA Soil DNA Kit D5625-01 (Omega Bio-Tek, Norcross, GA, USA) based on the manufacturer's protocol. The quality and quantity of the extracted DNA were measured by a Nanodrop 1000 spectrophotometer (NanoDrop Technologies, Wilmington, DE, USA).

For gene libraries construction, the V3–V4 regions of 16S rDNA genes were PCR-amplified using BAC319F/806R (5′-ACTCCTACGGGAGGCAGCAG-3′/5′-GGACTACHVGGGTWTCTAAT-3′) and ARC349F/806R (5′-GYGCASCAGKCGMGAAW-3′/5′-GGACTACHVGGGTWTCTAAT-3′), respectively. PCR reactions were conducted in a total volume of 25 μL mixture containing 12.5 μL Premix Ex TaqTM Hot Start Version (TaKaRa, Dalian, China), 2.5 μL of each primer (1 μM), 25 ng template DNA and $ddH_2O$. The bacterial PCR amplification was performed under the following conditions: initial denaturing at 98 °C for 30 s, 30 cycles of denaturing (98 °C for 30 s), annealing (56 °C for 30 s), and elongation (72 °C for 45 s), and a final extension step at 72 °C for 10 min. Archaeal community PCR amplification condition was similar to the bacterial community except that the cycle's number was 35. The PCR products were confirmed by 2% agarose gel electrophoresis and purified using the AxyPrepDNA Gel Extraction Kit (Axygen, Union City, CA, USA) following the manufacturer's protocol. Finally, the purified amplicons were pooled in equimolar and paired-end sequenced using the Illumina MiSeq platform (Illumina Inc., San Diego, CA, USA).

After sequencing, the raw sequences were filtered for quality, trimmed and processed based on the Mothur (version 1.30.1, http://www.mothur.org) analytic pipeline. Briefly, the adapters, barcodes and primers were trimmed. The sequence reads containing ambiguous base calls, with homopolymers >6 bp and shorter than 200 bp were removed. Chimeras detected by UCHIME (version 4.2.40, http://drive5.com/usearch/manual/uchime_algo.html) were filtered out. The resulting high quality sequences were clustered into operation taxonomic units (OTU) at 97% similarity level by Mothur. Representative sequences selected for each OTU were assigned taxonomy using a Ribosomal Database Project (RDP) classifier with a confidence threshold of 80%. The raw sequencing data were deposited to the National Center for Biotechnology Information (NCBI) Short Read Archive (Accession number: SRP101990).

## 3. Results and Discussion

### *3.1. Bioreactor Performance*

#### 3.1.1. COD Removal

To evaluate the performance of the AWFR system, the seasonal COD removal efficiencies were investigated at different HRTs for 364 days. During the whole operation period, the influent COD concentration ranged from approximately 157 to 469 mg/L (Figure 2). At the beginning of the start-up period (days 1–30), the AWFR was operated in the temperatures ranged from 8 °C to 17 °C, with an HRT of three days. The COD removal efficiency fluctuated in the range of 24.6–51.8%. Based on the COD removal efficiency in the start-up period, we concluded that microbial community adapted to the operation conditions and the AWFR was successfully initiated. The COD removal efficiency increased gradually in spring (days 31–90), thus resulting in an average COD removal efficiency of 61 ± 7.9% ($n$ = 60) at an HRT of two days, with average temperature variations of 21 ± 3.0 °C ($n$ = 60). COD removal efficiency can improve with temperature. Thus, the HRT was adjusted to one day when the average temperature increased to 31 ± 3.7 °C ($n$ = 90) in summer (days 91–180). Despite shortening the HRT, following temperature increase, the average effluent COD concentration decreased to 82.5 mg/L, and the average COD removal efficiency increased to 76 ± 7.2% ($n$ = 90). The change in COD removal efficiency might be due to higher activities of microbial communities in summer. In order to keep COD removal efficiency above 50%, the HRT was adjusted from two days (autumn) to three days (winter). Even though the OLR decreased from 135 ± 32.9 mgCOD/L/day ($n$ = 90) in autumn to 82 ± 12.9 mgCOD/L/day ($n$ = 94) in winter, the COD removal efficiency still decreased from 57 ± 9.4% (autumn, average temperature 25 ± 5.2 °C, $n$ = 90) to 52 ± 5.9% (winter, average temperature 10 ± 2.2 °C, $n$ = 94). These findings indicated that the temperature played an important role in the AWFR system operation. Low temperature might increase the wastewater viscosity, slow down the rate of biological reaction, and decrease COD removal efficiency. Compared to the previous study with elastic fiber carriers in two-stage anaerobic filter for domestic wastewater treatment [12], the COD removal efficiency in this study was better, which may be because the wool felt carrier enhanced the agglomeration of the microbial community in the AWFR system. Overall, the performance of the AWFR system was relatively stable and the COD removal efficiency was higher in summer than in winter. To achieve better performance of the AWFR system at the temperature ranging from 8 °C to 35 °C, an HRT of three days is recommended.

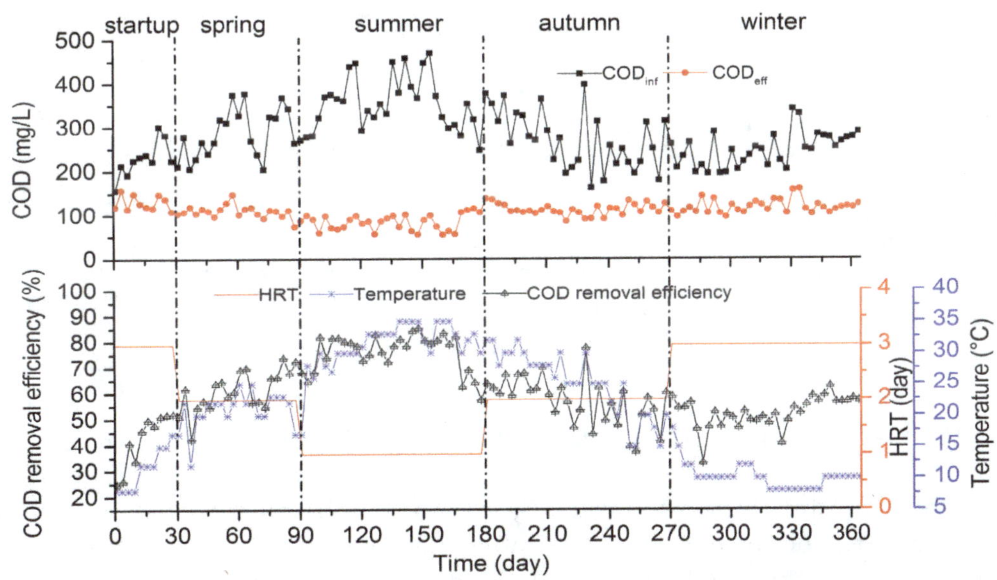


**Figure 2.** Influent and effluent concentrations of chemical oxygen demand (COD) and COD removal efficiency.

### 3.1.2. VFA Accumulation and Biogas Production

Since volatile fatty acids (VFA) is an important performance indicator of the anaerobic system, the variation of VFAs with respect to seasonal temperature variations was studied in the AWFR system. It was observed clearly that acetate and propionate were the main VFAs in the effluent as shown in Figure 3, indicating that organic matter was degraded to VFAs by both hydrolytic and acidogenic bacteria. However, no butyrate and valerate were detected, which might be due to their low concentrations below the detection threshold. During spring (days 31–90), the average effluent concentrations of acetate and propionate were 44.5 ± 8.2 mg/L ($n$ = 60) and 3.8 ± 1.6 mg/L ($n$ = 60), respectively. With the average temperature increasing to 31 ± 3.7 °C ($n$ = 90) in summer (days 91–180), the average effluent concentrations of acetate and propionate drastically dropped to 30.1 ± 5.0 mg/L ($n$ = 90) and 2.6 ± 0.7 mg/L ($n$ = 90), respectively. Nevertheless, the concentrations of acetate and propionate increased from 45.6 ± 15.6 mg/L ($n$ = 90) and 6.3 ± 2.6 mg/L ($n$ = 90) to 62.1 ± 8.5 mg/L ($n$ = 94) and 6.3 ± 3.2 mg/L ($n$ = 94) in winter. Similar results were also observed with AFBR to treat domestic wastewater in winter [36]. The effluent concentrations of both acetate and propionate were much lower than those in winter even though the OLR in summer (352 ± 61.8 mgCOD/L/day, $n$ = 90) was higher. This coincided with higher COD removal efficiency in summer than in winter, indicating that the conversion of VFAs to biogas was higher in summer than in winter because of high activities of the microbial community, especially for acetoclastic methanogens. Based on these findings, we further confirmed that the seasonal temperature variations affected the anaerobic metabolism of decentralized wastewater.

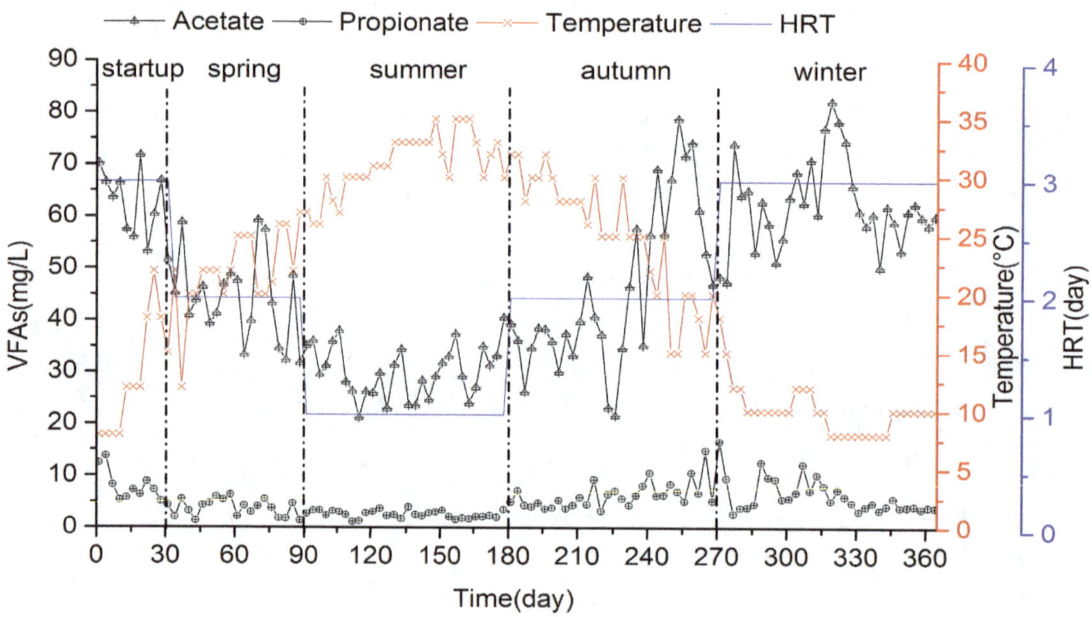


**Figure 3.** Volatile fatty acids (VFAs) production in the AWFR system during the whole operation.

The daily biogas production is shown in Figure 4. The daily biogas production in the start-up period was low, but after the formation of the stable anaerobic biofilm, it gradually increased with temperature increase, even though the HRT was reduced. The maximum daily biogas production of 10.7 L/day was achieved in summer, when the HRT was one day. This result corresponded well with a high reduction of COD concentration in effluent and a high COD removal efficiency during that period. However, the biogas production decreased with temperature decrease in winter, while acetate and propionate were accumulated. The lower biogas production in winter might be explained by poor degradation of organic matter and higher solubility of biogas at a lower temperature.

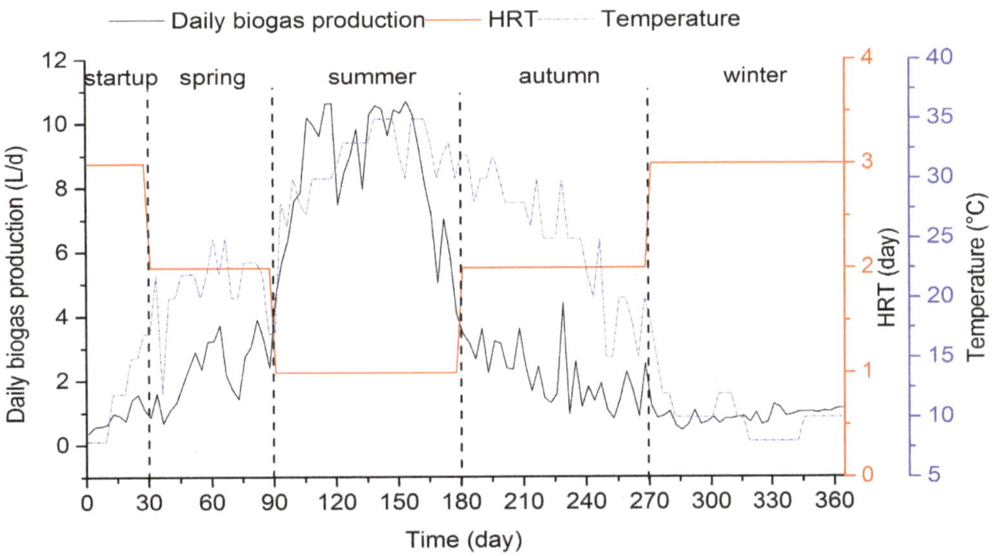


**Figure 4.** Biogas production in the system during the whole operation.

### *3.2. COD Mass Balance*

Based on Equations (1)–(3), a COD mass balance was used to assess the bioconversion process of organic matter during spring, summer, autumn and winter (Figure 5). Approximately 43.5–52.5% of influent COD was converted to methane (including gaseous methane and dissolved methane). Similar COD conversions were reported for an anaerobic filter treating on-site domestic wastewater [37]. In addition, 10.2–31.0% of influent COD was converted to VFAs, while 25.5–37.3% of influent COD was transformed to biomass, $CO_2$ and non-biodegradable organic matters. During the spring, about 28.3% of influent COD was converted to gaseous methane and 18.7% of influent COD was accounted for VFAs. With an increase of temperature in summer, more organic matter was converted to gaseous methane (41.1% of influent COD) than VFAs (10.2% of influent COD). In contrast, more VFAs were accumulated in winter. Meanwhile, the results demonstrated that dissolved methane represented about 17.7%, 11.4%, 16.4%, and 23.7% of COD during spring, summer, autumn, and winter, respectively, indicating that more methane dissolved in effluent with a decrease of temperatures. As a potential energy resource, future studies should focus on the recovery of dissolved methane in the AWFR system to avoid the loss of dissolved methane.

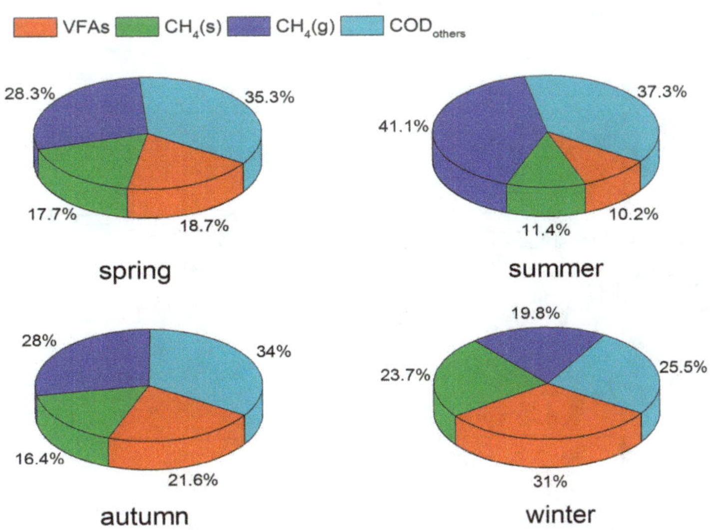


**Figure 5.** COD balance in different seasons.

### 3.3. Nutrient Removal

Nitrogen and phosphorous are two important nutrients for agricultural and landscaping reuse. The influent and effluent concentrations of TN and TP were monitored throughout the whole operation period. As shown in Figure 6a,b, the influent concentration of TN fluctuated from 19.6 to 54.2 mg/L throughout the operation period. In spring, with the temperature of 21 ± 3.0 °C ($n$ = 60), the TN concentration in the effluent of the AWFR system was 30.2 ± 5.9 mg/L ($n$ = 60), corresponding to a 15.7 ± 2.5% ($n$ = 60) TN removal efficiency at an HRT of two days. When the temperature increased to 31 ± 3.7 °C ($n$ = 90) in summer, the TN removal efficiency slightly increased to 19.0 ± 2.7% ($n$ = 90) at an HRT of one day. Despite an HRT increase from one day in summer to three days in winter, the average TN removal efficiency decreased to 12.9 ± 2.8% ($n$ = 94). Overall, the TN removal was low in this study because TN was probably removed by microbial assimilation only, and not via a nitrification–denitrification pathway. Meanwhile, a similar trend was observed for TP removal (Figure 6c,d). 25.0 ± 3.5% ($n$ = 90) of TP was removed in summer at an HRT of one day, corresponding to an average effluent TP concentration of 2.2 ± 0.48 mg/L ($n$ = 90). With the decrease of temperature from summer to winter, TP removal efficiency in winter was 16.1 ± 4.4% ($n$ = 94) at an HRT of three days, which corresponded to an average TP concentration in the effluent of 2.0 ± 0.44 mg/L ($n$ = 94). Most of the TN and TP remained in the effluent of the AWFR system. Therefore, future studies should focus on the ecological post-treatment of the AWFR system.

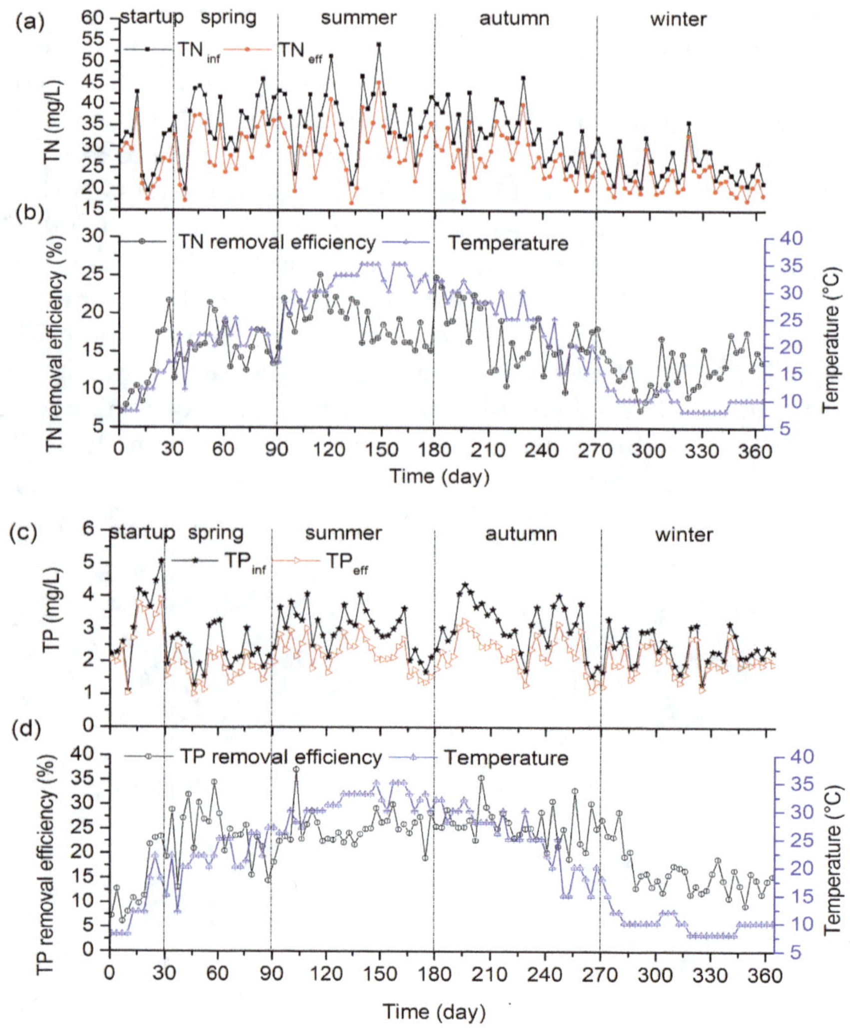


**Figure 6.** Total nitrogen (TN) (**a**,**b**) and total phosphorus (TP) (**c**,**d**) concentration in influent, effluent and the removal efficiency throughout the operation period.

### 3.4. Morphology and Structure of Anaerobic Biofilm Development

The biofilm in an anaerobic bioreactor comprises a complex microbial community. To observe the morphology and structures of the anaerobic biofilm, representative anaerobic biofilm samples (15, 90, 180, 270 and 340 days) were taken from the system for SEM analysis. Images of SEM revealed that the wool felt carrier had a highly porous and rough surface structure that played important roles in the anaerobic biofilm formation (Figure 7). Various types of microbes initially began to develop on the surface of the wool felt carrier primarily by Van der Waals and electrostatic forces (Figure 7a) [38,39]. From the anaerobic biofilm, cocci, bacillus, and filamentous bacteria were observed on the wool felt carrier. The anaerobic biofilm gradually became thicker (Figure 7c,e,g). A predominance of bacillus and filamentous bacteria deep into the biofilm matrix could be noticed. In addition, the filament-shaped *Methanosaeta-like* structures, long rod-shaped *Methanobacterium-like* structures, and *Methanolinea-like* structures were observed by SEM (Figure 7b,d,f,h,j), which were also observed in previous studies [31,40]. SEM observations were further confirmed by Illumina MiSeq sequencing analysis.

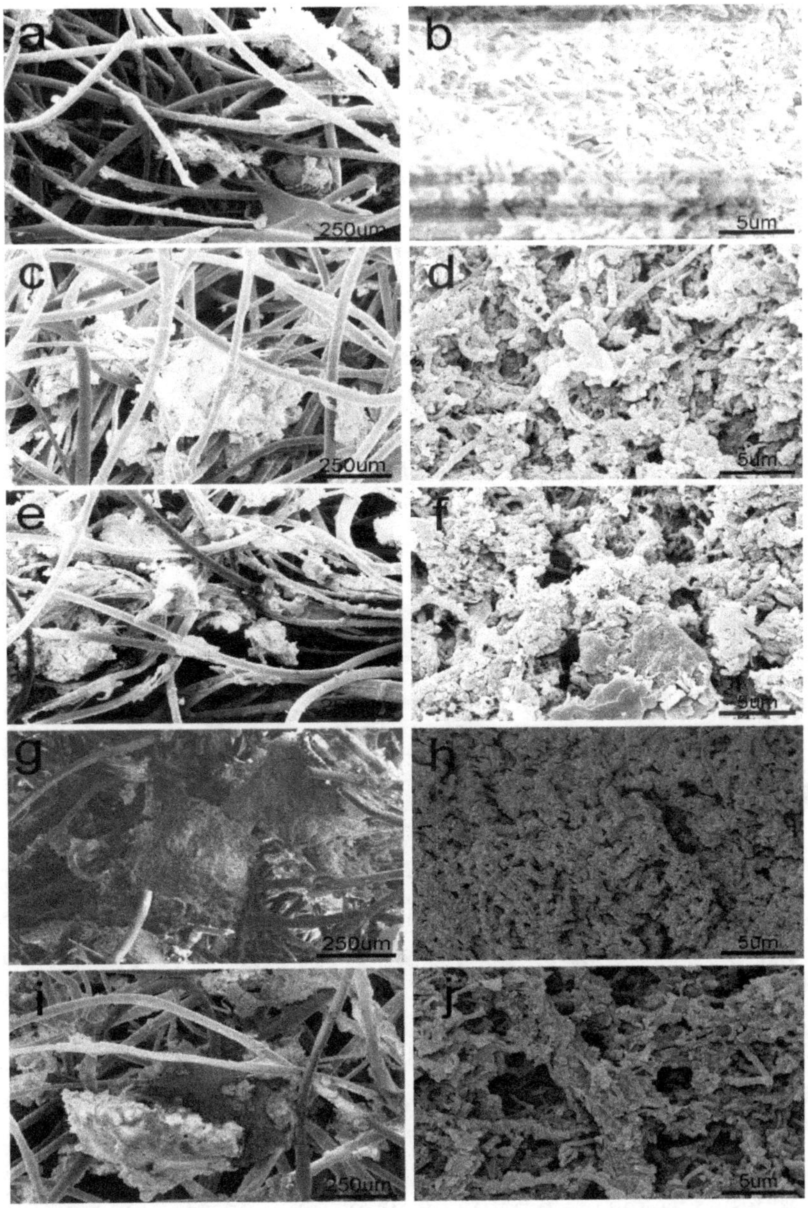


**Figure 7.** SEM of biofilm on the wool felt in the reactor: (**a**,**b**): start-up on day 15; (**c**,**d**): spring on day 90; (**e**,**f**): summer on day 180; (**g**,**h**): autumn on day 270; (**i**,**j**): winter on day 340.

### *3.5. Shifts in Microbial Community Structures with Seasonal Temperature*

High COD removal and biogas production were achieved by the AWFR system in summer, but not in winter. To better elucidate the changes of the microbial community and explore the COD removal mechanism during the whole operation period, the composition and structures of the microbial community were analyzed. Biofilm samples on days 15, 90, 180, 270 and 340, which represented five different periods of the AWFR, were chosen for high-throughput sequencing by the Illumina MiSeq platform.

The archaeal community had a lower diversity and most of the archaeal sequences were assigned to the three main orders *Methanosarcinales*, *Methanobacteriales* and *Methanomicrobiales* throughout the operation period (Figure 8a). These orders occupied 92.8–96.5% of the total community and were closely related to acetotrophic and hydrogenotrophic methanogen. During the start-up period, the predominant orders *Methanosarcinales*, *Methanobacteriales* and *Methanomicrobiales* accounted for 49.3%, 21.4%, and 22.2%, respectively, of the total community. However, the relative abundance of the order *Methanosarcinales* (34.3%) decreased in spring, whereas the relative abundances of the orders *Methanobacteriales* (35.0%) and *Methanomicrobiales* (26.0%) increased. In contrast, the order *Methanosarcinales* showed an increase in abundance during summer, but a reduction in abundance in winter. The relative abundances of the orders *Methanobacteriales* and *Methanomicrobiales* significantly decreased in summer but largely increased in winter.

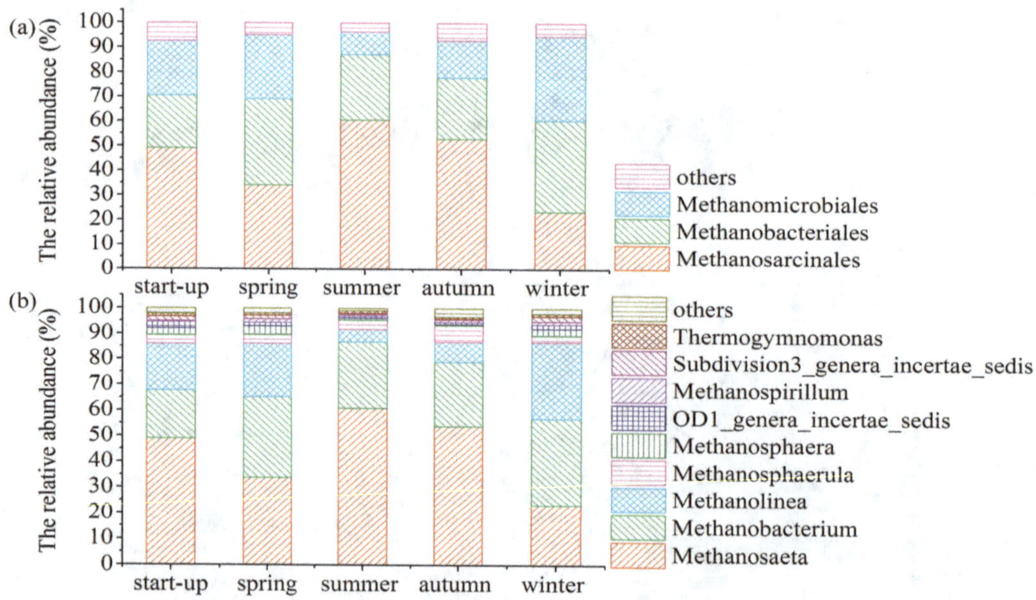


**Figure 8.** Archaeal community shifts at (**a**) order and (**b**) genus level.

To further validate the function of archaeal community, the microbial distribution at genus level was shown in Figure 8b. The results demonstrated that *Methanosaeta*, *Methanobacterium* and *Methanolinea* were the predominant genera throughout the operation period. Among these genera, *Methanosaeta* was affiliated with acetoclastic methanogen and the remaining genera *Methanobacterium* and *Methanolinea* were affiliated with hydrogenotrophic methanogens. *Methanosaeta* affiliated with the order *Methanosarcinales* had a high relative abundance (49.2%) in the start-up period, whereas the relative abundances of the genera *Methanobacterium* and *Methanolinea* were 18.7% and 18.3%, respectively. During summer, the genus *Methanosaeta* (61.0%) was the most abundant followed by the genera *Methanobacterium* (26.1%) and *Methanolinea* (4.8%). These findings indicated that acetoclastic methanogenesis was the main pathway, with some contribution from the hydrogenotrophic methanogens. These genera may have helped the AWFR system to maintain low VFA level, leading to relatively better COD removal efficiency and higher biogas production in summer. These results were in accordance with that in a previous study, achieving higher COD efficiency and biogas production in summer in a UASB treating domestic wastewater [8]. Nevertheless, during the winter period, the hydrogenotrophic genus *Methanobacterium* (33.8%) was the most dominant species followed by the hydrogenotrophic genus *Methanolinea* (29.9%) and the acetoclastic genus *Methanosaeta* (23.2%). Furthermore, other less abundant hydrogenotrophic genera were also detected in winter, including the genera *Methanospirllum*, *Methanosphaerula* and *Methanosphaera*, which are capable of using $H_2/CO_2$ to produce methane. Hydrogenotrophic genera (>63.7%) made up a large portion of the archaeal community in winter. These findings were in line with the previous observations of methanogen population shifts to favor hydrogen utilization under low temperatures [25,41–43]. It might be explained that lower temperature in winter affected the microbial membrane fluidity and inhibited the utilization of acetate than $H_2$ [44]. Although the relative abundance of the hydrogenotrophic genera was much higher than the acetoclastic genera in winter, COD removal and biogas production in winter were lower compared with in summer. The low temperature in winter may affect more the $H_2/CO_2$ methanogenesis than $H_2/CO_2$-dependent acetate production, and decrease the conversion rate of $H_2/CO_2$ to methane and increase the acetate accumulation [41,44].

In contrast to the archaeal community, the bacterial community in the AWFR system showed high diversity: 33 taxonomic categories at phyla level were identified (Table S1). This result was consistent with the previous studies [45–47]. This high diversity might suggest that a variety of bacterial communities participated in multiple metabolic pathways of organic matter degradation under seasonal temperature variation. The predominant bacteria in the AWFR system during the whole operation period were grouped into four phyla affiliated with *Firmicutes*, *Proteobacteria*, *Bacteroidetes*, and *Chloroflexi*, which occupied 89.3 $\pm$ 4.08% of the total phyla (Figure 9). These dominant phyla were also detected in the full-scale biogas digesters [47,48]. During the start-up period, the predominant microorganisms in the AWFR system belonged to the phylum *Proteobacteria*, *Firmicutes*, *Bacteroidetes* and *Chloroflexi*, with a relative abundance of 33.01%, 22.19%, 19.94% and 11.29%, respectively. With an increase of temperatures from spring to summer, the bacterial community structures remarkably changed. The abundance of the phylum *Firmicutes* increased from 37.26% to 64.97%, whereas the relative abundances of the phyla *Proteobacteria*, *Bacteroidetes*, and *Chloroflexi* decreased from 28.02% to 14.96%, 16.03% to 10.06%, and 6.00% to 3.87%, respectively. Interestingly, with a decrease of seasonal temperatures from summer to winter, the relative abundance of *Firmicutes* drastically decreased from 64.97% to 29.39%. In contrast, *Proteobacteria*, and *Bacteroidetes* obviously increased from 14.96% to 31.50% and 10.06% to 20.66%, respectively. Each phylum in our system likely possessed different tolerance and adaptation mechanisms to seasonal temperature variations. This was consistent with previous studies, stressing that temperature was the vital factor affecting the structures of the bacterial community in anaerobic digestion [48].

To gain further insight into the bacterial community structure in the AWFR system, some main bacteria at class level are shown in Figure 9b. During the start-up period, the dominant class was *Gammaproteobacteria* affiliated with the phylum *Proteobacteria*, with a relative abundance of 13.67%, followed by *Bacteroidia*, *Clostridia*, *Anaerolineae* and *Bacilli*, with a relative abundance of 12.87%, 11.88%, 10.96% and 9.89%, respectively. Among these predominant classes, *Bacteroidia* and *Clostridia* played significant roles in hydrolysis metabolism [47–49]. *Gammaproteobacteria* was the most representative class during the start-up period, which was also the main contributor to the phylum *Proteobacteria*. However, with an increased temperature during the summer period, *Bacilli* affiliated with the phylum *Firmicutes* was the most representative class, with a relative abundance of 57.17%, while the relative abundances of *Bacteroidia* and *Clostridia* were 6.61% and 7.43%, respectively. These results probably indicate that seasonal temperature in summer benefited the growth of *Bacilli* bacteria. *Bacteroidales* affiliated with the class *Bacteroidia* had been identified as hydrolyser and fermenter that participated in the conversation of cellulose and polysaccharide into VFAs [49]. *Clostridiales* affiliated with the class *Clostridia* was correlated with methane production, due to their diverse metabolism that utilized polysaccharide fermentation to produce VFAs for methanogenesis [47,48]. It was interesting to note that the class *Bacteroidales* was much more abundant in winter (12.45%) than in summer (6.61%), corresponding to an increase of VFAs in the winter. These results also corresponded with the decrease of COD removal efficiency and biogas production in the winter. Furthermore, the class *Deltaproteobacteria* was much less abundant in summer (1.94%) than in winter (7.81%). A similar tendency was shown by the class *Alphaproteobacteria* (4.92% in summer, and 9.57% in winter) and *Betaproteobacteria* (3.84% in summer, and 7.63% in winter).

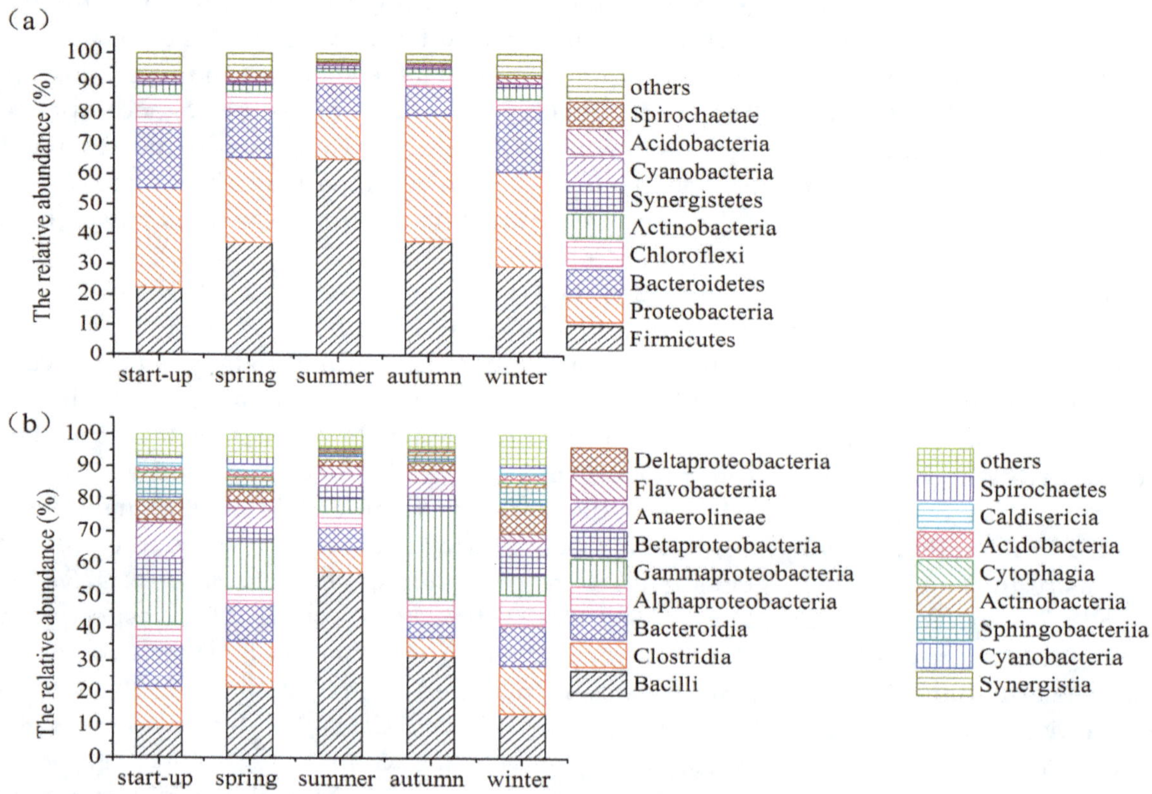


**Figure 9.** Bacterial community shifts at (**a**) phylum and (**b**) class level.

In order to better evaluate the structures of bacterial communities, the top 10 dominant genera in different periods of the system are presented in Figure 10. Among these genera, although the relative abundance of the bacteria at genus level differed depending on the operation conditions, it appeared that *Bacillus* was the most abundant genus throughout the whole operation period. *Bacillus* has been reported to possess the abilities of degrading various organic compounds [50–53]. The genus *Pseudomonas*, with a chemolithoautotrophic and heterotrophic functionality, has been reported to play an important role in organic compound degradation [54,55]. The presence of *Bacillus* and *Pseudomonas* in summer might be responsible for the degradation of organic matters by the AWFR system. This finding was consistent with the high COD removal efficiency and low effluent concentration of COD in summer. In addition, *VadinBC27_wastewater-sludge_group* was also observed throughout the operation period. It might be participating in the degradation of amino acids and some refractory organic matters [56–58]. The genus *Smithella*, which could degrade propionate to acetate [42,59], was dominant in start-up (3.13%) and winter (3.5%), and this was probably one of the reasons of acetate accumulation. The genus *Acinetobacter* was observed in autumn and winter, and was probably involved in the oxidation of organic matter or sulfides [13,60].

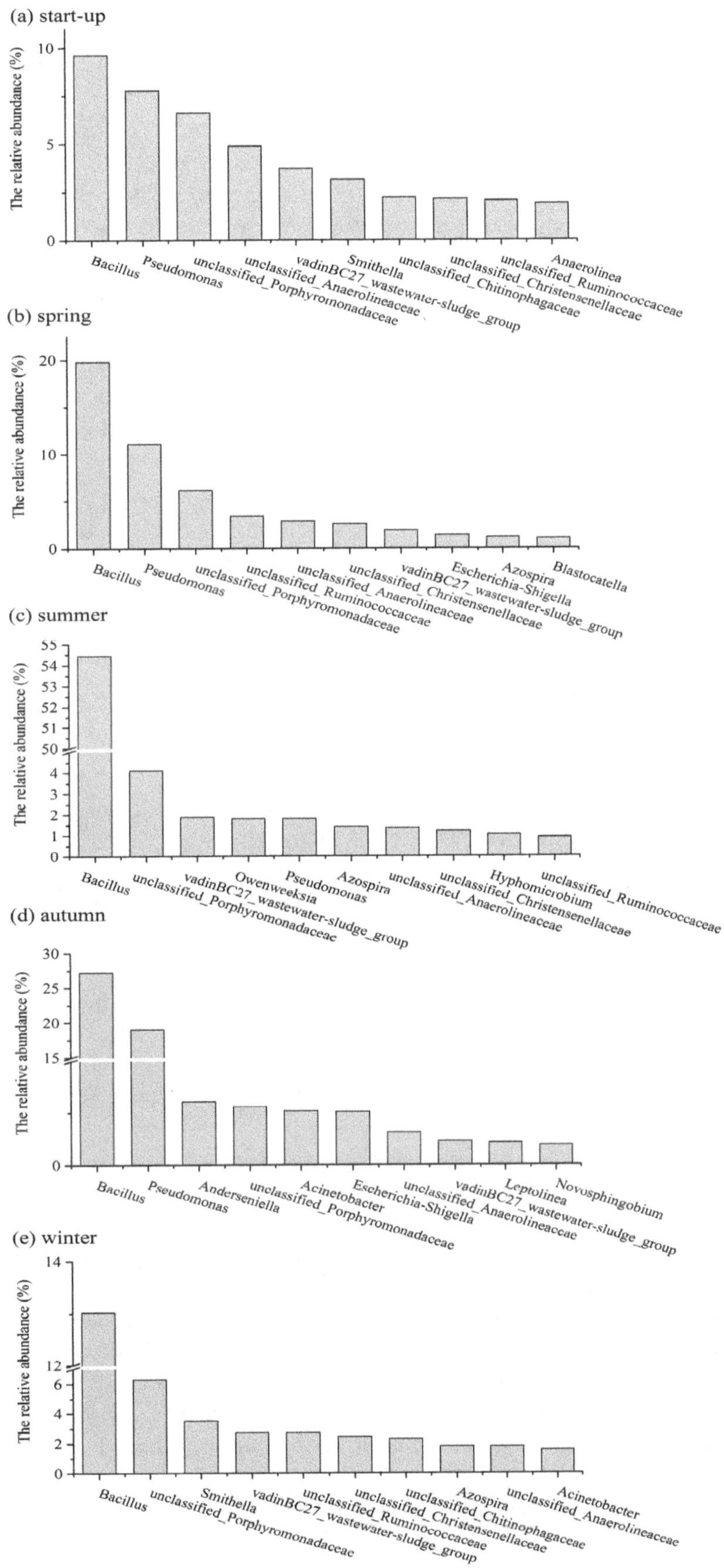


**Figure 10.** The top 10 bacterial genera during (**a**) start-up; (**b**) spring; (**c**) summer; (**d**) autumn; (**e**) winter period.

## 4. Conclusions

This study assesses the performance of the three-stage AWFR system for the treatment of decentralized domestic wastewater under seasonal variation of temperature. The COD removal efficiency changed with temperature: the average COD removal efficiency in summer and winter were 76 ± 7.2% (1-day HRT) and 52 ± 5.9% (3-day HRT), respectively. Although COD removal was lower in winter, approximately 43.5% of the influent COD was still converted to methane during that period. Miseq sequencing results suggested that seasonal temperature had a strong impact on the microbial community composition. The genera *Methanosaeta*, *Methanobacterium*, and *Methanolinea* were the predominant methanogens, whereas *Bacillus* was always the most abundant genus, which probably contributed to the fermentation processes throughout the whole operation period. Most of the nutrients, i.e., N and P, remained in the effluent, which could be treated by wetland or used for irrigation for agriculture. The AWFR system appears to be a sustainable option for the pretreatment of the decentralized domestic wastewater. However, longer HRT needs to be applied during winter.

**Acknowledgments:** This research was supported by China National Water Pollution Control and Management Technology Major Projects (2012ZX07101-005). The authors would like to thank the editor and the anonymous reviewers for editing and review. We thank Liwei Sun, Ran Yu, Haq Nawaz Abbasi and John Leju Celestino Ladu for useful comments and manuscript polishing.

**Author Contributions:** Juanhong Li and Xiwu Lu conceived and designed the experiments, analyzed the data, and wrote the paper.

## References

1. Yu, R.; Wu, Q.; Lu, X. Constructed wetland in a compact rural domestic wastewater treatment system for nutrient removal. *Environ. Eng. Sci.* **2012**, *29*, 751–757. [CrossRef]
2. Dong, H.-Y.; Qiang, Z.-M.; Wang, W.-D.; Jin, H. Evaluation of rural wastewater treatment processes in a county of eastern China. *J. Environ. Monit.* **2012**, *14*, 1906–1913. [CrossRef] [PubMed]
3. McCarty, P.L.; Bae, J.; Kim, J. Domestic wastewater treatment as a net energy producer—Can this be achieved? *Environ. Sci. Technol.* **2011**, *45*, 7100–7106. [CrossRef] [PubMed]
4. Chernicharo, C.; Van Lier, J.; Noyola, A.; Ribeiro, T.B. Anaerobic sewage treatment: State of the art, constraints and challenges. *Rev. Environ. Sci. Bio/Technol.* **2015**, *14*, 649–679. [CrossRef]
5. López-López, A.; Albarrán-Rivas, M.G.; Hernández-Mena, L.; León-Becerril, E. An assessment of an anaerobic filter packed with a low-cost material for treating domestic wastewater. *Environ. Technol.* **2013**, *34*, 1151–1159. [CrossRef] [PubMed]
6. Smith, A.L.; Skerlos, S.J.; Raskin, L. Membrane biofilm development improves COD removal in anaerobic membrane bioreactor wastewater treatment. *Microb. Biotechnol.* **2015**, *8*, 883–894. [CrossRef] [PubMed]
7. Mei, X.; Wang, Z.; Miao, Y.; Wu, Z. Recover energy from domestic wastewater using anaerobic membrane bioreactor: Operating parameters optimization and energy balance analysis. *Energy* **2016**, *98*, 146–154. [CrossRef]
8. Bandara, W.M.; Kindaichi, T.; Satoh, H.; Sasakawa, M.; Nakahara, Y.; Takahashi, M.; Okabe, S. Anaerobic treatment of municipal wastewater at ambient temperature: Analysis of archaeal community structure and recovery of dissolved methane. *Water Res.* **2012**, *46*, 5756–5764. [CrossRef] [PubMed]
9. Chong, S.; Sen, T.K.; Kayaalp, A.; Ang, H.M. The performance enhancements of upflow anaerobic sludge blanket (UASB) reactors for domestic sludge treatment–a state-of-the-art review. *Water Res.* **2012**, *46*, 3434–3470. [CrossRef] [PubMed]
10. Feng, J.; Wang, Y.; Ji, X.; Yuan, D.; Li, H. Performance and bioparticle growth of anaerobic baffled reactor (ABR) fed with low-strength domestic sewage. *Front. Environ. Sci. Eng.* **2014**, *9*, 352–364. [CrossRef]
11. Smith, A.L.; Stadler, L.B.; Love, N.G.; Skerlos, S.J.; Raskin, L. Perspectives on anaerobic membrane bioreactor treatment of domestic wastewater: A critical review. *Bioresour. Technol.* **2012**, *122*, 149–159. [CrossRef] [PubMed]

12. Gouveia, J.; Plaza, F.; Garralon, G.; Fdzpolanco, F.; Pena, M. Long-term operation of a pilot scale anaerobic membrane bioreactor (AnMBR) for the treatment of municipal wastewater under psychrophilic conditions. *Bioresour. Technol.* **2015**, *185*, 225–233. [CrossRef] [PubMed]
13. Jo, Y.; Kim, J.; Hwang, S.; Lee, C. Anaerobic treatment of rice winery wastewater in an upflow filter packed with steel slag under different hydraulic loading conditions. *Bioresour. Technol.* **2015**, *193*, 53–61. [CrossRef] [PubMed]
14. Gannoun, H.; Khelifi, E.; Omri, I.; Jabari, L.; Fardeau, M.-L.; Bouallagui, H.; Godon, J.-J.; Hamdi, M. Microbial monitoring by molecular tools of an upflow anaerobic filter treating abattoir wastewaters. *Bioresour. Technol.* **2013**, *142*, 269–277. [CrossRef] [PubMed]
15. Gannoun, H.; Othman, N.B.; Bouallagui, H.; Moktar, H. Mesophilic and thermophilic anaerobic co-digestion of olive mill wastewaters and abattoir wastewaters in an upflow anaerobic filter. *Ind. Eng. Chem. Res.* **2007**, *46*, 6737–6743. [CrossRef]
16. Couto, E.D.A.D.; Calijuri, M.L.; Assemany, P.P.; Santiago, A.D.F.; Lopes, L.S. Greywater treatment in airports using anaerobic filter followed by uv disinfection: An efficient and low cost alternative. *J. Clean. Prod.* **2015**, *106*, 372–379. [CrossRef]
17. Wu, Y.; Zhu, W.; Lu, X. Identifying key parameters in a novel multistep bio-ecological wastewater treatment process for rural areas. *Ecol. Eng.* **2013**, *61*, 166–173. [CrossRef]
18. Tonon, D.; Tonetti, A.L.; Coraucci Filho, B.; Bueno, D.A.C. Wastewater treatment by anaerobic filter and sand filter: Hydraulic loading rates for removing organic matter, phosphorus, pathogens and nitrogen in tropical countries. *Ecol. Eng.* **2015**, *82*, 583–589. [CrossRef]
19. Lew, B.; Tarre, S.; Beliavski, M.; Green, M. Anaerobic degradation pathway and kinetics of domestic wastewater at low temperatures. *Bioresour. Technol.* **2009**, *100*, 6155–6162. [CrossRef] [PubMed]
20. Dutta, K.; Tsai, C.-Y.; Chen, W.-H.; Lin, J.-G. Effect of carriers on the performance of anaerobic sequencing batch biofilm reactor treating synthetic municipal wastewater. *Int. Biodeterior. Biodegrad.* **2014**, *95*, 84–88. [CrossRef]
21. Han, Z.; Chen, F.; Zhong, C.; Zhou, J.; Wu, X.; Yong, X.; Zhou, H.; Jiang, M.; Jia, H.; Wei, P. Effects of different carriers on biogas production and microbial community structure during anaerobic digestion of cassava ethanol wastewater. *Environ. Technol.* **2016**, 1–10. [CrossRef] [PubMed]
22. Bodkhe, S. Development of an improved anaerobic filter for municipal wastewater treatment. *Bioresour.Technol.* **2008**, *99*, 222–226. [CrossRef] [PubMed]
23. De Oliveira Cruz, L.M.; Stefanutti, R.; Coraucci Filho, B.; Tonetti, A.L. Coconut shells as filling material for anaerobic filters. *SpringerPlus* **2013**, *2*, 655. [CrossRef] [PubMed]
24. Esmaeili, A.; Beni, A.A. Novel membrane reactor design for heavy-metal removal by alginate nanoparticles. *J. Ind. Eng. Chem.* **2015**, *26*, 122–128. [CrossRef]
25. Seib, M.D.; Berg, K.J.; Zitomer, D.H. Influent wastewater microbiota and temperature influence anaerobic membrane bioreactor microbial community. *Bioresour. Technol.* **2016**, *216*, 446–452. [CrossRef] [PubMed]
26. Mckeown, R.M.; Hughes, D.; Collins, G.; Mahony, T.; Flaherty, V.O. Low-temperature anaerobic digestion for wastewater treatment. *Curr. Opin. Biotechnol.* **2012**, *23*, 444–451. [CrossRef] [PubMed]
27. Enright, A.-M.; Collins, G.; O'Flaherty, V. Temporal microbial diversity changes in solvent-degrading anaerobic granular sludge from low-temperature (15 °C) wastewater treatment bioreactors. *Syst. Appl. Microbiol.* **2007**, *30*, 471–482. [CrossRef] [PubMed]
28. Gomec, C.Y.; Letsiou, I.; Ozturk, I.; Eroglu, V.; Wilderer, P.A. Identification of archaeal population in the granular sludge of an uasb reactor treating sewage at low temperatures. *J. Environ. Sci. Heal. A.* **2008**, *43*, 1504–1510. [CrossRef] [PubMed]
29. Gao, W.; Leung, K.; Qin, W.; Liao, B. Effects of temperature and temperature shock on the performance and microbial community structure of a submerged anaerobic membrane bioreactor. *Bioresour. Technol.* **2011**, *102*, 8733–8740. [CrossRef] [PubMed]
30. Regueiro, L.; Carballa, M.; Lema, J.M. Outlining microbial community dynamics during temperature drop and subsequent recovery period in anaerobic co-digestion systems. *J. Biotechnol.* **2014**, *192*, 179–186. [CrossRef] [PubMed]
31. Abubakkar, S.; Kundu, K.; Sreekrishnan, T.R. Comparative study of the performance of an anaerobic rotating biological contactor and its potential to enrich hydrogenotrophic methanogens. *J. Chem. Technol. Biotechnol.* **2015**, *90*, 398–406. [CrossRef]

32. Takai, K.; Horikoshi, K. Rapid detection and quantification of members of the archaeal community by quantitative pcr using fluorogenic probes. *Appl. Environ. Microbiol.* **2000**, *66*, 5066–5072. [CrossRef] [PubMed]
33. McElhoe, J.A.; Holland, M.M.; Makova, K.D.; Su, M.S.-W.; Paul, I.M.; Baker, C.H.; Faith, S.A.; Young, B. Development and assessment of an optimized next-generation DNA sequencing approach for the mtgenome using the Illumina Miseq. *Forensic Sci. Int. Genet.* **2014**, *13*, 20–29. [CrossRef] [PubMed]
34. American Public Health Association (APHA). *Standard Methods for the Examination of Water and Wastewater*; American Public Health Association: Washington, DC, USA, 2005.
35. Krishna, G.V.T.G.; Kumar, P.; Kumar, P. Treatment of low strength complex wastewater using an anaerobic baffled reactor (ABR). *Bioresour. Technol.* **2008**, *99*, 8193–8200. [CrossRef] [PubMed]
36. Shin, C.; Mccarty, P.L.; Kim, J.; Bae, J. Pilot-scale temperate-climate treatment of domestic wastewater with a staged anaerobic fluidized membrane bioreactor (SAF-MBR). *Bioresour. Technol.* **2014**, *159*, 95–103. [CrossRef] [PubMed]
37. Sharma, M.K.; Khursheed, A.; Kazmi, A.A. Modified septic tank-anaerobic filter unit as a two-stage onsite domestic wastewater treatment system. *Environ. Technol.* **2014**, *35*, 2183–2193. [CrossRef] [PubMed]
38. Hermansson, M. The DLVO theory in microbial adhesion. *Colloids Surf. B: Biointerfaces* **1999**, *14*, 105–119. [CrossRef]
39. Fang, H.H.; Chan, K.-Y.; Xu, L.-C. Quantification of bacterial adhesion forces using atomic force microscopy (AFM). *J. Microbiol. Methods* **2000**, *40*, 89–97. [CrossRef]
40. Fernández, N.; Díaz, E.E.; Amils, R.; Sanz, J.L. Analysis of microbial community during biofilm development in an anaerobic wastewater treatment reactor. *Microb. Ecol.* **2008**, *56*, 121–132. [CrossRef] [PubMed]
41. Lettinga, G.; Rebac, S.; Zeeman, G. Challenge of psychrophilic anaerobic wastewater treatment. *Trends Biotechnol.* **2001**, *19*, 363–370. [CrossRef]
42. Mckeown, R.M.; Scully, C.; Enright, A.; Chinalia, F.A.; Lee, C.; Mahony, T.; Collins, G.; Oflaherty, V. Psychrophilic methanogenic community development during long-term cultivation of anaerobic granular biofilms. *ISME J.* **2009**, *3*, 1231–1242. [CrossRef] [PubMed]
43. Bialek, K.; Kumar, A.; Mahony, T.; Lens, P.N.L.; Flaherty, V.O. Microbial community structure and dynamics in anaerobic fluidized-bed and granular sludge-bed reactors: Influence of operational temperature and reactor configuration. *Microb. Biotechnol.* **2012**, *5*, 738–752. [CrossRef] [PubMed]
44. Fey, A.; Conrad, R. Effect of temperature on carbon and electron flow and on the archaeal community in methanogenic rice field soil. *Appl. Environ. Microbiol.* **2000**, *66*, 4790–4797. [CrossRef] [PubMed]
45. Smith, A.L.; Skerlos, S.J.; Raskin, L. Psychrophilic anaerobic membrane bioreactor treatment of domestic wastewater. *Water Res.* **2013**, *47*, 1655–1665. [CrossRef] [PubMed]
46. Mchugh, S.; Carton, M.W.; Collins, G.; Oflaherty, V. Reactor performance and microbial community dynamics during anaerobic biological treatment of wastewaters at 16–37 °C. *FEMS Microbiol. Ecol.* **2004**, *48*, 369–378. [CrossRef]
47. Sundberg, C.; Al-Soud, W.A.; Larsson, M.; Alm, E.; Yekta, S.S.; Svensson, B.H.; Sørensen, S.J.; Karlsson, A. 454 pyrosequencing analyses of bacterial and archaeal richness in 21 full-scale biogas digesters. *FEMS Microbiol. Ecol.* **2013**, *85*, 612–626. [CrossRef] [PubMed]
48. Lee, S.H.; Kang, H.J.; Lee, Y.H.; Lee, T.J.; Han, K.; Choi, Y.; Park, H.D. Monitoring bacterial community structure and variability in time scale in full-scale anaerobic digesters. *J. Environ. Monit.* **2012**, *14*, 1893–1905. [CrossRef] [PubMed]
49. Goux, X.; Calusinska, M.; Lemaigre, S.; Marynowska, M.; Klocke, M.; Udelhoven, T.; Benizri, E.; Delfosse, P. Microbial community dynamics in replicate anaerobic digesters exposed sequentially to increasing organic loading rate, acidosis, and process recovery. *Biotechnol. Biofuels* **2015**, *8*, 122. [CrossRef] [PubMed]
50. Chebbi, A.; Mnif, S.; Mhiri, N.; Jlaiel, L.; Sayadi, S.; Chamkha, M. A moderately thermophilic and mercaptan-degrading bacillus licheniformis strain can55 isolated from gas-washing wastewaters of the phosphate industry, Tunisia. *Int. Biodeterior. Biodegrad.* **2014**, *94*, 207–213. [CrossRef]
51. Nakkabi, A.; Sadiki, M.; Fahim, M.; Ittobane, N.; Ibnsoudakoraichi, S.; Barkai, H.; Abed, S.E. Biodegradation of Poly(ester urethane)s by Bacillus subtilis. *Int. J. Environ. Res.* **2015**, *9*, 157–162.
52. Patowary, K.; Saikia, R.R.; Kalita, M.C.; Deka, S. Degradation of polyaromatic hydrocarbons employing biosurfactant-producing Bacillus pumilus KS2. *Ann. Microbiol.* **2014**, *65*, 225–234. [CrossRef]

53. Xiao, Y.; Chen, S.; Gao, Y.; Hu, W.; Hu, M.; Zhong, G. Isolation of a novel beta-cypermethrin degrading strain Bacillus subtilis BSF01 and its biodegradation pathway. *Appl. Microbiol. Biotechnol.* **2014**, *99*, 2849–2859. [CrossRef] [PubMed]
54. Dhall, P.; Kumar, R.; Kumar, A. Biodegradation of sewage wastewater using autochthonous bacteria. *Sci. World J.* **2012**, *2012*, 861903. [CrossRef] [PubMed]
55. Antwi, P.; Li, J.; Boadi, P.O.; Meng, J.; Shi, E.; Xue, C.; Zhang, Y.; Ayivi, F. Functional bacterial and archaeal diversity revealed by 16S rRNA gene pyrosequencing during potato starch processing wastewater treatment in an UASB. *Bioresour. Technol.* **2017**, *235*, 348–357. [CrossRef] [PubMed]
56. Xie, Z.; Wang, Z.; Wang, Q.; Zhu, C.; Wu, Z. An anaerobic dynamic membrane bioreactor (AnMBR) for landfill leachate treatment: Performance and microbial community identification. *Bioresour. Technol.* **2014**, *161*, 29–39. [CrossRef] [PubMed]
57. Sun, W.; Yu, G.; Louie, T.S.; Liu, T.; Zhu, C.; Xue, G.; Gao, P. From mesophilic to thermophilic digestion: The transitions of anaerobic bacterial, archaeal, and fungal community structures in sludge and manure samples. *Appl. Microbiol. Biotechnol.* **2015**, *99*, 10271–10282. [CrossRef] [PubMed]
58. Tang, Y.; Shigematsu, T.; Morimura, S.; Kida, K. Microbial community analysis of mesophilic anaerobic protein degradation process using bovine serum albumin (BSA)-fed continuous cultivation. *J. Biosci. Bioeng.* **2005**, *99*, 150–164. [CrossRef] [PubMed]
59. Regueiro, L.; Lema, J.M.; Carballa, M. Key microbial communities steering the functioning of anaerobic digesters during hydraulic and organic overloading shocks. *Bioresour. Technol.* **2015**, *197*, 208–216. [CrossRef] [PubMed]
60. Resende, J.A.; Silva, V.L.D.; De Oliveira, T.L.R.; Fortunato, S.; Carneiro, J.D.C.; Otenio, M.H.; Diniz, C.G. Prevalence and persistence of potentially pathogenic and antibiotic resistant bacteria during anaerobic digestion treatment of cattle manure. *Bioresour. Technol.* **2014**, *153*, 284–291. [CrossRef] [PubMed]

# Quantitative PCR Detection of Enteric Viruses in Wastewater and Environmental Water Sources by the Lisbon Municipality

**Pedro Teixeira [1,2,3,*], Sílvia Costa [1], Bárbara Brown [1], Susana Silva [4], Raquel Rodrigues [3] and Elisabete Valério [3]**

1 Câmara Municipal de Lisboa, Direcção Municipal do Ambiente, Estrutura Verde, Clima e Energia, Laboratório de Bromatologia e Águas, Avenida Cidade do Porto S/N-1700-111 Lisboa, Portugal; silvia.a.costa@cm-lisboa.pt (S.C.); barbara.brown@cm-lisboa.pt (B.B.)
2 Centro de Estudos do Ambiente e do Mar (CESAM), Faculdade de Ciências da Universidade de Lisboa, Campo Grande, 1749-016 Lisboa, Portugal
3 Departamento de Saúde Ambiental, Instituto Nacional de Saúde Doutor Ricardo Jorge, Avenida Padre Cruz, 1649-016 Lisboa, Portugal; raquel.rodrigues@insa.min-saude.pt (R.R.); elisabete.valerio@insa.min-saude.pt (E.V.)
4 Departamento de Epidemiologia, Instituto Nacional de Saúde Doutor Ricardo Jorge, Avenida Padre Cruz, 1649-016 Lisboa, Portugal; susana.pereira@insa.min-saude.pt
* Correspondence: pedro.teixeira@cm-lisboa.pt

**Abstract:** Current regulations and legislation require critical revision to determine safety for alternative water sources and water reuse as part of the solution to global water crisis. In order to fulfill those demands, Lisbon municipality decided to start water reuse as part of a sustainable hydric resources management, and there was a need to confirm safety and safeguard for public health for its use in this context. For this purpose, a study was designed that included a total of 88 samples collected from drinking, superficial, underground water, and wastewater at three different treatment stages. Quantitative Polimerase Chain Reaction (PCR) detection (qPCR) of enteric viruses Norovirus (NoV) genogroups I (GI) and II (GII) and Hepatitis A (HepA) was performed, and also FIB (*E. coli*, enterococci and fecal coliforms) concentrations were assessed. HepA virus was only detected in one untreated influent sample, whereas NoV GI/ NoV GI were detected in untreated wastewater (100/100%), secondary treated effluent (47/73%), and tertiary treated effluent (33/20%). Our study proposes that NoV GI and GII should be further studied to provide the support that they may be suitable indicators for water quality monitoring targeting wastewater treatment efficiency, regardless of the level of treatment.

**Keywords:** norovirus; water reuse; water quality

## 1. Introduction

Monitoring every pathogenic microorganism potentially present in water, namely viruses, bacteria, protozoa, or fungi, is unrealistic given the number of resources necessary for that purpose. As an alternative, microbiological water quality assessment has been focused essentially on detecting fecal indicator bacteria (FIB), namely *Escherichia coli* (*E. coli*) and *Enterococcus* spp. [1–6]. FIB are used as surrogates for enteric pathogens in particular for monitoring fecal contamination in environmental waters, relying on the principle that FIB existence is concomitant with pathogen presence, as demonstrated in several studies [7,8]. However, there are also numerous studies showing that pathogens do not correlate significantly with FIB [9–18]. Since FIB, like *E. coli* and *Enterococcus* spp. are shed in most animal feces [4,19,20], the lack of suitability between FIB levels and human health outcomes may be related to the FIB source. Amplifying the disconnection between FIB and pathogens is the ability for FIB not only to persist but to grow in environmental habitats like terrestrial soils, aquatic sediments, and aquatic vegetation [16,21–26]. Further studies are thus essential to assess the suitability of FIB as sole indicators not only in environmental waters but in an area that has been gaining more importance worldwide: treated wastewater use, as demand for water reuse, is increasing worldwide - whether by necessity in developing countries or by environmental objectives in developed countries.

An efficient and sustainable hydric resources management allows non-potable uses for treated wastewater, such as irrigation, industrial processes, firefighting, recreational, or municipal services. Besides the existence of heavy metals, chemicals, hormones and endocrine disruptors in wastewater, it is still necessary to deal with the expected presence of resistant pathogenic microorganisms, many of which are not tested or included in current standards or legislation for water quality assessment. These microorganisms include viruses, bacteria, protozoa, and helminths, responsible for a significant number of potentially dangerous pathologies. There are some reports showing that wastewater treatment processes do not completely remove enteric viruses [27–29], even from effluents with adequate chlorine concentrations or UV treatment [30]. With an increase in reclaimed water use, the potential health impacts resultant from microbial contamination need to be further explored, as outbreaks of viral infectious diseases have been linked to insufficient treatment [31–33]. Several viruses, including Norovirus (NoV) and Hepatitis A (HepA), are listed in United States Environmental Protection Agency (USEPA's) drinking water Contaminant Candidate List (4-CCL 4), heightening waterborne viruses as a research priority [34]. According to a recent review by Teixeira et al. ([35], in press), numerous studies have focused on detecting enteric viruses in water samples, including treated and untreated wastewater, as well as environmental waters. Data on NoV and HepA viruses' concentrations, however, particularly in tertiary-treated wastewater, are scarce ([35], in press). Therefore, further studies are necessary to evaluate the adequate indicators for water reuse quality evaluation and support their application for regulatory purposes.

Globally, NoV is responsible for nearly 20% of all acute gastroenteritis cases [36], with 677 million cases per year and over 213,000 deaths [37]. In risk groups comprising children, elderly, or immunocompromised individuals, morbidity, and mortality rates of NoV infection are significant [38–42]. NoV exposure, and possible outbreaks, have been reported in schools, hospitals, cruise ships, nursing homes, swimming pools, and restaurants [43–47]. Transmission of the virus primarily occurs via fecal-oral contamination route, direct contact with an infected individual, and contaminated water or food consumption [47–51]. Contact with the virus may occur through drinking [52], recreational [44], or irrigation water [53], leading to waterborne outbreaks [54–56]. The vast contaminated aqueous sources linked to NoV indicate a ubiquitous distribution of the virus [57]. HepA virus transmission occurs mainly by the oral-fecal route (about 95%), and ingestion of contaminated water and food [58–62]. HepA virus has been detected in drinking water, surface water, groundwater, treated, and untreated wastewater [63–67]. With global distribution, HepA is a major causal agent of acute viral hepatitis, with approximately 1.4 million cases reported annually globally [60′68], and a high endemicity of hepatitis A in regions with low sanitation [60′69′70].

Taking benefit from an existing standardized method for detecting NoV and HepA viruses in food and bottled water using real-time quantitative PCR detection (RT-qPCR) (ISO/TS, 15216-1:2017), the aim of our study was to develop a method based on this standardized method and assess the suitability of these enteric viruses as water quality indicators for water use and reuse. Moreover, the method developed aimed to be suitable for several water matrices—drinking, underground, superficial, treated, and untreated wastewater, thus enabling its use in routine water quality testing for detecting and quantifying NoV GI/GII and HepA viruses, and allowing managing entities to evaluate more accurately potential public health risks.

## 2. Materials and Methods

### 2.1. Sampling

Fieldwork was carried out from February 2018 to December 2018, with a total of 88 samples collected. Sampling included: 1) underground water samples collected at two different sites in Lisbon (n = 19); 2) superficial water intended for drinking water production (n = 10); 3) drinking water from Lisbon´s supply system (n = 11); 4) wastewater collected at a Wastewater Treatment Plant in the Lisbon district, at three different stages – untreated influent (n = 15), effluent with secondary treatment (n = 15) and effluent with tertiary, sand filtration and UV treatment (n = 15). Additionally, blank assays were performed with distilled, sterilized water samples (n = 3). All samples were collected in sterile polyethylene containers and stored at 4 °C for less than 2 h until chemical and microbiological determinations were initiated. 0.5 mL of a 10% dechlorination agent $Na_2S_2O_3$ was added to containers before sampling in order to neutralize possible existing residual chlorine from drinking water samples. Water samples for enteric viruses' determinations were collected in sterile glass containers. Enteric viruses were determined in different volumes according to the water source (25–5000 mL). Microbiological indicators were determined in 100 mL, and chemical indicators were determined according to standard procedures, from a total of 1000 mL [71]. The time lapse between sample collection and laboratory processing did not exceed 24 h.

### 2.2. Detection and Quantification of Enteric Viruses

The procedure established in this study was based on the international standard method for the determination of viruses in foods ISO/TS 15216-1:2017 [72]. Initially, 10 µL of Mengo virus strain vMC0 (ceeramTOOLS®, Biomérieux, France) were added to each sample, to be used as an internal process control to assure the RNA extraction efficiency. Samples with extraction efficiency, ≥1% were considered valid, as established in Norm ISO/TS 15216-1:2017 [72]. Viral particles were captured through filtration with a 0.45 µm pore size (47 mm diameter) positively charged nylon membrane (Amersham Hybond N+, GE Healthcare Life Sciences, UK), from 25 mL (untreated wastewater), 1000 mL (blank assay, secondary, tertiary treated wastewater and superficial water) and 5000 mL (drinking and underground water) samples. It should be noted that there are existing alternative protocols for virus collection [73], which stipulate significantly high volumes for filtration and virus collection in surface and groundwater—300 L and 1500 L, respectively. For practical reasons, namely to avoid filter clogging and the possible inexistence of such water volumes in some groundwater sources, and more importantly, aiming to establishing a reasonably achievable method for detecting and quantification enteric viruses in diverse water samples for laboratories that perform the routine water monitoring, in this study different volumes were tested and filtered according to the sample matrix. Filters were then transferred into a sterile tube, and 4 mL of tris/glycine/beef extract (TGBE) buffer were added and shaken at approximately 50 oscillations $min^{-1}$ for 20 ± 5 min. An additional 4 mL of TGBE buffer was added, and the eluates pooled into a single clean tube. The pH was adjusted to 7.0 ± 0.5 with HCl (≥5 mol/L). Subsequently, samples were concentrated using Amicon® Ultra-15 Centrifugal Filter Devices with a 100 kDa molecular weight cut-off (Merck Millipore, Darmstadt, Germany), through centrifugation 4000 g, at 4 °C, for 15 min. For RNA isolation, 500 µL TRIzol® reagent (Thermo Fisher Scientific, Waltham, MA, USA) was added to the samples and mixed for 5 min, 30 °C at 350 rpm in a thermomixer (Eppendorf, Germany). Afterward, 200 µL of chloroform were added and samples were mixed in a thermomixer (Eppendorf, Germany) for 3 min, 30 °C at 350 rpm. Centrifugation was then performed for 15 min, at 4 °C and 12,000 g. RNA was extracted from 140 µL final volume (approx.) of the concentrated sample (aqueous phase) using a QIAamp Viral RNA Mini kit (QIAGEN, Hilden, Germany), according to the manufacturer's instructions. Samples were stored at −80 °C until analysis. Quantitative Real-Time PCR (RT-qPCR) was performed (Applied Biosystems AB7500 qPCR) for the specific detection of NoV GI, NoV GII, and HepA viruses. An initial screening for the selected viruses was performed, in addition to confirmation for the absence of inhibitors with

10-fold dilutions. RT-qPCR was performed for NoV GI, NoV GII, HepA, and Mengo virus detection and quantification, with the use of commercial kits (ceeramTOOLS®, Biomérieux, France), according to the manufacturer´s specifications.

### 2.3. Microbiological Analysis

Detection of total coliforms, E. coli, enterococci, and fecal coliforms was performed through the use of Colilert and Enterolert with Quanti-Tray (IDDEX Laboratories, Westbrook, ME, USA). Samples were processed according to manufacturer´s instructions. For heterotrophic plate count at 22 °C and 37 °C, agar inclusion technique was performed using 1 mL aliquots of the water sample (after the necessary dilutions were performed) and adding 15 mL of Yeast Extract agar (VWR Chemicals, Radnor, PA, USA), and incubated at 36 °C for 44 h and 22 °C for 68 h. After the incubation period at each temperature, all colonies were quantified for each case.

### 2.4. Physical and Chemical Assessment

Temperature and pH were both determined (Thermo Scientific™ Orion™3-Star Benchtop pH Meter, Thermo Fisher Scientific, Waltham, MA, USA) according to standard methods, as well as free chlorine and total organic carbon (TOC) [71]. Conductivity was measured according to Standard Guideline NP EN 27888:1996 (MeterLab CDM 210) and Mohr´s Method was performed to assess chlorides.

### 2.5. Statistical Analysis

A descriptive analysis was made for numeric and categorical variables. For comparing different measures among the three wastewater treatment phases, a Kruskal–Wallis non-parametric or Fisher's Exact Test was used. When differences were found, multiple comparisons with Bonferroni correction was also performed. Correlation in each wastewater treatment phase was measured with the Pearson correlation coefficient. Statistical tests were performed bilaterally at a significance level of 5%. The statistical analysis of the data was performed using statistical software R, version 3.4.3. [74].

## 3. Results

### 3.1. Environmental and Drinking Waters

A statistical summary for virus, microbiological, physical and chemical results (median) is presented in Table 1, concerning blank assays (BA), superficial water intended for drinking water production (DW1), drinking water from Lisbon's public supply system (DW2) and underground water samples collected at two different sites in Lisbon (GW1 and GW2). None of the enteric viruses targeted in this study—NoV GI, NoV GII, and HepA—were detected in samples from superficial water intended for drinking water production (n = 10), drinking water from Lisbon's supply system (n = 11) and underground water samples (n = 19).

Microbiological, physical, and chemical results for drinking water samples (n = 11) were all in compliance with national legislation—Law Decree No. 306/2007, 27th August, and Law Decree No. 236/1998, 1st August, with no microbial contamination detected. Environmental - superficial and underground bodies of water - presented similar and reduced levels of fecal contamination, with no significant differences between them (Table 1). Apart from conductivity and TOC, which presented lower and higher values, respectively, in the superficial water samples. The physical and chemical results were similar between the superficial and underground water samples.

**Table 1.** Microbiological, physical, and chemical results (median) for blank assays (BA), superficial water intended for drinking water production (DW1), drinking water from Lisbon's supply system (DW2), and underground water samples collected at two different sites in Lisbon (GW1 and GW2). MNP—Most Probable Number, CFU—Colony Forming Units, GC—Genomic Copies. ND—Not detected. n = total analyzed samples. LD (limit of detection) = 1 MPN/100 mL.

| | **BA** | **DW1** | **DW2** | **GW1** | **GW2** |
|---|---|---|---|---|---|
| | **(n = 3)** | **(n = 10)** | **(n = 11)** | **(n = 8)** | **(n = 11)** |
| Total Coliforms (MPN/100 mL) | <LD | $7.58 \times 10^2$ | <LD | $5.6 \times 10^1$ | $2.01 \times 10^2$ |
| *E. coli* (MPN/100 mL) | <LD | 2 | <LD | 3 | <LD |
| Fecal Coliforms (MPN/100 mL) | <LD | 1 | <LD | 2 | 1 |
| Enterococcus (MPN/100 mL) | <LD 0 | 4 | <LD | 5 | $2.6 \times 10^1$ |
| HPC37 °C (CFU/mL) | ND | $8.4 \times 10^1$ | ND | $1.4 \times 10^1$ | $1.57 \times 10^2$ |
| HPC22 °C (CFU/mL) | ND | $1.86 \times 10^2$ | ND | $8.9 \times 10^1$ | $3.01 \times 10^2$ |
| TOC (mg/L) | 0.0 | 3.9 | 0.0 | 0.0 | 0.0 |
| Chlorides (mg $Cl^-$/L) | 1 | 7 | 14 | 30 | 69 |
| Free chlorine (mg $Cl_2$/L) | 0.06 | 0.06 | 0.60 | 0.08 | 0.07 |
| Conductivity (μS/cm at 20 °C) | 2 | 71 | 173 | 552 | 827 |
| pH (25°) | 6.0 | 7.9 | 7.9 | 7.8 | 7.1 |
| NoV GI (GC/100 mL) | ND | ND | ND | ND | ND |
| NoV GII (GC/100 mL) | ND | ND | ND | ND | ND |
| HepA (GC/100 mL) | ND | ND | ND | ND | ND |

### 3.2. Wastewater

Untreated wastewater influent (WW1), effluent with secondary treatment (WW2) and effluent with tertiary treatment (WW3) statistical results summary for virus, microbiological, physical and chemical results (median) is presented in Table 2. HepA virus was only detected in one untreated influent sample, with a $3.99 \times 10^6$ gc/100 mL concentration. NoV GI was detected in all wastewater treatment stages - untreated wastewater influent (n = 15), effluent with secondary treatment (n = 7) and effluent with tertiary treatment (n = 5). Similar results were obtained for NoV GII with the following positive samples: untreated wastewater influent (n = 15), effluent with secondary treatment (n = 11) and effluent with tertiary treatment (n = 3). It can be noted that NoV GI median concentration reductions are significant between stages; a $6.67 \times 10^7$ gc/100 mL initial concentration is reduced to $8.34 \times 10^6$ gc/100 mL (WW2) and to $8.81 \times 10^5$ gc/100 mL final concentration. NoV GII initial $4.86 \times 10^7$ gc/100 mL median concentration is also significantly reduced with secondary treatment to $2.29 \times 10^6$ gc/100 mL (WW2). With UV and sand filtration, the effluent's concentration is also diminished ($9.69 \times 10^5$ gc/100 mL), although with no observed statistical significance.

Significant differences were found between all treatment wastewater phases for FIB and HPC at 37 °C and 22 °C concentrations. Notably, we can observe a significant FIB decay not only with secondary treatment but also with tertiary sand filtration and UV treatment—after which FIB have been eliminated upon tertiary treatment (results < limit of detection). According to current Portuguese national legislations—Law Decree No. 119/2019, 21st August—these effluents (with tertiary treatment) can be adequate for non-potable uses including irrigation, industrial uses, firefighting or street cleaning. Excluding chlorides, with no significant variations between wastewater treatments, the majority of the physical and chemical results, display a statistically significant reduction with secondary treatment - activated sludge.

**Table 2.** Microbiological, physical, and chemical results (median) for untreated wastewater influent (WW1), effluent with secondary treatment (WW2) and effluent with tertiary treatment (WW3). * Kruskall–Wallis non parametric test. ** Multiple comparisons tests with Bonferroni correction. a—Statistically significant difference between untreated influent and secondary treated effluent. b—Statistically significant difference between untreated influent and tertiary treated effluent. c—Statistically significant difference between secondary and tertiary treated effluent. MNP—Most Probable Number, CFU—Colony Forming Units, GC—Genomic Copies. ND—Not Detected. n = total analyzed samples. LD (limit of detection) = 1 MPN/100 mL.

| | WW1 | WW2 | WW3 | *P*-Value * | Pairwise Test ** |
|---|---|---|---|---|---|
| | (n = 15) | (n = 15) | (n = 15) | | |
| Total Coliforms (MPN/100 mL) | $7.70 \times 10^7$ | $9.80 \times 10^4$ | <LD | <0.001 | a, b, c |
| *E. coli* (MPN/100 mL) | $1.66 \times 10^7$ | $2.14 \times 10^4$ | <LD | <0.001 | a, b, c |
| Fecal Coliforms (MPN/100 mL) | $1.19 \times 10^7$ | $1.58 \times 10^4$ | <LD | <0.001 | a, b, c |
| Enterococcus (MPN/100 mL) | $3.99 \times 10^6$ | $8.80 \times 10^3$ | <LD | <0.001 | a, b, c |
| HPC37 °C (CFU/mL) | $3.10 \times 10^6$ | $1.50 \times 10^4$ | 4 | <0.001 | a, b, c |
| HPC22 °C (CFU/mL) | $4.20 \times 10^6$ | $2.04 \times 10^4$ | 2 | <0.001 | a, b, c |
| TOC (mg/L) | 169.0 | 7.90 | 7.70 | <0.001 | a, b |
| Chlorides (mg $Cl^-$/L) | 173 | 136 | 129 | 0.250 | - |
| Free chlorine (mg $Cl_2$/L) | 0.06 | 0.10 | 0.09 | <0.001 | a, b |
| Conductivity (μS/cm at 20 °C) | 1262 | 839 | 860 | <0.001 | a, b |
| pH (25°) | 8.0 | 7.0 | 7.5 | <0.001 | a, b, c |
| NoV GI (GC/100 mL) | $6.67 \times 10^7$ | $8.34 \times 10^6$ | $8.81 \times 10^5$ | <0.001 | a, b, c |
| NoV GII (GC/100 mL) | $4.86 \times 10^7$ | $2.29 \times 10^6$ | $9.69 \times 10^5$ | <0.001 | a, b |
| HepA (GC/100 mL) | $3.99 \times 10^6$ | ND | ND | - | - |

Correlations between the microbiological parameters analyzed in each wastewater treatment phase (WW1, WW2 and WW3) were measured (Figure 1A–C). A positive correlation was observed between NoV GI and NoV GII in untreated influent (ρ = 0.54), and between NoV GI and fecal coliforms (ρ = 0.61). One of the highest correlations on this phase was found between Total Coliforms and Enterococcus (ρ = 0.68). Several positive correlations were detected for secondary treated effluent samples in microbiological parameters but none significant concerning NoV GI/GII. In tertiary treated effluents, only one moderate correlation was observed for HPC 22 °C/37 °C (ρ = 0.60).

A)

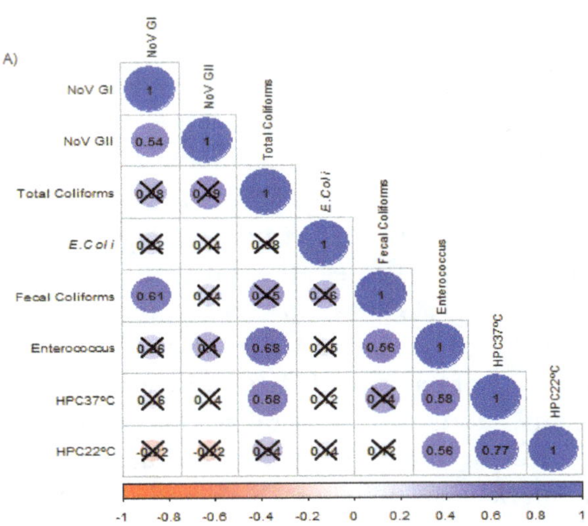


**Figure 1.** *Cont.*

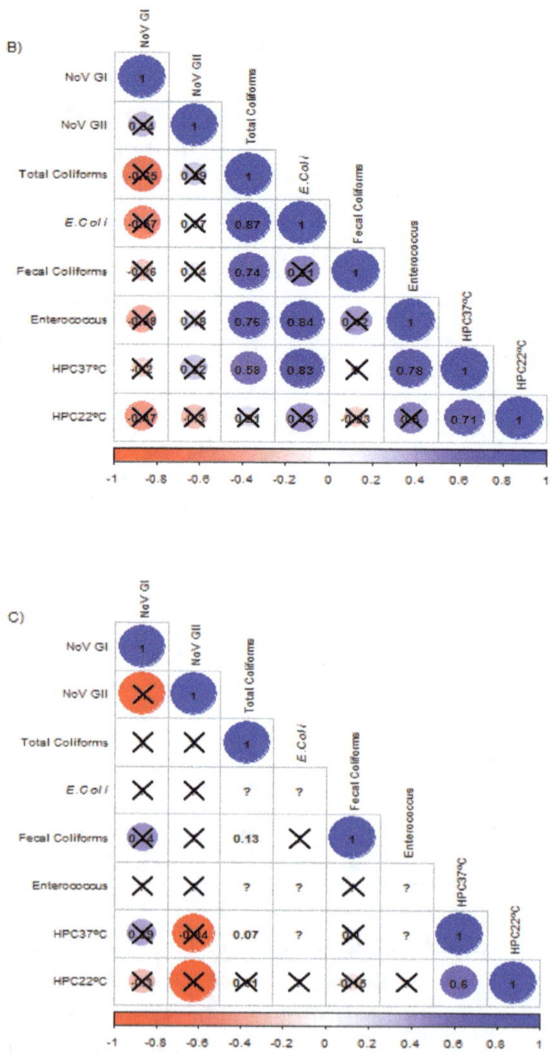


**Figure 1.** Graphical display of the correlation matrix of the microbiological parameters tested in each wastewater treatment phase (**A**) WW1, (**B**) WW2, and (**C**) WW3. The "?"—means "Not enough finite observations".

## 4. Discussion and Conclusions

In the case of environmental and drinking waters, enteric viruses NoV GI, NoV GII, and HepA virus were not detected in any of the samples in drinking, superficial, or underground water, suggesting an absence of fecal contamination from a human source in these types of water. However, the presence of total coliforms, *E. coli*, enterococci, and fecal coliforms were detected in samples from underground and superficial samples, and though in reduced concentrations, it suggests a natural and expected fecal contamination from animal origin. Studies showed that NoV, could be detected in particularly high concentrations in feces - up to $10^{11}$ GC/g [75]. Consequently, the occurrence of NoV in surface waters can be linked to contaminated water sources (i.e., fecal contamination) and with environmental conditions determining the survival of the virus [14,76]. Since these viruses are highly infectious [49] and demonstrate a high resistance to environmental degradation in water [77–79], its presence in surface and groundwater intended for human consumption raises as a potential public health risk, by which frequent monitoring of enteric viruses such as NoV besides FIB, could be desirable and prudent.

On what concerns the treated wastewaters, our results demonstrate that the absence of FIB does not imply the absence of pathogenic microorganisms, namely NoV GI and GII in tertiary treated wastewater (obtained after sand filtration and UV). FIB monitoring and their correlations with enteric viruses have recently been questioned [17,80,81], hence the necessity for more adequate indicators in water quality monitoring, especially for wastewater use. Growing water resources management concerns arise, not only in areas affected by droughts but also importantly in urban areas, like the one targeted in this study—Lisbon, where water reuse represents one of the municipality's commitments as part of a sustainable hydric resources management, with self-evident safeguard for public health. The results here obtained for Nov GI and GII concentrations are in accordance with previous studies targeting untreated, secondary, and tertiary treated wastewater [82–85].

The results of our study support the hypothesis that NoV GI and GII might be suitable indicators for water quality monitoring regarding wastewater treatment efficiency, regardless of the level of the treatment—secondary or tertiary. A correlation between NoV (GI) and FIB (fecal coliforms) is observed in untreated effluent, but in subsequent treatment phases, there was no correlation between the targeted enteric viruses and FIB. While positive correlations observed between FIB and NoV in untreated wastewater could be expected and were indeed observed, the absence of correlations between FIB and secondary and tertiary treatment stages points out to a lack of a link between FIB and the targeted enteric viruses concentrations. Moreover, in several samples with an apparent total FIB elimination with UV and sand filtration treatment, NoV was still detectable. Our results point out to the need of further studies and expansion to different WWTPs, in diverse locations and including a higher number of samples, not only to further reinforce the results of this study and validate our hypothesis but also to expand detection methods to other viral pathogens and evaluate the best and most comprehensive indicators to use for water reuse situations. While cell culture methods have been the gold standard for the detection of infectious viruses, qPCR methods became essential for enteric virus detection in water samples due to shorter detection times, high sensitivity and specificity, and the ability to detect viruses that are not easily or at all culturable [86]. A significant limitation of the method applied in our study is the inability to differentiate infective viruses from non-infective viruses. Nonetheless, viral genomes are present in water samples, particularly in reclaimed wastewater. Considering an estimated 50% infectious dose (ID50) between 18 and 1,015 genomic equivalents [49], public health risks may arise even from non-potable uses like irrigation or street cleaning, as incorrect usage of treated wastewater has caused outbreaks of viral infectious diseases worldwide [31–33]. To this fact, it must enhance while qPCR delivers quantitative data with high accuracy; results obtained by this method should be interpreted prudently because of potential losses during the sample concentration/extraction/purification procedure, which may result in an underestimation in the detection/quantification process [86]. The whole process control used in our study – Mengo virus – allowed evaluating the efficiency for virus recovery in the different stages – from sample concentration and RNA extraction, to RT-qPCR. However, even within the established ISO values (recovery $\geq$1%), the obtained recovery rates for this control were usually low (<10%), which can underestimate NoV GI and GII concentrations, particularly in reclaimed wastewater. Importantly, this study shows that a procedure based on the international standard method for the determination of viruses in foods ISO/TS 15216-1 [72] could be applied for routine water quality monitoring of NoV and HepA, in different water matrices.

The importance of FIB is undoubtedly as they are responsible for a significant improvement in water quality assessment and safety management for many decades [1]. Nevertheless, current regulations and legislation require critical revision to determine safety in particular for water reuse, as we can observe samples determined compliant for bacterial indicators that are positive for NoV GI and GII presence. Representing a key to the global water crisis, water reuse should also be determined safe considering the different treatments applied, as enteric viruses' infections also represent a worldwide economic concern.

**Author Contributions:** Conceptualization, E.V. and P.T.; methodology, E.V. and P.T.; validation, E.V. and P.T; formal analysis, P.T., E.V. and S.S.; investigation, P.T., S.C., B.B. and E.V.; resources, P.T., R.R. and E.V.; data curation, P.T. and E.V.; writing—original draft preparation, P.T.; writing—review and editing, E.V., P.T., S.C., S.S., B.B and R.R.; visualization, P.T., S.S. and E.V.; supervision, E.V.; project administration, E.V. and P.T.; funding acquisition, P.T. and E.V. All authors have read and agreed to the published version of the manuscript.

**Acknowledgments:** The authors are grateful to Ângelo Mesquita and José Canêdo for their contributions to the implementation of this project. We would also like to thank Helena Rebelo and João Brandão for all the help provided. Any opinions, findings and conclusions expressed are those of the authors and do not necessarily reflect the views of the Lisbon Municipality.

## References

1. Tallon, P.; Magajna, B.; Lofranco, C.; Leung, K.T. Microbial Indicators of Faecal Contamination in Water: A Current Perspective. *Water. Air. Soil Pollut.* **2005**, *166*, 139–166. [CrossRef]
2. José Figueras, M.; Borrego, J.J. New perspectives in monitoring drinking water microbial quality. *Int. J. Environ. Res. Public Health* **2010**, *7*, 4179–4202. [CrossRef] [PubMed]
3. Saxena, G.; Bharagava, R.N.; Kaithwas, G.; Raj, A. Microbial indicators, pathogens and methods for their monitoring in water environment. *J. Water Health* **2015**, *13*, 319–339. [CrossRef] [PubMed]
4. Leclerc, H.; Mossel, D.A.A.; Edberg, S.C.; Struijk, C.B. Advances in the Bacteriology of the Coliform Group: Their Suitability as Markers of Microbial Water Safety. *Annu. Rev. Microbiol.* **2001**, *55*, 201–234. [CrossRef]
5. WHO. Guidelines for safe recreational water environments—Coastal and Fresh waters. *Guidel. Safe Recreat. Water.* **2003**, *1*, 253.
6. EPA. Recreational Water Quality Criteria. Available online: https://www.epa.gov/sites/production/files/2015-10/documents/rwqc2012.pdf (accessed on 13 February 2020).
7. Wade, T.J.; Sams, E.; Brenner, K.P.; Haugland, R.; Chern, E.; Beach, M.; Wymer, L.; Rankin, C.C.; Love, D.; Li, Q.; et al. Rapidly measured indicators of recreational water quality and swimming-associated illness at marine beaches: A prospective cohort study. *Environ. Heal. A Glob. Access Sci. Source* **2010**, *9*, 1–14. [CrossRef]
8. Zmirou, D.; Pena, L.; Ledrans, M.; Letertre, A. Risks associated with the microbiological quality of bodies of fresh and marine water used for recreational purposes: Summary estimates based on published epidemiological studies. *Arch. Environ. Health* **2003**, *58*, 703–711. [CrossRef]
9. Lund, V. Evaluation of E. Coli as an indicator for the presence of Campylobacter jejuni and Yersinia enterocolitica in chlorinated and untreated oligotrophic lake water. *Water Res.* **1996**, *30*, 1528–1534. [CrossRef]
10. Bonadonna, L.; Briancesco, R.; Ottaviani, M.; Veschetti, E. Occurrence of Cryptosporidium oocysts in sewage effluents and correlation with microbial, chemical and physical water variables. *Environ. Monit. Assess.* **2002**, *75*, 241–252. [CrossRef]
11. Lemarchand, K.; Lebaron, P. Occurrence of Salmonella spp. and Cryptosporidium spp. in a French coastal watershed: Relationship with fecal indicators. *FEMS Microbiol. Lett.* **2003**, *218*, 203–209. [CrossRef]
12. Harwood, V.J.; Levine, A.D.; Scott, T.M.; Chivukula, V.; Lukasik, J.; Farrah, S.R.; Rose, J.B. Validity of the indicator organism paradigm for pathogen reduction in reclaimed water and public health protection. *Appl. Environ. Microbiol.* **2005**, *71*, 3163–3170. [CrossRef]
13. Anderson, K.L.; Whitlock, J.E.; Valerie, J.; Harwood, V.J. Persistence and Differential Survival of Fecal Indicator Bacteria in Subtropical Waters and Sediments. *Appl. Environ. Microbiol.* **2005**, *71*, 3041–3048. [CrossRef]
14. Hörman, A.; Rimhanen-finne, R.; Maunula, L.; Von Bonsdorff, C.; Torvela, N. Indicator Organisms in Surface Water in and Indicator Organisms in Surface Water in Southwestern Finland, 2000–2001. *Appl. Environ. Microbiol.* **2004**, *70*, 87–95. [CrossRef]
15. Zhang, Q.; Eichmiller, J.J.; Staley, C.; Sadowsky, M.J.; Ishii, S. Correlations between pathogen concentration and fecal indicator marker genes in beach environments. *Sci. Total Environ.* **2016**, *15*, 826–830. [CrossRef]
16. Bradshaw, J.K.; Snyder, B.J.; Oladeinde, A.; Spidle, D.; Berrang, M.E.; Meinersmann, R.J.; Oakley, B.; Sidle, R.C.; Sullivan, K.; Molina, M. Characterizing relationships among fecal indicator bacteria, microbial source tracking markers, and associated waterborne pathogen occurrence in stream water and sediments in a mixed land use watershed. *Water Res.* **2016**, *101*, 498–509. [CrossRef]
17. McQuaig, S.; Griffith, J.; Harwood, V.J. Association of fecal indicator bacteria with human viruses and microbial source tracking markers at coastal beaches impacted by nonpoint source pollution. *Appl. Environ. Microbiol.* **2012**, *78*, 6423–6432. [CrossRef]
18. Arnold, B.F.; Schiff, K.C.; Griffith, J.F.; Gruber, J.S.; Yau, V.; Wright, C.C.; Wade, T.J.; Burns, S.; Hayes, J.M.;

Mcgee, C.; et al. Swimmer illness associated with marine water exposure and water quality indicators: Impact of widely used assumptions. *Epidemiology* **2013**, *24*, 845–853. [CrossRef]
19. Harwood, V.J.; Butler, J.; Parrish, D.; Wagner, V. Isolation of fecal coliform bacteria from the diamondback terrapin (Malaclemys terrapin centrata). *Appl. Environ. Microbiol.* **1999**, *65*, 865–867. [CrossRef]
20. Field, K.G.; Samadpour, M. Fecal source tracking, the indicator paradigm, and managing water quality. *Water Res.* **2007**, *41*, 3517–3538. [CrossRef]
21. Solo-Gabriele, H.M.; Wolfert, M.A.; Desmarais, T.R.; Palmer, C.J. Sources of Escherichia coli in a coastal subtropical environment. *Appl. Environ. Microbiol.* **2000**, *66*, 230–237. [CrossRef]
22. Byappanahalli, M.N.; Shively, D.A.; Nevers, M.B.; Sadowsky, M.J.; Whitman, R.L. Growth and survival of Escherichia coli and enterococci populations in the macro-alga Cladophora (Chlorophyta). *FEMS Microbiol. Ecol.* **2003**, *46*, 203–211. [CrossRef]
23. Jeng, H.C.; England, A.J.; Bradford, H.B. Indicator organisms associated with stormwater suspended particles and estuarine sediment. *J. Environ. Sci. Heal. Part A Toxic/Hazardous Subst. Environ. Eng.* **2005**, *40*, 779–791. [CrossRef] [PubMed]
24. Badgley, B.D.; Thomas, F.I.M.; Harwood, V.J. Quantifying environmental reservoirs of fecal indicator bacteria associated with sediment and submerged aquatic vegetation. *Environ. Microbiol.* **2011**, *13*, 932–942. [CrossRef] [PubMed]
25. Eifan, S.A. Enteric Viruses And Aquatic Environment. *Internet J. Microbiol.* **2013**, *12*, 1–7.
26. He, X.; Wei, Y.; Cheng, L.; Zhang, D.; Wang, Z. Molecular detection of three gastroenteritis viruses in urban surface waters in Beijing and correlation with levels of fecal indicator bacteria. *Environ. Monit. Assess.* **2012**, *184*, 5563–5570. [CrossRef]
27. Blatchley, E.R.; Gong, W.-L.; Alleman, J.E.; Rose, J.B.; Huffman, D.E.; Otaki, M.; Lisle, J.T. Effects of Wastewater Disinfection on Waterborne Bacteria and Viruses. *Water Environ. Res.* **2007**, *79*, 81–92. [CrossRef]
28. El-Senousy, W.M.; Guix, S.; Abid, I.; Pintó, R.M.; Bosch, A. Removal of astrovirus from water and sewage treatment plants, evaluated by a competitive reverse transcription-PCR. *Appl. Environ. Microbiol.* **2007**, *73*, 164–167. [CrossRef]
29. Lazarova, V.; Levine, B.; Sack, J.; Cirelli, G.; Jeffrey, P.; Muntau, H.; Salgot, M.; Brissaud, F. Role of water reuse for enhancing integrated water management in Europe and Mediterranean countries. *Water Sci. Technol.* **2001**, *43*, 25–33. [CrossRef]
30. Sano, D.; Amarasiri, M.; Hata, A.; Watanabe, T.; Katayama, H. Risk management of viral infectious diseases in wastewater reclamation and reuse: Review. *Environ. Int.* **2016**, *91*, 220–229. [CrossRef]
31. Bernard, H.; Faber, M.; Wilking, H.; Haller, S.; Höhle, M.; Schielke, A.; Ducomble, T.; Siffczyk, C.; Merbecks, S.; Fricke, G.; et al. Large multistate outbreak of norovirus gastroenteritis associated with frozen strawberries, Germany, 2012. *Eurosurveillance* **2014**, *19*, 20719. [CrossRef]
32. Okoh, A.I.; Sibanda, T.; Gusha, S.S. Inadequately treated wastewater as a source of human enteric viruses in the environment. *Int. J. Environ. Res. Public Health* **2010**, *7*, 2620–2637. [CrossRef] [PubMed]
33. Sinclair, R.G.; Jones, E.L.; Gerba, C.P. Viruses in recreational water-borne disease outbreaks: A review. *J. Appl. Microbiol.* **2009**, *107*, 1769–1780. [CrossRef] [PubMed]
34. United States Environmental Protection Agency. Contaminant Candidate List (CCL) and Regulatory Determination. Available online: https://www.epa.gov/ccl/microbial-contaminants-ccl-4 (accessed on 30 January 2020).
35. Teixeira, P.; Salvador, D.; Brandão, J.; Ahmed, W.; Sadowsky, M.J.; Valério, E. Environmental and Adaptive Changes Necessitate a Paradigm Shift for Indicators of Fecal Contamination. *Microbiol Spectr.*. in press.
36. Ahmed, S.M.; Hall, A.J.; Robinson, A.E.; Verhoef, L.; Premkumar, P.; Parashar, U.D.; Koopmans, M.; Lopman, B.A. Global prevalence of norovirus in cases of gastroenteritis: A systematic review and meta-analysis. *Lancet Infect. Dis.* **2014**, *14*, 725–730. [CrossRef]
37. Pires, S.M.; Fischer-Walker, C.L.; Lanata, C.F.; Devleesschauwer, B.; Hall, A.J.; Kirk, M.D.; Duarte, A.S.R.; Black, R.E.; Angulo, F.J. Aetiology-specific estimates of the global and regional incidence and mortality of diarrhoeal diseases commonly transmitted through food. *PLoS ONE* **2015**, *10*, e0142927. [CrossRef]
38. Harris, J.P.; Edmunds, W.J.; Pebody, R.; Brown, D.W.; Lopman, B.A. Deaths from norovirus among the elderly, England and Wales. *Emerg. Infect. Dis.* **2008**, *14*, 1546–1552. [CrossRef]
39. Hutson, A.M.; Atmar, R.L.; Estes, M.K. Norovirus disease: Changing epidemiology and host susceptibility factors. *Trends Microbiol.* **2004**, *12*, 279–287. [CrossRef]

40. Payne, D.C.; Vinjé, J.; Szilagyi, P.G.; Edwards, K.M.; Staat, M.A.; Weinberg, G.A.; Hall, C.B.; Chappell, J.; Bernstein, D.I.; Curns, A.T.; et al. Norovirus and medically attended gastroenteritis in U.S. children. *N. Engl. J. Med.* **2013**, *368*, 1121–1130. [CrossRef]
41. Siebenga, J.J.; Vennema, H.; Zheng, D.; Vinjé, J.; Lee, B.E.; Pang, X.; Ho, E.C.M.; Lim, W.; Choudekar, A.; Broor, S.; et al. Norovirus Illness Is a Global Problem: Emergence and Spread of Norovirus GII.4 Variants, 2001–2007. *J. Infect. Dis.* **2009**, *200*, 802–812. [CrossRef]
42. Tian, G.; Jin, M.; Li, H.; Li, Q.; Wang, J.; Duan, Z.J. Clinical characteristics and genetic diversity of noroviruses in adults with acute gastroenteritis in Beijing, China in 2008–2009. *J. Med. Virol.* **2014**, *86*, 1235–1242. [CrossRef]
43. Fankhauser, R.L.; Monroe, S.S.; Noel, J.S.; Humphrey, C.D.; Bresee, J.S.; Parashar, U.D.; Ando, T.; Glass, R.I. Epidemiologic and Molecular Trends of "Norwalk-like Viruses" Associated with Outbreaks of Gastroenteritis in the United States. *J. Infect. Dis.* **2002**, *186*, 1–7. [CrossRef] [PubMed]
44. Podewils, L.J.; Zanardi Blevins, L.; Hagenbuch, M.; Itani, D.; Burns, A.; Otto, C.; Blanton, L.; Adams, S.; Monroe, S.S.; Beach, M.J.; et al. Outbreak of norovirus illness associated with a swimming pool. *Epidemiol. Infect.* **2007**, *135*, 827–833. [CrossRef] [PubMed]
45. White, P.A. Evolution of norovirus. *Clin. Microbiol. Infect.* **2014**, *20*, 741–745. [CrossRef] [PubMed]
46. Xerry, J.; Gallimore, C.I.; Iturriza-Gómara, M.; Allen, D.J.; Gray, J.J. Transmission events within outbreaks of gastroenteritis determined through analysis of nucleotide sequences of the P2 domain of genogroup II noroviruses. *J. Clin. Microbiol.* **2008**, *46*, 947–953. [CrossRef] [PubMed]
47. Widdowson, M.A.; Sulka, A.; Bulens, S.N.; Beard, R.S.; Chaves, S.S.; Hammond, R.; Salehi, E.D.P.; Swanson, E.; Totaro, J.; Woron, R.; et al. Norovirus and foodborne disease, United States, 1991–2000. *Emerg. Infect. Dis.* **2005**, *11*, 95–102. [CrossRef]
48. Kotwal, G.; Cannon, J.L. Environmental persistence and transfer of enteric viruses. *Curr. Opin. Virol.* **2014**, *4*, 37–43. [CrossRef]
49. Teunis, P.F.M.; Moe, C.L.; Liu, P.; Miller, S.E.; Lindesmith, L.; Baric, R.S.; Le Pendu, J.; Calderon, R.L. Norwalk virus: How infectious is it? *J. Med. Virol.* **2008**, *80*, 1468–1476. [CrossRef]
50. Bitler, E.J.; Matthews, J.E.; Dickey, B.W.; Eisenberg, J.N.S.; Leon, J.S. Norovirus outbreaks: A systematic review of commonly implicated transmission routes and vehicles. *Epidemiol. Infect.* **2013**, *141*, 1563–1571. [CrossRef]
51. Glass, R.I.; Parashar, U.D.; Estes, M.K. Current Concepts—Norovirus Gastroenteritis. *N. Engl. J. Med.* **2009**, *361*, 1776–1785. [CrossRef]
52. Kukkula, M.; Maunula, L.; Silvennoinen, E.; Von Bonsdorff, C. Outbreak of Viral Gastroenteritis Due to Drinking Water Contaminated by Norwalk-like Viruses. *J. Infect. Dis.* **1999**, *180*, 1771–1776. [CrossRef]
53. Doyle, M.P.; Erickson, M.C. Summer meeting 2007—The problems with fresh produce: An overview. *J. Appl. Microbiol.* **2008**, *105*, 317–330. [CrossRef] [PubMed]
54. Maunula, L.; Miettinen, I.T.; Von Bonsdorff, C.H. Norovirus outbreaks from drinking water. *Emerg. Infect. Dis.* **2005**, *11*, 1716–1721. [CrossRef] [PubMed]
55. Riera-Montes, M.; Brus Sjölander, K.; Allestam, G.; Hallin, E.; Hedlund, K.O.; Löfdahl, M. Waterborne norovirus outbreak in a municipal drinking-water supply in Sweden. *Epidemiol. Infect.* **2011**, *139*, 1928–1935. [CrossRef] [PubMed]
56. Matthews, J.E.; Dickey, B.W.; Miller, R.D.; Felzer, J.R.; Dawson, B.P.; Lee, A.S.; Rocks, J.J.; Kiel, J.; Montes, J.S.; Moe, C.L.; et al. The epidemiology of published norovirus outbreaks: A review of risk factors associated with attack rate and genogroup. *Epidemiol. Infect.* **2012**, *140*, 1161–1172. [CrossRef]
57. Robilotti, E.; Deresinski, S.; Pinsky, B.A. Norovirus. *Clin. Microbiol. Rev.* **2015**, *28*, 134–164. [CrossRef]
58. Yezli, S.; Otter, J.A. Minimum Infective Dose of the Major Human Respiratory and Enteric Viruses Transmitted Through Food and the Environment. *Food Environ. Virol.* **2011**, *3*, 1–30. [CrossRef]
59. Ciocca, M. Clinical course and consequences of hepatitis A infection. *Vaccine* **2000**, *18*, S71–S74. [CrossRef]
60. Yong, H.T.; Son, R. Review Article Hepatitis A virus—A general overview. *Int. Food Res. J.* **2009**, *467*, 455–467.
61. Jacobsen, K.H.; Wiersma, S.T. Hepatitis A virus seroprevalence by age and world region, 1990 and 2005. *Vaccine* **2010**, *28*, 6653–6657. [CrossRef]
62. Yeh, H.Y.; Hwang, Y.C.; Yates, M.V.; Mulchandani, A.; Chen, W. Detection of hepatitis A virus by using a combined cell culture-molecular beacon assay. *Appl. Environ. Microbiol.* **2008**, *74*, 2239–2243. [CrossRef]
63. Borchardt, M.A.; Spencer, S.K.; Kieke, B.A.; Lambertini, E.; Loge, F.J. Viruses in nondisinfected drinking

water from municipal wells and community incidence of acute gastrointestinal illness. *Environ. Health Perspect.* **2012**, *120*, 1272–1279. [CrossRef] [PubMed]

64. Hellmér, M.; Paxéus, N.; Magnius, L.; Enache, L.; Arnholm, B.; Johansson, A.; Bergström, T.; Norder, H. Detection of pathogenic viruses in sewage provided early warnings of hepatitis A virus and norovirus outbreaks. *Appl. Environ. Microbiol.* **2014**, *80*, 6771–6781. [CrossRef] [PubMed]
65. Ouardani, I.; Manso, C.F.; Aouni, M.; Romalde, J.L. Efficiency of hepatitis A virus removal in six sewage treatment plants from central Tunisia. *Appl. Microbiol. Biotechnol.* **2015**, *99*, 10759–10769. [CrossRef] [PubMed]
66. Chigor, V.N.; Sibanda, T.; Okoh, A.I. Assessment of the Risks for Human Health of Adenoviruses, Hepatitis A Virus, Rotaviruses and Enteroviruses in the Buffalo River and Three Source Water Dams in the Eastern Cape. *Food Environ. Virol.* **2014**, *6*, 87–98. [CrossRef] [PubMed]
67. Dias, J.; Pinto, R.N.; Vieira, C.B.; de Abreu Corrêa, A. Detection and quantification of human adenovirus (HAdV), JC polyomavirus (JCPyV) and hepatitis A virus (HAV) in recreational waters of Niterói, Rio de Janeiro, Brazil. *Mar. Pollut. Bull.* **2018**, *133*, 240–245. [CrossRef] [PubMed]
68. Jacobsen, K.H. *The Global Prevalence of Hepatitis A Virus Infection and Susceptibility: A Systematic Review (No. WHO/IVB/10.01)*; WHO Press: Geneva, Switzerland, 2009.
69. Gupta, E.; Ballani, N. State of the globe: Hepatitis A virus—Return of a water devil. *J. Glob. Infect. Dis.* **2014**, *6*, 57–58. [CrossRef]
70. Osuolale, O.; Okoh, A. Incidence of human adenoviruses and Hepatitis A virus in the final effluent of selected wastewater treatment plants in Eastern Cape Province, South Africa. *Virol. J.* **2015**, *12*, 1–8. [CrossRef]
71. APHA/AWWA/WEF. Standard Methods for the Examination of Water and Wastewater. 2012. Available online: https://beta-static.fishersci.com/content/dam/fishersci/en_US/documents/programs/scientific/technical-documents/white-papers/apha-water-testing-standard-methods-white-paper.pdf (accessed on 13 February 2020).
72. ISO. Microbiology of the Food Chain—Horizontal Method for Determination of Hepatitis A Virus and Norovirus Using Real-Time RT-PCR—Part 1: Method for Quantification. ISO/TS 15216-1. Available online: https://www.iso.org/standard/65681.html (accessed on 13 February 2020).
73. Fout, G.S.; Cashdollar, J.L.; Varughese, E.A.; Parshionikar, S.U.; Grimm, A.C. EPA method 1615. Measurement of enterovirus and norovirus occurrence in water by culture and RT-qPCR. I. collection of virus samples. *J. Vis. Exp.* **2015**, e52067. [CrossRef]
74. R Core Team. R: A Language and Environment for Statistical Computing. Available online: https://www.r-project.org/ (accessed on 16 July 2019).
75. Atmar, R.L.; Opekun, A.R.; Gilger, M.A.; Estes, M.K.; Crawford, S.E.; Neill, F.H.; Graham, D.Y. Norwalk virus shedding after experimental human infection. *Emerg. Infect. Dis.* **2008**, *14*, 1553–1557. [CrossRef]
76. Lee, S.G.; Jheong, W.H.; Suh, C.I.; Kim, S.H.; Lee, J.B.; Jeong, Y.S.; Ko, G.P.; Jang, K.L.; Lee, G.C.; Paik, S.Y. Nationwide groundwater surveillance of noroviruses in South Korea, 2008. *Appl. Environ. Microbiol.* **2011**, *77*, 1466–1474. [CrossRef] [PubMed]
77. Bae, J.; Schwab, K.J. Evaluation of murine norovirus, feline calicivirus, poliovirus, and MS2 as surrogates for human norovirus in a model of viral persistence in surface water and groundwater. *Appl. Environ. Microbiol.* **2008**, *74*, 477–484. [CrossRef] [PubMed]
78. Ngazoa, E.S.; Fliss, I.; Jean, J. Quantitative study of persistence of human norovirus genome in water using TaqMan real-time RT-PCR. *J. Appl. Microbiol.* **2008**, *104*, 707–715. [CrossRef] [PubMed]
79. Seitz, S.R.; Leon, J.S.; Schwab, K.J.; Lyon, G.M.; Dowd, M.; McDaniels, M.; Abdulhafid, G.; Fernandez, M.L.; Lindesmith, L.C.; Baric, R.S.; et al. Norovirus infectivity in humans and persistence in water. *Appl. Environ. Microbiol.* **2011**, *77*, 6884–6888. [CrossRef]
80. Wyer, M.D.; Wyn-Jones, A.P.; Kay, D.; Au-Yeung, H.K.C.; Gironés, R.; López-Pila, J.; de Roda Husman, A.M.; Rutjes, S.; Schneider, O. Relationships between human adenoviruses and faecal indicator organisms in European recreational waters. *Water Res.* **2012**, *46*, 4130–4141. [CrossRef]
81. Henry, R.; Schang, C.; Kolotelo, P.; Coleman, R.; Rooney, G.; Schmidt, J.; Deletic, A.; McCarthy, D.T. Effect of environmental parameters on pathogen and faecal indicator organism concentrations within an urban estuary. *Estuar. Coast. Shelf Sci.* **2016**, *174*, 18–26. [CrossRef]
82. Da Silva, A.K.; Le Saux, J.C.; Parnaudeau, S.; Pommepuy, M.; Elimelech, M.; Le Guyader, F.S. Evaluation of

removal of noroviruses during wastewater treatment, using real-time reverse transcription-PCR: Different behaviors of genogroups I and II. *Appl. Environ. Microbiol.* **2007**, *73*, 7891–7897. [CrossRef]

83. Iwai, M.; Hasegawa, S.; Obara, M.; Nakamura, K.; Horimoto, E.; Takizawa, T.; Kurata, T.; Sogen, S.I.; Shiraki, K. Continuous presence of noroviruses and sapoviruses in raw sewage reflects infections among inhabitants of Toyama, Japan (2006 to 2008). *Appl. Environ. Microbiol.* **2009**, *75*, 1264–1270. [CrossRef]
84. López-Gálvez, F.; Truchado, P.; Sánchez, G.; Aznar, R.; Gil, M.I.; Allende, A. Occurrence of enteric viruses in reclaimed and surface irrigation water: Relationship with microbiological and physicochemical indicators. *J. Appl. Microbiol.* **2016**, *121*, 1180–1188. [CrossRef]
85. Haramoto, E.; Fujino, S.; Otagiri, M. Distinct behaviors of infectious F-specific RNA coliphage genogroups at a wastewater treatment plant. *Sci. Total Environ.* **2015**, *520*, 32–38. [CrossRef]
86. Haramoto, E.; Kitajima, M.; Hata, A.; Torrey, J.R.; Masago, Y.; Sano, D.; Katayama, H. A review on recent progress in the detection methods and prevalence of human enteric viruses in water. *Water Res.* **2018**, *135*, 168–186. [CrossRef] [PubMed]

9

# An Innovative Dual-Column System for Heavy Metallic Ion Sorption by Natural Zeolite

**Amanda L. Ciosek * and Grace K. Luk**

Department of Civil Engineering, Faculty of Engineering and Architectural Science, Ryerson University, Toronto, ON M5B 2K3, Canada; gluk@ryerson.ca
* Correspondence: amanda.alaica@ryerson.ca.

**Abstract:** This study investigates the design and performance of a novel sorption system containing natural zeolite. The apparatus consists of packed, fixed-bed, dual-columns with custom automated controls and sampling chambers, connected in series and stock fed by a metering pump at a controlled adjustable distribution. The purpose of the system is to remove heavy metallic ions predominately found in acid mine drainage, including lead ($Pb^{2+}$), copper ($Cu^{2+}$), iron ($Fe^{3+}$), nickel ($Ni^{2+}$) and zinc ($Zn^{2+}$), combined in equal equivalence to form an acidified total 10 meq/L aqueous solution. Reported trends on the zeolite's preference to these heavy metallic ions is established in the system breakthrough curve, as $Pb^{2+} >> Fe^{3+} > Cu^{2+} > Zn^{2+} >> Ni^{2+}$. Within a 3-h contact period, $Pb^{2+}$ is completely removed from both columns. Insufficient $Ni^{2+}$ removal is achieved by either column with the promptest breakthrough attained, as zeolite demonstrates the least affinity towards it; however, a 48.97% removal is observed in the cumulative collection at the completion of the analysis period. The empty bed contact times for the first and second columns are 20 and 30 min, respectively; indicating a higher bed capacity at breakthrough and a lower usage rate of the zeolite mineral in the second column. This sorption system experimentally demonstrates the potential for industrial wastewater treatment technology development.

**Keywords:** zeolite; sorption; packed fixed-bed columns; heavy metallic ions; automated sampling design

---

## 1. Introduction

Acid mine drainage (AMD) generated by industrial mines contains highly acidic wastewater and toxic heavy metallic ions (HMIs). These HMIs are a serious threat to human health and the environment [1–5], with their high solubility [6], as well as non-biodegradable and bio-accumulative properties [7,8]. If mines are abandoned, there is a risk of AMD contamination to both surface and groundwater, causing catastrophic damage to the ecosystem [4,9]. Therefore, the HMIs must be removed by advanced treatment methods prior to discharge [10–12], to abide by the effluent maximum allowable limits [7,8]. Due to the high site-to-site variability present in mines, mitigation feasibility can become a challenge [3], and a simple, resilient and cost-effective strategy must be developed [1].

Various industrial wastewater treatment methods include precipitation, electro-chemical remediation, oxidation and hydrolysis, neutralization, ion-exchange and solvent extraction, ion-exchange and precipitation, titration, bio-sorption, adsorption, reverse osmosis [4,9], and ultrafiltration [13]. Among these, sorption is a simple but promising treatment method [5,14], based on demonstrated industrial viability and effectiveness, cost efficiency and environmental sustainability [15,16]. The removal of HMIs is attributed to the mechanisms of both adsorption (on the surface of the sorbents' micropores) and ion-exchange (through the sorbents' framework pores and channels) [17]; and culminate in the unified process of sorption [18,19].

Zeolites hold great potential as naturally occurring minerals [1], and are recognized as effective materials for the removal of HMIs from industrial wastewater by sorption [20]. They have attracted a lot of researchers' interests [11], due to its coexisting molecular sieve action, ion-exchange and catalytic properties [14,21]. The mineral zeolite is a hydro-aluminosilicate with a crystalline structure comprised of $SiO_4$ and $AlO_4$ tetrahedras linked by oxygen atoms, which form an open, homogeneous microporous three-dimensional framework creating voids and channels [21]. Clinoptilolite, a globally abundant and well-documented form of zeolite [13,20,22,23], is used in this research. Two of the most significant properties of zeolite are its high cation exchange capacity [23], and its selectivity of certain metals [24]. In addition to wide deposit distribution and low exploitation costs [7,25], zeolite is considered as a strong candidate for the removal of wastewater contaminants [26], especially those containing high HMIs from mine waste.

Currently, there is limited comprehension of the sorption by zeolite for waste with multi-component HMIs [24,27], in order to fully benefit from using it in tertiary treatment processes [1]. In particular, the composition of AMD waste contains numerous contaminants, which include HMIs and other pollutants, and the presence of these in solution affect the overall removal potential [3,18]. Research is needed on the simultaneous sorption of the multi-metallic components that are prevalent in industrial effluent, and to quantify uptake interference of these HMIs in combination [16]. Helfferich [18] (p. 201) points out that for multi-component systems, the exchange rate may vary for the various counter-ions of HMIs in solution, with the possibility that the concentrations of certain species in either the sorbent or solution may fluctuate prior to attaining its balanced state. The performance of columns or fixed-bed reactors (FBR) is convenient for industrial scale applications [7], which requires less investment and operational costs, and is more economically feasible than its discontinuous batch mode counterparts [8]. FBR columns have demonstrated performance efficiency in treating large volumes, and are frequently implemented in sorption studies. However, the removal of multi-component solutions of HMIs is complex due to ion competition, different affinity sequences, and zeolite selectivity. Its operations are affected by equilibrium (isotherm and capacity), kinetic (diffusion and convection coefficients) and hydraulic (liquid holdup, geometric analogies and mal-distribution) factors [28]. In practice, the influence of operative conditions on the overall system performance is not experimentally verified [17,29,30], but they are extremely important to large-scale development. Although the FBR system is highly valuable, its analysis is unpredictably multi-faceted [28,31] and even more so with the presence of numerous interfering ions. Complications due to ion competition and solute-surface interactions [4], as well as the unique affinity sequences and sorbent material selectivity [25], have been reported.

The authors have conducted a four-phase research project, consisting of the analysis of: (1) the effects of preliminary parameters and operative conditions (particle size, sorbent-to-sorbate dosage, influent concentration, contact time, set-temperature, heat pre-treatment), (2) heavy metallic ions component system combinations and selectivity order [32], (3) kinetic modelling trends [33], and finally in this current study, (4) the design of a packed fixed-bed, dual-column sorption system. The first three phases are conducted in batch mode, to which reveal a key trend among the HMIs selected as lead ($Pb^{2+}$) >> iron ($Fe^{3+}$) > copper ($Cu^{2+}$) > zinc ($Zn^{2+}$) > nickel ($Ni^{2+}$). The findings of these preliminary phases have established a platform for the design of the sorption system in continuous mode, presented in this paper.

Existing column experimental designs involve various limitations, including:

1. The evaluation of predominantly single- or dual-component HMI system combinations;
2. The implementation of primarily slender column aspect ratios (i.e., bed depth/particle diameter, column height/diameter), causing a challenge to eventual scale-up design;
3. The use of inconsistent and/or vague sorbent compaction techniques, and;
4. The application of simple, idealized flow patterns (i.e., set single and continuous flow rate).

The objective of this paper is to develop a novel dual-column sorption system to overcome some of these shortcomings. Important design factors such as the zeolite compaction, column dimensions and aspect ratios, flow control, sampling and analytical procedure, will be taken into consideration. The exclusivity of this prototype is attributed to an automated, variable-flow configuration with a custom sampling technique. In contrast to most previous single-component sorption set-ups, this study evaluates the simultaneous sorption process by natural zeolite of the five most commonly occurring HMIs found in AMD, including $Pb^{2+}$, $Cu^{2+}$, $Fe^{3+}$, $Ni^{2+}$ and $Zn^{2+}$. This paper will demonstrate the effectiveness and the removal efficiency in a continuous-flow FBR system over a 3-h duration from the dual columns, providing insights on HMIs selectivity and treatment system breakthroughs. It is envisaged that this research will provide much-needed information to the wastewater treatment industry for the design and implementation of innovative sorption technologies.

## 2. Experimental Design

### *2.1. Packed Fixed-Bed Column Design Considerations*

When the concentration of the effluent reaches 5%–10% of the influent, this point on a typical S-shaped breakthrough curve is commonly referred to as the 'breakthrough point' or 'breakpoint' (BP). The point of column exhaustion (EP) is when the effluent reaches maximum capacity to 90%–95% of its influent value [12,34]. The efficiency of the column performance is related to the bed capacity at breakthrough and at exhaustion, represented by the following relationship [12,34]:

$$\eta = \frac{C_{BP}}{C_{EP}} \tag{1}$$

where $\eta$ is the column efficiency (degree of saturation), $C_{BP}$ is the breakthrough capacity of the bed (in meq/g), and $C_{EP}$ is the maximum capacity at exhaustion of the bed, defined by the total amount of HMI ions bound in the zeolite (or exchanged in the packed fixed-bed) (in meq/g).

The breakthrough capacity and equilibrium capacity are further expressed in Equations (2) and (3), respectively [12,34]:

$$C_{BP} = \frac{\int_0^{V_{BP}} (C_0 - C)dV}{\rho HA} = \frac{C_0 \cdot V_{BP}}{m} = \frac{\eta_{BP}}{m} \tag{2}$$

$$C_{EP} = \frac{\int_0^{V_{EP}} (C_0 - C)dV}{\rho HA} = \frac{\int_0^{V_{EP}} (C_0 - C)dV}{m} = \frac{\eta_{EP}}{m} \tag{3}$$

where $V_{BP}$ is the effluent volume collected up to breakthrough point (BP) (in L), $V_{EP}$ is the effluent volume at which the exhaustion point (EP) is reached in the zeolite bed (in L), $C_0$ is the influent concentration (in meq/L), C is the effluent concentration (in meq/L), $\rho$ is the packing density of the bed (in $g/cm^3$), H is the bed depth (in cm), A is the bed cross-sectional area (in $cm^2$), m is the zeolite mass (in g); where $\eta_{BP}$ and $\eta_{EP}$ is the total amount of HMI ions removed up to BP and EP (in meq), respectively.

Empty bed contact time (EBCT) is the residence time (in min) a fluid element is in contact with the bed, and is related to the systems' removal kinetics [35]. This is represented by the relationship between the bed depth (H) in the column and the feed solution velocity (v) [25,34], as given by:

$$EBCT = \frac{H}{v} = \frac{H}{(Q/A)} = \frac{d^2 \pi H}{4Q} \tag{4}$$

Research conducted by Vukojevic Medvidovic et al. [34] demonstrates that the breakthrough curve results reveal that the flow through the column determines the EBCT; with the same initial concentration, the increase in flow rate decreases the contact time and increases the mass transfer zone

(MTZ) height. The MTZ is the restricted area where the exchange process occurs, and is defined as the zeolite layer height between the equilibrium zone and the unused bed zone [34]; where the effluent concentration varies from 5% to 95% of the influent concentration [25]. As the HMI solution is fed through the packed fixed-bed, the MTZ moves in the direction of flow and eventually reaches the exit [12,34]. Peric et al. [36] distinctly demonstrates the importance of the column bed depth on the removal of lead from aqueous solutions. The results show that as the bed depth increases, a delay in breakthrough and exhaustion occurs, with an increase of the MTZ height. The higher the bed depth, the longer the service time at various breakthrough points due to the increase in binding sites on the sorbent material (zeolite mineral) [9]. Adequate wetting of the zeolite, and ideal contact time between the zeolite and solution interface are important for mass transfer and equilibrium conditions based on the selection of the flow rate and particle size. To minimize possible wall and axial dispersion effects in the fixed-bed column, the bed depth-to-particle diameter ratio ($H/d_p$) must be greater than 20. At a higher $H/d_p$ ratio, the breakthrough point appears later and the curve is steeper.

The usage rate ($v_U$, in g/L) determines the rate at which the sorbent would be exhausted and how often it must be replaced or regenerated, and is expressed in the following relationship [4,25]:

$$v_U = \frac{m_Z}{V_{BP}} \tag{5}$$

where $m_Z$ is the zeolite mass in the bed (in g) and $V_{BP}$ is the volume of the effluent treated at breakthrough (BP) (in L) [25]. Inglezakis [28] states that it is extremely difficult to model multi-component system interactions, as numerous time-consuming data are required and the process involves significant mathematical complexity. Breakthrough and exhaustion thresholds of specific HMIs within a fixed-bed are important for experimental specific conditions. In order to optimize the liquid-solid contact time and removal capacity, it is necessary to develop these relationships, between EBCT and usage rate [28].

### 2.2. *Natural Zeolite Mineral*

This study employs a natural zeolite mineral sample composed primarily of 85%–95% clinoptilolite (CAS No. 12173-10-3) and is sourced from a deposit located in Preston, Idaho [37]. The natural zeolite sample specifications are provided in Table 1, where typical elemental analysis indicates the presence of various elements, including $Na^+$, $Ca^{2+}$, $Mg^{2+}$ and $K^+$, as well as lead, copper, iron, and zinc. No significant concentrations of toxic trace elements are present in its composition, nor are trace metal elements water soluble. The low-clay content unique to this sample ensures good hydraulic conductivity, low dust content, and a harder and more resistant structure [37]. The zeolite mineral sample is applied in its natural state, without any chemical modifications, to minimize all associated costs and environmental impacts of this study. As suggested by the laboratory-scale packed bed system investigations by Inglezakis et al. [38], a zeolite fraction of 0.8 mm to 1 mm nominal diameter is recommended for the columns to ensure the full exploitation of the material but also to prevent considerable pressure drop during the analysis period. Therefore, the zeolite sample used for this research is obtained from standard mechanical mesh sieves set to a range of 0.841 mm to 1.19 mm (standard mesh $-16 + 20$) [27], resulting in a geometric mean diameter of 1.00 mm [39]. The sieved zeolite is exposed to a cleaning cycle, which involves rinsing in deionized distilled water to remove residual debris and dust, and drying at $80 \pm 3$°C for 24 h to remove any residual moisture [30].

**Table 1.** Natural zeolite specifications [32,37].

| | | | | |
|---|---|---|---|---|
| **Chemical Composition** | Mineral Component | 85%–95% Clinoptilolite (non-crystalline silica opaline balance) | | |
| | Cation Exchange Capacity (CEC) | 180–220 meq/100 g (as ammonium, -N) (high) | | |
| | Maximum Water Retention | >55 wt % (hydrophilic) | | |
| | Overall Surface Area | 24.9 $m^2/g$ (large) | | |
| | Bulk Density | approx. 55–60 lb/$f^3$ | | |
| | Hardness | Moh's No. 4 (high) | | |
| | pH | 7–8.64 | | |
| | Colour | Pale Green | | |
| **MSDS Composition Information** | | **Chemical** | **wt %** | **CAS No.** |
| | | Clinoptilolite | 90–97 | 12173-10-3 |
| | | Water | 3–10 | 7732-18-5 |
| **Analytical Rock Data** | | $SiO_2$ | | 67.4% |
| | | $Al_2O_3$ | | 10.6% |
| | | MgO | | 0.45% |
| | | $K_2O$ | | 4.19% |
| | | MnO | | <0.01% |
| | | CaO | | 2.23% |
| | | $TiO_2$ | | 0.27% |
| | | $Fe_2O_3$ | | 1.70% |
| | | $Na_2O$ | | 0.59% |
| | | $P_2O_5$ | | 0.10% |
| | | Loss-On-Ignition (LOI) 925 °C | | 11.40% |
| **Major Cation Range** | | Ca | | 1.60%–2.0% |
| | | K | | 2.93%-3.47% |
| | | Na | | <0.5% |

### 2.3. *Heavy Metallic Ion Solution*

Due to a greater presence in various Ontario mine waste streams as presented by Wilson [40] and the strict limitations required by the Canadian Government [41], this study focuses on the presence of the heavy metallic ions of $Pb^{2+}$, $Cu^{2+}$, $Fe^{3+}$, $Ni^{2+}$ and $Zn^{2+}$ [32,42]. The cations present in the sorbent have valences that differ from those in solution. Consequently, as the dilution increases, the selectivity of the adsorbent for the ion with a higher valence also increases. Accordingly, comparative analysis of various metal ions should be conducted at the same normality and temperature, in order to minimize the changes observed in isotherm configuration with dilution [43]. The synthetic metallic ion solutions are prepared from analytical grade nitrate salts in deionized distilled water, namely $Pb(NO_3)_2$ (CAS No. 10099-74-8), $Cu(NO_3)_2{\cdot}3H_2O$ (CAS No. 10031-43-3), $Fe(NO_3)_3{\cdot}9H_2O$ (CAS No. 7782-61-8), $Ni(NO_3)_2{\cdot}6H_2O$ (CAS No. 13478-00-7), and $Zn(NO_3)_2{\cdot}6H_2O$ (CAS No. 10196-18-6), respectively, and combined equally to maintain a total normality of 0.01 N (10 meq/L). The five metals in a multi-component system of 2.0 meq/L per metal correspond to concentrations of approximately 207 mg/L for $Pb^{2+}$, 64 mg/L for $Cu^{2+}$, 37 mg/L for $Fe^{3+}$, 59 mg/L for $Ni^{2+}$, and 65 mg/L for $Zn^{2+}$, respectively. The $NO_3^-$ anions in the aqueous solution do not influence the ion-exchange process, since they do not form any metal-anion complexes and do not hydrolyze in solution [13,44].

The uptake of multiple HMIs from aqueous solutions on natural zeolite is a complex process consisting of predominately ion-exchange and adsorption. At high initial concentrations, this process could be accompanied by precipitation and the metal ion hydroxo-complexes formed can be sorbed on zeolite surface sites that encompass different sorption affinity [13]. Research has demonstrated [26,43,45] that the sorbate solution acidity level affects the uptake of metals, and this is particularly the case for the HMIs that have low preference by zeolite. Such factors include the metal ion speciation and natural stability, as well as the electro-kinetic properties of zeolite in aqueous solutions. At a low pH level, the hydrogen cation ($H^+$) is considered as a competitive ion to the HMI during the ion-exchange process [43]; evidently, the process is preferred at higher pH levels, which should be lower than the minimum pH of precipitation [10]. The pH level of the effluent solution decreases, depending on the metal removed. Therefore, the pH range under which sorption takes place should be specified [11].

The Canada-Wide Survey of Acid Mine Drainage [40] reports a seasonal average of a majority of the mines surveyed to have documented pH values ranging from 2 to 5. This present study is conducted in the conservative end of this range, with the influent stock acidified to a pH level of around 2 with concentrated nitric ($HNO_3$) acid (CAS No. 7697-37-2) [46], to prevent precipitation of the metal ions [14,43]. With this low pH, however, the $H^+$ ion competition is significant, and so the removals obtained are on the conservative side and are lower than would normally be expected in field installations.

### 2.4. Analytical Procedure and Quality Control

There are various atomic spectrometry techniques, which include Flame Atomic Adsorption (AA), Graphite Furnace AA, and Inductively Coupled Plasma Mass Spectrometry (ICP-MS). In particular, the Inductively Coupled Plasma—Atomic Emission Spectroscopy (ICP-AES) technique permits the complete atomization of the elements in a sample. This feature minimizes the potential for chemical interferences. It is considered as a true multi-element technique with exceptional sample throughput, and with a very wide range of analytical signal intensity [47]. Therefore, the HMIs are analyzed in their aqueous phase by ICP-AES technology (Optima 7300 DV, Part No. N0770796, Serial No. 077C8071802, Firmware Version 1.0.1.0079; Perkin Elmer Inc.; Waltham, MA, USA); with corresponding WinLab32 Software (Version 4.0.0.0305). The analyte primary wavelengths of each HMI element targeted were selected as 327.393 for Cu, 238.204 for Fe, 231.604 for Ni, 220.353 for Pb, and 206.200 for Zn, respectively; on the basis that these wavelengths have the strongest emission and provide the best quantifiable detection limits (QDL). Analysis is conducted with a plasma setting in radial view (to concentrations of greater than 1 mg/L), with QDLs of 0.05 μg/mL for Cu, Fe, Ni, and Zn, and 0.10 μg/mL for Pb. The spectrometer settings involve auto sampling of 45 s normal time at a rate of 1.5 mL/min, and a processing setting of 3 to 5 points per peak with 2 point spectral corrections. The calibration curve is generated 'through zero' by applying a stock blank and a multi-element Quality Control Standard 4 with 1, 10, 50, 90, and 100 mg/L concentrations (as per Standard Methods Part 3000) [46]. In comparison to the 'linear calculated intercept' calibration method, only a 0.15 mg/L (or 0.66%) maximum discrepancy is observed among all average concentration values, demonstrating an accurate overall calibration. In addition, the median 50 mg/L calibration standard is applied as a check parameter, with the intent to ensure a higher accuracy of all experimental measurements. Triplicate readings and their mean concentrations in calibration units are generated in mg/L by the corresponding WinLab32 Software. The sorbed amount of HMI is calculated from the difference between the starting concentration and its concentration in the 0.45 μm filtered samples' supernatant.

During every ICP-AES analytical session, several quality control methods are applied, and evaluated by three check parameters to assess the calibration quality [48]. Firstly, the percent relative standard deviation (%RSD) reports an average of 0.433%, which is well within the <3% limit recommended. The triplicate concentration of the median standard has an average value of 49.26 mg/L, and is within 5% of the known value. Finally, the correlation coefficient of each HMI analyte primary wavelength reports an average of 0.999977, which is very close to unity. Therefore, the data is relatively accurate, highly reproducible, and the experimental replicates are reliable based on the calibration relationship established.

The multi-component stock is created by diluting the respective HMI nitrate salts of three 1-L stock solutions, acidified to a pH of 2.0 ± 0.1, and then re-combined. These 3 stocks (denoted as X, Y, Z) are diluted by one 50% step to be within the 0–100 mg/L calibration range, analyzing each separately and combined (denoted as M). The consistency in stock preparation is demonstrated in Table 2. The average diluted concentrations of the X, Y, and Z influent stocks for $Cu^{2+}$, $Fe^{3+}$, $Ni^{2+}$, $Pb^{2+}$, and $Zn^{2+}$ are 70.15, 39.25, 61.18, 216.02, and 66.54 mg/L, respectively. The diluted concentration of the M influent stock for $Cu^{2+}$, $Fe^{3+}$, $Ni^{2+}$, $Pb^{2+}$, and $Zn^{2+}$ are 70.22, 39.21, 60.48, 213.22, and 65.98 mg/L, respectively. A maximum difference of 2.80 mg/L, equivalent to 1.3%, is detected for the $Pb^{2+}$ stock. Also, the corresponding HMI concentrations in mg/L are comparable to the theoretically expected values, based on the selected total 10 meq/L initial concentration; only a 0.05% difference between the average of all initial concentrations of the theoretical and combined M stock is detected. Overall, this demonstrates that strong quality control has been implemented.

**Table 2.** Inductively Coupled Plasma—Atomic Emission Spectroscopy (ICP-AES) generated multi-component stock concentration.

| Sample ID | Analyte | Int (Corr) | RSD (Corr Int) | Conc (Calib) (mg/L) |
|---|---|---|---|---|
| M-X | Cu 327.393 | 188,070.71 | 0.27 | 34.26 |
| | Fe 238.204 | 79,641.94 | 0.39 | 19.18 |
| | Ni 231.604 | 41,071.22 | 0.50 | 29.81 |
| | Pb 220.353 | 32,330.16 | 0.52 | 105.28 |
| | Zn 206.200 | 55,015.91 | 0.38 | 32.31 |
| M-Y | Cu 327.393 | 186,885.03 | 0.74 | 34.04 |
| | Fe 238.204 | 79,083.95 | 0.90 | 19.04 |
| | Ni 231.604 | 40,721.53 | 0.48 | 29.55 |
| | Pb 220.353 | 31,973.87 | 0.31 | 104.12 |
| | Zn 206.200 | 54,758.60 | 1.09 | 32.16 |
| M-Z | Cu 327.393 | 202,742.71 | 0.91 | 36.93 |
| | Fe 238.204 | 85,771.53 | 1.02 | 20.65 |
| | Ni 231.604 | 44,652.28 | 3.73 | 32.41 |
| | Pb 220.353 | 35,199.76 | 3.77 | 114.63 |
| | Zn 206.200 | 60,176.84 | 4.12 | 35.34 |
| MM | Cu 327.393 | 192,776.82 | 0.63 | 35.11 |
| | Fe 238.204 | 81,419.80 | 0.80 | 19.60 |
| | Ni 231.604 | 41,667.28 | 0.38 | 30.24 |
| | Pb 220.353 | 32,738.45 | 0.40 | 106.61 |
| | Zn 206.200 | 56,170.16 | 0.94 | 32.99 |

### *2.5. Sorption System Design*

Based on qualitative observations, the uptake of counter-ions in a continuous column system is favoured by various factors, including: a strong preference of the zeolite for the HMI counter-ions in solution, low concentration of HMI counter-ions, small and uniform particle size, high volume capacity and low degree of cross-linking, elevated temperature and low flow rate, as well as a high column height or aspect ratio [18] (p. 427). With this is mind, the apparatus development considers an extensive material and equipment selection process, with numerous stages of optimization in order to maintain flow continuity and repeatability. The final design was adopted in consideration of the following factors:

- Zeolite Compaction Technique
    - Regulated Layers of Dry Mass
    - Systematic Tampered Compaction
- Column Dimensions
    - Modular Design
    - Internal Diameter (1 in)
    - Sorption Column Height (1 ft)
- Flow Configuration
    - Upflow Distribution
    - Dual-Column Series Connection
    - Methodical Flow Rate Variability
- Pump Type
    - Diaphragm Metering

- Sampling Method
  - Automated Mode Controls
  - Customized Sampling Chambers
  - Modes' Interchange in Five (5)-minute Intervals
- Analysis Period
  - Three (3)-hour Contact Period

Based on these critical parameters, the sorption system design is finalized. Figure 1 is a schematic representation of the constructed prototype, detailing the flow paths through the system. The fundamental components include:

- HMI Multi-Component Influent Stock
- Metering Pump
- Silicon Tubing and Polyvinyl chloride (PVC) Connections
- Check Valves
- Automatable Solenoid Valves (symbol S)
- Packed Fixed-Bed Sorption Columns
- Custom Sampling Chambers
- Sampling Ports
- Effluent Collection Basin

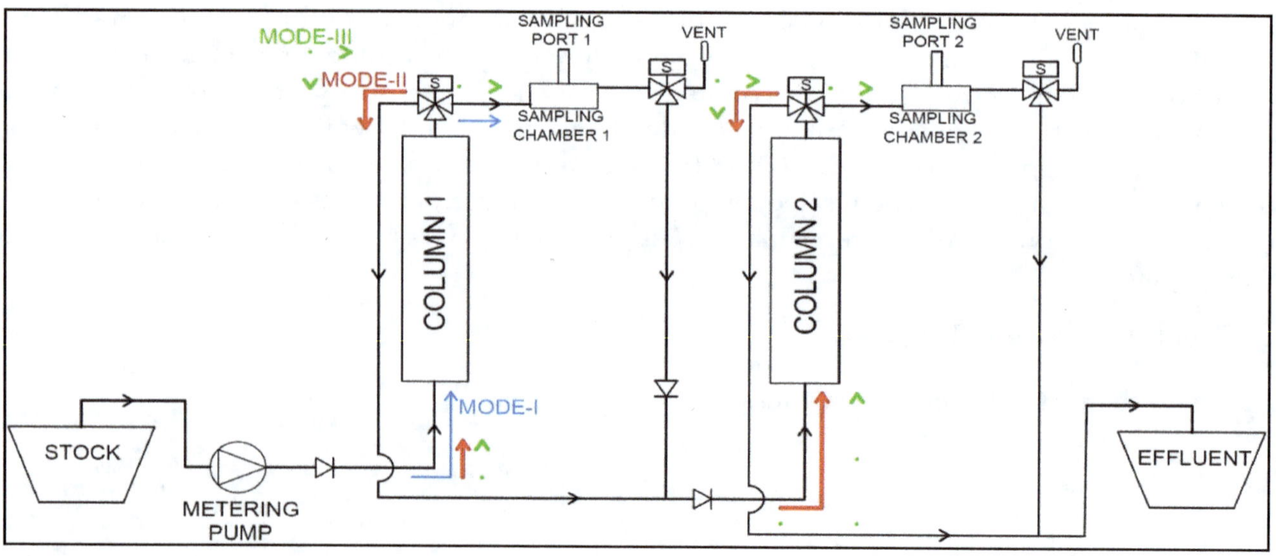


**Figure 1.** Schematic representation of automated sorption system prototype flow path layout.

2.5.1. Column Dimensions

The column is made of a circular section of clear PVC SCH-40 pipe (Part No. r4-1000; Fabco Plastics; Maple, ONT, Canada), 30.48 cm in height with 2.61 cm internal diameter. In order to minimize potential effects of wall and axial dispersion in the columns, the bed depth-to-particle diameter ratio should be kept greater than 20 [36]. Using the average nominal zeolite diameter of 1.00 mm as a reference, this ratio works out to be over 300 for the design. The cleaned and dry zeolite particles are added to the column at nine layers applied at 20-mL or 16.9 g amounts. Each layer is compacted with medium force, pounding six times with a customized PVC plunger of a diameter equal to the internal diameter of the column; such that the column height and the zeolite bed depth are equivalent. Inert plastic mesh with a smaller size than the minimum zeolite particle gradation of 0.841 mm is used to contain the zeolite material, and to permit sample flow through the columns. This mesh is set at each end of the column, within the two halves of a PVC SCH-80 socket union fitting incorporating a viton o-ring (Part No. 897010; Fabco Plastics; Maple, ONT, Canada), connected to a nominal 1 × 1/2-inch PVC SCH-40 reducer bushing (Part No. 438130 (slip × FPT); Fabco Plastics; Maple, ONT, Canada). All components are connected to 1/4-inch silicon tubing and with corresponding adapters and nipples fittings.

### 2.5.2. Flow Rate and Configuration

As a dual-stage system, the two columns are connected in a serial-flow arrangement, such that the first column receives the original stock with a higher HMI concentration and the second column receives the effluent from the first column. The upflow configuration ensures an overall better quality of flow, with a low liquid hold-up and a good stock feed distribution across the column cross sectional area. In contrast, for the downflow mode, an increase in pressure drop and flooding of the column bed is more probable [28]. Consequently, the stock is fed in an upflow direction to ensure proper and thorough distribution to the column beds and to minimize the need for backwashing and head loss effects.

A critical parameter in the design process is the flow rate. Existing research demonstrates that lower flow rates result in high detention times in the column, which is needed due to the relatively slow uptake rate of zeolite [24,27]. The HMI solution volume element is in contact with a given zeolite bed layer for only a limited period of time. Consequently, equilibrium is not usually achieved and thereby results in a lower overall uptake of HMIs from the influent stock solution. Preliminary testing involved a peristaltic pump, using the corresponding silicon rubber tubing. Significant back pressure was observed and the capacity of the peristaltic pump was hindered. Consequently, the required flow rate was unachievable; the rotational speed and strength decreased for the feed to completely traverse through the entire system. Subsequently a diaphragm-type metering pump (No. 950218125-C Plus, max 45-LPD, 80-psi, 125-AC, 50/60-Hz; PULSAtron; Punta Gorda, FL, USA) is employed in the final design, which mechanically facilitates the desired stock feeding rate. Based on the 45-LPD (31.25 mL/min) capacity of the metering pump, preliminary flow rate testing of the pump set to 100% stroke (mechanically pumped volume) and 50% rate established an initial, repeatable, point-of-reference flow rate of 6.36 ± 0.32 mL/min. This stroke-rate setting is maintained and is comparable to the lower end of the 6–18 mL/min range recommended by Inglezakis et al. [24,27], to provide sufficient detention time in the system.

### 2.5.3. Sampling Method

Another critical component to the design is the sampling method, and how to maintain continuous flow through the system while sampling the effluent of both columns. Due to the relatively slow feeding rate, the time to collect the desired sample volume for dilution and ICP-AES analysis would require residual sample volume and minutes of valuable contact time. Three-way solenoid valves (No. 00457979, 0124-C, 1/8-FKM-PP, NPT-1/4, max 145-psi, 24-V, 60-Hz, 8-W, 38-mL; burkert; Ingelfingen, Germany) are implemented to ensure that while a sample volume is collected at the desired sampling time, both columns would still be fed continuously. The MODE valves and custom fabricated rotating 30-mL sampling chambers are attached to the top exit of each sorption column, with accessible sampling ports. A second three-way solenoid VENT valve is included at the exit of each sampling chamber to introduce an air vent to assist in rapid sample extraction by preventing vacuum pockets within the sample chamber and discharge tubing. A multi-turn valve is included at the exit of the vent for the first column to introduce minor back pressure similar to that of the second column, so as not to alter the flow characteristics through the first column. Check valves are placed at critical locations throughout the hydraulic circuit to prevent back flows.

### 2.5.4. System Modes of Operation

The sorption system presented in this paper is comprised of three distinct modes of operation that are controlled by the MODE valves for each column:

- **Mode-I**
  - Sorption System Activation
  - Fill Sorption Column 1 and Sample Chamber 1
- **Mode-II**
  - Flow Circulation through Sorption Columns 1 (C1) and 2 (C2)
  - Detour of flow to Sampling Chambers 1 (SC1) and 2 (SC2)
  - VENT Valve Activation for Sample Collection
- **Mode-III**
  - Flow Rate Division
  - Concurrently 'Pulse' Fill Sampling Chambers 1 and 2

Figure 2 presents the arrangement of the prototype components, including an adjustable bi-stable timer which determines the time division modulation of the MODE and VENT solenoid valves.

Activating the process in Mode-I, the fluid element is mechanically pumped from the acidified 3-L multi-component influent stock. Once the pump is turned on, the inlet tubing is primed with the influent stock and passes the column check valve at the system inlet. The fluid element passes through the mesh-union fitting and reaches the base of the first column (C1), and traverses up through the sample chamber entry solenoid valve to the first sampling chamber. Once the 30-mL sample chamber is filled, the fluid element begins to drip at its exit against the multi-turn valve, which is an indication to switch the sample chamber entry solenoid MODE valves to Mode-II using the automated mode controls.

In Mode-II, the fluid element by-passes the first sampling chamber (SC1), continues to traverse through column 1 (C1) and begins to fill column 2 (C2). The fluid element does not cross-circuit back towards the exits of first column, due to the additional check valves connection located at the entry of the second column. While the fluid element traverses up both columns C1 and C2, the sampling chamber exit solenoid VENT valve is switched from closed to open. The sampling port tube is uncapped, twisted using the custom rotating handle and inverted to draw a 30-mL sample. The VENT valve is then turned off (closed from atmosphere). It is important to note that the inlet-outlet offset of the sampling chambers guarantees a highly repeatable sample volume. It is designed to minimize cross-contamination, for when the chamber is rotated from the vertical upward (sample collection in Mode-III) to downward (sample dispense in Mode-II and VENT) position, the chamber contents are completely void.

Once C2 is filled, both MODE valves of the sampling chambers are switched from Mode-II to Mode-III. The fluid element now simultaneously traverses through C1 and C2, while filling SC1 and SC2, dividing the flow rate and maintaining a continuous flow through the system. Once both sampling chambers are filled, the MODE valves are switched from Mode-III back to Mode-II, such that the fluid element by-passes the sample chambers and only traverses through the columns. At this time, the VENT valves are switched from closed to open, and the samples are taken from the sampling ports of both SC1 and SC2. Once both samples are collected, the VENT valves are closed and the MODE valves are once again switched back to pulse in Mode-III until SC1 and SC2 are filled. This sequence is repeated at approximate 5-min increments between Mode-II and Mode-III, for a total analysis period of just over 3-h. The prototype is secured to a sturdy, level frame that includes supporting clamps for the packed fixed-bed columns and a removable sampling chamber lock mechanism for maintenance accessibility.

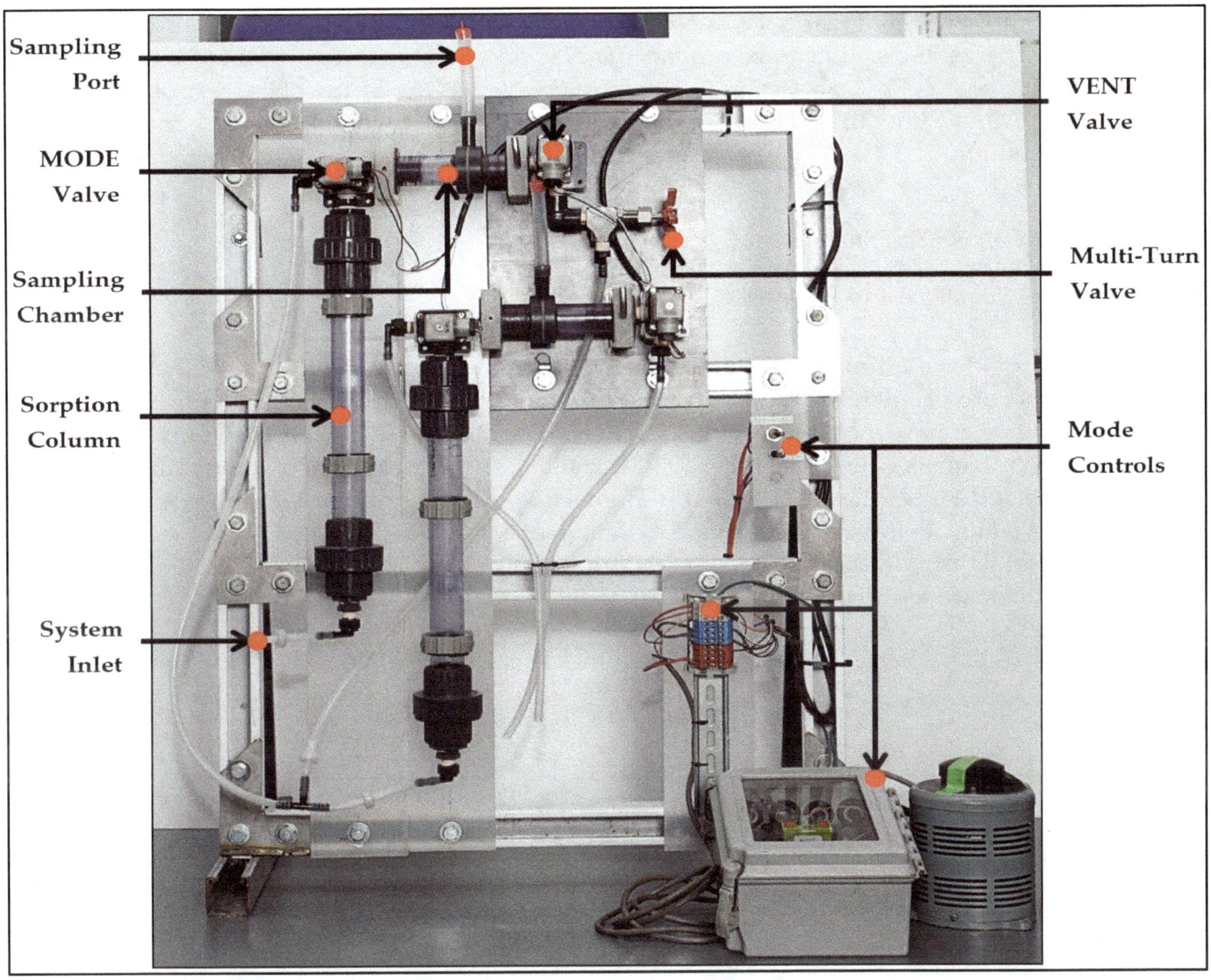


**Figure 2.** Image of automated sorption system prototype design.

It is important to note that the sample chambers are fabricated to a 30-mL capacity, to ensure that the 25-mL required volume is attained, to be filtered for dilution in preparation of ICP-AES analysis. This influences the time to collect the sample volume, based on the selected pump flow rate of this study. Also, the spacing of the prototype components influences the tubing connection lengths. The dual-column sorption system design presented in Figures 1 and 2 provides the opportunity to analyse higher flow rates and/or prolonged sample collection in Mode-III in future research endeavours.

## 3. Results and Discussion

### 3.1. Preliminary Batch Mode Results

Detailed analysis on the selected HMIs of this study was conducted by Ciosek and Luk [32,33] in batch-mode configuration, consisting of a synthetic nitrate salt solution at 10 meq/L total concentration, acidified to a pH of 2 by concentration $HNO_3$ acid, with a zeolite dosage of 4 g per 100-mL HMI solution. The aqueous solution is agitated within a contact period 180 min by means of a triple-eccentric drive orbital shaker operating at 400 r/min set to 22°C. The five (5) HMIs were methodically combined in single-, dual-, triple-, and multi-component systems. Elemental analysis by ICP-AES concludes that after 3 contact h, a total HMI uptake of 0.0986 meq/g is achieved the multi-component system.

The percent removal of $Pb^{2+}$, $Cu^{2+}$, $Fe^{3+}$, $Ni^{2+}$ and $Zn^{2+}$ are 94.0%, 21.9%, 56.2%, 9.10%, and 16.5%, respectively. The zeolite's preference among the HMIs is demonstrated by the selectivity series, which is established as $Pb^{2+} >> Fe^{3+} > Cu^{2+} > Zn^{2+} > Ni^{2+}$. One of the objectives of this current study is to investigate how these HMIs interact and affect the removal uptake in a continuous flow, dual-column settling.

### *3.2. Automated Column Sorption System*

#### 3.2.1. Sampling Sequence and Flow Rate

Table 3 provides the timeline of modes in the system set-up sequence. Once the inlet tubing and check valve are primed, the pump starts to fill the inlet connection cavity. At full flow rate in Mode-I, it requires approximately 8:39 min:s to travel from the base to the top of column 1 (C1). After approximately 3:40 min:s, sample chamber 1 (SC1) is filled, and Mode-II (circulation) is initiated while the first sample (C1-A) is collected. In the continuous flow of Mode-II, and it requires approximately 8:27 min:s for the flow to travel from the base to the top of column 2 (C2). The flow is then switched to Mode-III (pulse), which divides the flow to fill both sample chambers SC1 and SC2. Once the 30-mL volumes are filled, Mode-III is switched back to Mode-II and the samples C1-B and C2-B are collected at 42:50 min:s.

**Table 3.** Sorption system set-up sequence.

| MODE | Function | Flow Description | Time (min:s) |
|---|---|---|---|
| I | Fill C1 | Primed Inlet to C1 Base | 2:26 |
| | | C1 Base to C1 Top | 11:05 |
| | | C1 Top to SC1 Drip | 15:10 |
| II | Fill SC1 | | |
| | | Sample C1-A | 18:50 |
| | Fill C2 | C2 Inlet to C2 Base | 24:08 |
| | | C2 Base to C2 Top | 32:35 |
| | | C2 Top to SC2 Drip | 36:14 |
| III | Fill SC1 and SC2 | | |
| II | | Sample C1-B and C2-B | 42:50 |
| III | Fill SC1 and SC2 | | 48:04 |
| II | | Sample C1-1 and C2-1 | 54:27 |

Once the system is set-up, there is an orderly switch between Mode-II (circulation) and Mode-III (pulse). Table 4 summarizes this sampling sequence. Altogether, there are twenty-nine 30-mL samples collected throughout the analysis period. During the system set-up, the collection of the first sample (C1-A) is followed by the second column 1 sample (C1-B) and first column 2 sample (C2-B). The orderly sequence begins at the collection of Cx-1 (48:04 min:s), for a total of two samples for each of the thirteen (13) runs. A total waste (TW) sample in the collection basin of the sorption system is also collected half-way through sampling (115:45 min:s) and at the end of the analysis period (195:00 min:s). The final influent stock and total effluent volumes are approximately 1.45-L and 550-mL, respectively.

**Table 4.** Sorption system sampling sequence.

| Sample | MODE | Start Time (min:s) | End Time (min:s) | |
|---|---|---|---|---|
| | | | SC1 | SC2 |
| C1-A | I<br>II | 15:10<br>18:50 | 18:50 | - |
| Cx-B | III<br>II | 36:40<br>42:50 | 42:42 | 42:49 |
| Cx-1 | III<br>II | 48:04<br>54:27 | 54:10 | 54:27 |
| Cx-2 | III<br>II | 59:39<br>66:34 | 66:15 | 66:34 |
| Cx-3 | III<br>II | 71:37<br>78:20 | 77:55 | 78:19 |
| Cx-4 | III<br>II | 83:24<br>89:57 | 89:56 | 89:56 |
| Cx-5 | III<br>II | 95:05<br>101:46 | 101:32 | 101:45 |
| Cx-6 | III<br>II | 106:52<br>113:11 | 112:53 | 113:10 |
| TW1 | | 115:45 | | |
| Cx-7 | III<br>II | 118:10<br>123:57 | 123:45 | 123:56 |
| Cx-8 | III<br>II | 129:00<br>136:20 | 136:07 | 136:19 |
| Cx-9 | III<br>II | 141:22<br>147:35 | 147:29 | 147:34 |
| Cx-10 | III<br>II | 152:40<br>158:46 | 158:35 | 158:45 |
| Cx-11 | III<br>II | 163:45<br>169:46 | 169:40 | 169:45 |
| Cx-12 | III<br>II | 174:50<br>181:10 | 181:06 | 181:04 |
| Cx-13 | III<br>II | 186:11 | 192:08<br>PUMP OFF | 192:36 |
| TW2 | | 195:00 | | |

The flow patterns are continuous and methodically kept consistent throughout the analysis period. Once samples C1-B and C2-B are collected, an average time of 6:26 min:s passes to switch from Mode-III to Mode-II, and 5:05 min:s from Mode-II to Mode-III. When the flow is divided in Mode-III, the average sampling acquisition time of 6:19 min:s is required to fill the 30-mL chambers, which is then collected for the filtering and dilution of the 25-mL sub-sample. The adjustable bi-stable timer at an approximate 50% duty setting automatically toggles the pulsing in Mode-III, to maintain a relatively consistent division of flow between the two columns, creating partial diversion to the two sampling chambers. This is demonstrated in relation to the start and end times of the sampling sequence.

It is important to note that the first sorption column (C1) receives a continuous inlet flow rate, as observed by the Mode-I filling rate of 8.18 mL/min for SC1 sample C1-A. Immediately after SC1 is filled and before 30-mL collection, the switch to Mode-II diverts the flow to begin filling the second sorption column (C2). Once both columns are filled, the flow is divided in Mode-III at the top outlet of C1, between SC1 and SC2, while maintaining consistent contact throughout the system. Again, during

Mode-III, C1 receives the same inlet flow rate, but the sampling chambers SC1 and SC2 receives this division of flow. It is column C2 that receives a variable flow rate during the analysis period, set by the adjustable division timer. Based on the filling start time (36:40 min:s) of C1-B and C2-B, and the end time of collection (192:36 min:s) for C1-13 and C2-13, a geometric mean flow rate between Mode-II and Mode-III in C2 is established as 5.39 mL/min. With these unique flow rates recognized for both columns, their corresponding EBCTs are established by Equation (4) to yield:

$$EBCT_1 = \frac{d^2\pi H}{4(Q_{C1})} = \frac{(2.61\ cm)^2\pi(30.48\ cm)}{4(8.18\ cm^3/min)} \cong 20:00\ min:s$$

$$EBCT_2 = \frac{d^2\pi H}{4(Q_{C2,AVG})} = \frac{(2.61\ cm)^2\pi(30.48\ cm)}{4(5.39\ cm^3/min)} \cong 30:20\ min:s$$

Due to the relatively slow kinetics of zeolites, long residence times are required. Any solution volume-element in contact with a given zeolite bed layer is for only a limited time period, which is usually insufficient to reach the equilibrium state. The failure of zeolite to attain local equilibrium causes a lower uptake of HMIs from solution [11]. The detention time that the fluid element is in contact with the fixed-bed per sorption column is a result of the flow rate selected in this present study. This trend between the columns provides insight into the overall treatment availability of the zeolite material in this very unique configuration.

### 3.2.2. Acidity Levels

Natural zeolites are known to raise the pH level in acidic aqueous solutions, which is due to: (1) the ion-exchange of $H^+$ ions, (2) the binding of $H^+$ ions to the Lewis basic sites linked to the oxygen atoms in the zeolite framework, and (3) the $OH^-$ ions in solution deriving from hydrolysis of some species present in the zeolite [11]. The pH level of the aqueous solution controls the overall sorption process; adsorption of the HMI at the solid–water interfaces as well as the ion-exchange of cations within the zeolite structure. Stylianou et al. [10] points out that for all minerals, a decrease in the ion-exchange capacity of HMIs occurs for a pH range of 1 to 2. However, very low pH levels may positively influence the sorption process with the hydrolysis of the HMIs in solution [10]. Table 5 presents the pH levels of the effluent for both columns, of equally distributed selected time-step checkpoints of Cx-3, Cx-6, Cx-9, and Cx-13. When the acidified influent stock is combined to a 3-L volume, the average multi-component MM pH level has a value of 1.90. By maintaining a very low initial pH level and the use of highly soluble nitrate salts, the precipitation of the HMIs is avoided. Additional trials verified that the filtered and unfiltered HMI influent stock concentrations are the same, indicating both effective dilution practices and complete solubility.

**Table 5.** The pH levels of selected sorption column samples.

| Sample | pH Level | |
|---|---|---|
| | SC1 | SC2 |
| C1-A | 6.34 | - |
| Cx-3 | 4.79 | 6.84 |
| Cx-6 | 3.99 | 6.72 |
| TW1 | 6.05 | |
| Cx-9 | 3.86 | 6.33 |
| Cx-13 | 3.60 | 5.76 |
| TW2 | 5.44 | |

As the sample traverses through the first column C1, the $H^+$ ions are captured by the zeolite, resulting in an increase in the pH level to 6.34 from the first sample C1-A. There is an interesting observation between the columns' pH levels, which is a direct reflection of the zeolite's removal capacity for both the HMIs of interest and the competitive $H^+$ ions in solution. The pH level gradually decreases in both columns, with the levels of C2 being slightly greater than that of C1. The total waste (TW) collects in the effluent basin throughout the analysis period, and its pH level decreases from the half-way check point of 6.05 (115:45 min:s) to 5.44 at the final collection (195:00 min:s). This is a clear indication that the zeolite capacity is becoming exhausted for the competing $H^+$ ions, as well as hindering the sorption process of the HMIs.

Research conducted by Vukojevic Medvidovic et al. [34] also demonstrated that the pH values changed during the uptake process, following the opposite shape of the typical breakthrough curves. At breakpoint, a drastic change in the pH value occurred, which corresponded with a rapid $Pb^{2+}$ concentration increase. The maximum pH level is reached at the breakthrough point, due to the absence of HMIs in the effluent. The minimum pH level is reached at the exhaustion point, due to the increase of the concentration of HMIs in the effluent and due to their hydrolysis in solution. After the exhaustion point, the pH level is constant [12]. These findings suggest that the continuous monitoring of pH levels is important and considerably contributes to the prediction of breakthrough and exhaustion points [11,12,34]; in order to monitor the progress of the service life and inevitably regeneration (adsorption/desorption cycles), both of which are very significant for practical industrial applications [34].

Also, the pH level may influence the ionization degree of the sorbate (HMI solution) and the surface property of the sorbent (zeolite mineral) [44]. The structural stability of the sorbent should not be compromised; for once the pH level reaches below 1, the structure of clinoptilolite breaks down in a process termed 'dealumination'. Precipitation should be avoided, for once the ions of interest have precipitated they cannot be sorbed [43]. It should be noted that while low pH levels prevent precipitation, the competitive $H^+$ ions present would hinder the sorption of HMIs. Therefore, it is to be expected that future field installations for the treatment of AMD (with typical pH range of 2 to 5 [40]) should potentially demonstrate even higher removal efficiencies. However, care should be taken in the design of industrial applications to incorporate pre-treatment processes to reduce particulates prior to applying the waste to any sorption system, to avoid flow obstruction in the sorption columns.

### 3.2.3. Hydraulic Conductivity Considerations

Following HMI uptake analysis, the columns are drained to sit overnight. The standard test method for permeability of granular soils (constant head) (ASTM 2434-68) [49] is adopted to determine the variance in the overall hydraulic conductivity between the sorption columns. The hydraulic coefficient of permeability is given by adapting the standard test in the following relationship:

$$k_T = \frac{V_C \cdot L}{A \cdot H_C \cdot T} \tag{6}$$

where $k_T$ (in cm/s) is the coefficient of permeability, $V_C$ (in $cm^3$) is the quantity of water that has discharged from the column and collected, L (in cm) is the column height, A (in $cm^2$) is the column cross-sectional area, $H_C$ (in cm) is the constant head of water on the column or the vertical distance between the feed head level and the column overflow level, and T (in s) is the time required to collect the $V_C$ volume. With a plumb tank clamp support system, the water is fed in upflow mode from its base. The collection volume ($V_C$) is set to 50-mL by a graduated cylinder, with a column height (zeolite bed depth) and cross-sectional area of 30.48 cm and 5.37 $cm^2$, respectively. Based on an 18 °C detected water temperature, the viscosity correction factor of $n_T/n_{20} = 1.0508$ is applied to reveal the hydraulic conductivity of columns C1 and C2 as $4.08 \times 10^{-4}$ m/s and $3.89 \times 10^{-4}$ m/s, respectively. With a 4.84% difference to the average between the columns, this demonstrates a consistency in the overall executed compaction method.

### 3.2.4. Heavy Metallic Ion Concentration Analysis

Table 6 provides the results of the HMI concentrations ($C_t$) in both sorption columns based on triplicate readings obtained by the ICP-AES software. The percent removal (%R) values are presented with respect to the 2.0 meq/L influent concentrations of each HMI. During the experimental sequence of sample collection, C2-11 is lost due to improper handling when transferring from the sampling port to filtration at 169:45 min:s. However, this sample is of a lower HMI concentration and the overall removal trend has been well-established by the time this sample is collected. The first major observation is that throughout the analysis period, $Pb^{2+}$ is not detected in both column effluents as well as the total waste, indicating a complete removal of the ion. The C1-13 sample for $Cu^{2+}$, $Fe^{3+}$, and $Zn^{2+}$ reports a removal of 18.09%, 82.54%, and 10.71%, respectively, from the first column. The dual-column configuration provides a substantial improvement on the removal as observed with the second pass in sample C2-13, to achieve a final removal of $Cu^{2+}$, $Fe^{3+}$, and $Zn^{2+}$ of 80.07%, 99.98%, and 51.53%, respectively. This improvement is also attributed to the unique feeding rate and design of the second column C2; the additional EBCT of approximately 10-minutes is available for the sorption process to occur as well as the slightly higher pH levels (and therefore lower presence of competitive $H^+$ ions). The final total waste (TW2) effluent concentrations report very good removal for all ions except for $Ni^{2+}$, with a removal of 48.97%. This removal trend is also consistent with the batch analyses conducted by Ciosek and Luk [32,33]. This is significant, as it proves that results from complex experimental batch studies, which are in high abundance, are useful in providing information on the sorption performance (i.e., removal efficiency, selectivity, and kinetics) in industrial applications where the process is run in a continuous flow-feeding configuration. In summary, the results demonstrate for the first time the effectiveness of multiple HMIs sorption by zeolite in a dual-column system with continuous flow.

**Table 6.** The heavy metallic ion (HMI) concentration (meq/L) and percent removal (%R) in the sorption columns.

| Sample | | HMI | | | | | | | | | |
|---|---|---|---|---|---|---|---|---|---|---|---|
| | | $Cu^{2+}$ | | $Fe^{3+}$ | | $Ni^{2+}$ | | $Pb^{2+}$ | | $Zn^{2+}$ | |
| | | meq/L | %R | meq/L | %R | meq/L | %R | meq/L | %R | meq/L | %R |
| | C1-A | 0.000 | 100.00 | 0.000 | 99.98 | 0.007 | 99.67 $^{BP}$ | 0.0003 | 99.98 | 0.002 | 99.92 $^{BP}$ |
| | C1-B | 0.129 | 93.55 $^{BP}$ | 0.000 | 99.99 | 0.911 | 54.47 | 0.0006 | 99.97 | 0.541 | 72.96 |
| | C1-1 | 0.517 | 74.17 | 0.000 | 99.99 | 1.483 | 25.85 | 0.0004 | 99.98 | 0.974 | 51.30 |
| | C1-2 | 0.938 | 53.09 | 0.001 | 99.93 | 1.906 | 4.71 | 0.0006 | 99.97 | 1.320 | 34.02 |
| | C1-3 | 1.221 | 38.93 | 0.011 | 99.46 | 2.116 | 0.00 | 0.0003 | 99.98 | 1.507 | 24.66 |
| | C1-4 | 1.369 | 31.54 | 0.030 | 98.49 | 2.231 | 0.00 | 0.0006 | 99.97 | 1.622 | 18.91 |
| | C1-5 | 1.431 | 28.47 | 0.052 | 97.42 | 2.269 | 0.00 | 0.0004 | 99.98 | 1.671 | 16.43 |
| C1 | C1-6 | 1.468 | 26.60 | 0.072 | 96.40 | 2.273 | 0.00 | 0.0005 | 99.97 | 1.703 | 14.83 |
| | C1-7 | 1.584 | 20.78 | 0.102 | 94.90 $^{BP}$ | 2.316 | 0.00 | 0.0005 | 99.98 | 1.816 | 9.20 |
| | C1-8 | 1.563 | 21.86 | 0.118 | 94.08 | 2.199 | 0.00 | 0.0005 | 99.98 | 1.751 | 12.43 |
| | C1-9 | 1.543 | 22.84 | 0.138 | 93.11 | 2.174 | 0.00 | 0.0004 | 99.98 | 1.730 | 13.51 |
| | C1-10 | 1.571 | 21.44 | 0.167 | 91.64 | 2.134 | 0.00 | 0.0004 | 99.98 | 1.739 | 13.04 |
| | C1-11 | 1.598 | 20.12 | 0.209 | 89.55 | 2.123 | 0.00 | 0.0004 | 99.98 | 1.752 | 12.38 |
| | C1-12 | 1.604 | 19.79 | 0.268 | 86.59 | 2.096 | 0.00 | 0.0001 | 100.00 | 1.750 | 12.49 |
| | C1-13 | 1.638 | 18.09 | 0.349 | 82.54 | 2.130 | 0.00 | 0.0002 | 99.99 | 1.786 | 10.71 |
| | C2-B | 0.00 | 100.00 | 0.0004 | 99.98 | 0.002 | 99.90 | 0.0003 | 99.99 | 0.002 | 99.88 |
| | C2-1 | 0.00 | 100.00 | 0.0003 | 99.98 | 0.002 | 99.88 | 0.0003 | 99.99 | 0.001 | 99.94 |
| | C2-2 | 0.00 | 100.00 | 0.0003 | 99.98 | 0.012 | 99.42 | 0.0004 | 99.98 | 0.001 | 99.95 |
| | C2-3 | 0.00 | 100.00 | 0.0002 | 99.99 | 0.046 | 97.68 | 0.0006 | 99.97 | 0.001 | 99.96 |
| | C2-4 | 0.00 | 100.00 | 0.0002 | 99.99 | 0.131 | 93.43 $^{BP}$ | 0.0002 | 99.99 | 0.001 | 99.93 |
| | C2-5 | 0.00 | 100.00 | 0.0002 | 99.99 | 0.285 | 85.75 | 0.0002 | 99.99 | 0.009 | 99.55 |
| C2 | C2-6 | 0.00 | 100.00 | 0.0003 | 99.98 | 0.505 | 74.77 | 0.0003 | 99.99 | 0.049 | 97.56 |
| | C2-7 | 0.00 | 100.00 | 0.0002 | 99.99 | 0.835 | 58.26 | 0.0003 | 99.98 | 0.155 | 92.23 $^{BP}$ |
| | C2-8 | 0.004 | 99.80 | 0.0002 | 99.99 | 1.163 | 41.85 | 0.0003 | 99.99 | 0.312 | 84.39 |
| | C2-9 | 0.029 | 98.57 | 0.0002 | 99.99 | 1.444 | 27.82 | 0.0004 | 99.98 | 0.455 | 77.24 |
| | C2-10 | 0.085 | 95.77 $^{BP}$ | 0.0003 | 99.99 | 1.675 | 16.26 | 0.0004 | 99.98 | 0.597 | 70.17 |
| | C2-12 | 0.289 | 85.57 | 0.0002 | 99.99 | 2.126 | 0.00 | 0.0006 | 99.97 | 0.895 | 55.23 |
| | C2-13 | 0.399 | 80.07 | 0.0003 | 99.98 | 2.198 | 0.00 | 0.0004 | 99.98 | 0.969 | 51.53 |
| TW1 | | 0.0514 | 97.43 | 0.0004 | 99.98 | 0.3077 | 84.61 | 0.0004 | 99.98 | 0.1107 | 94.46 |
| TW2 | | 0.1659 | 91.71 | 0.0057 | 99.72 | 1.0207 | 48.97 | 0.0000 | 100.00 | 0.4088 | 79.56 |

### 3.2.5. Breakthrough Curve, Capacity and Usage Rate Analysis

The breakthrough curve is displayed in Figure 3, as a plot of the solute outlet concentration ($C_t$) from Table 6 normalized to the inlet concentration ($C_o$) [34]. This normalized ratio trend over the service time of analysis [12,25] at which the sampling chambers (SC1, SC2) are extracted is presented for each of the five (5) HMIs combined in the multi-component solution; for both the first sorption column (C1) and second sorption column (C2). The breakthrough point (BP) and exhaustion point (EP) of each HMI in each column are indicated. The first observation to be had is that the breakthrough curves of the first column C1 do not have a defined S-shape. However, in the second sorption column C2, the curves take on this typical shape. Vukojevic Medvidovic et al. [34] points out that the shape-change may be attributed to an improved solid-solution phase contact for sorption to take place.

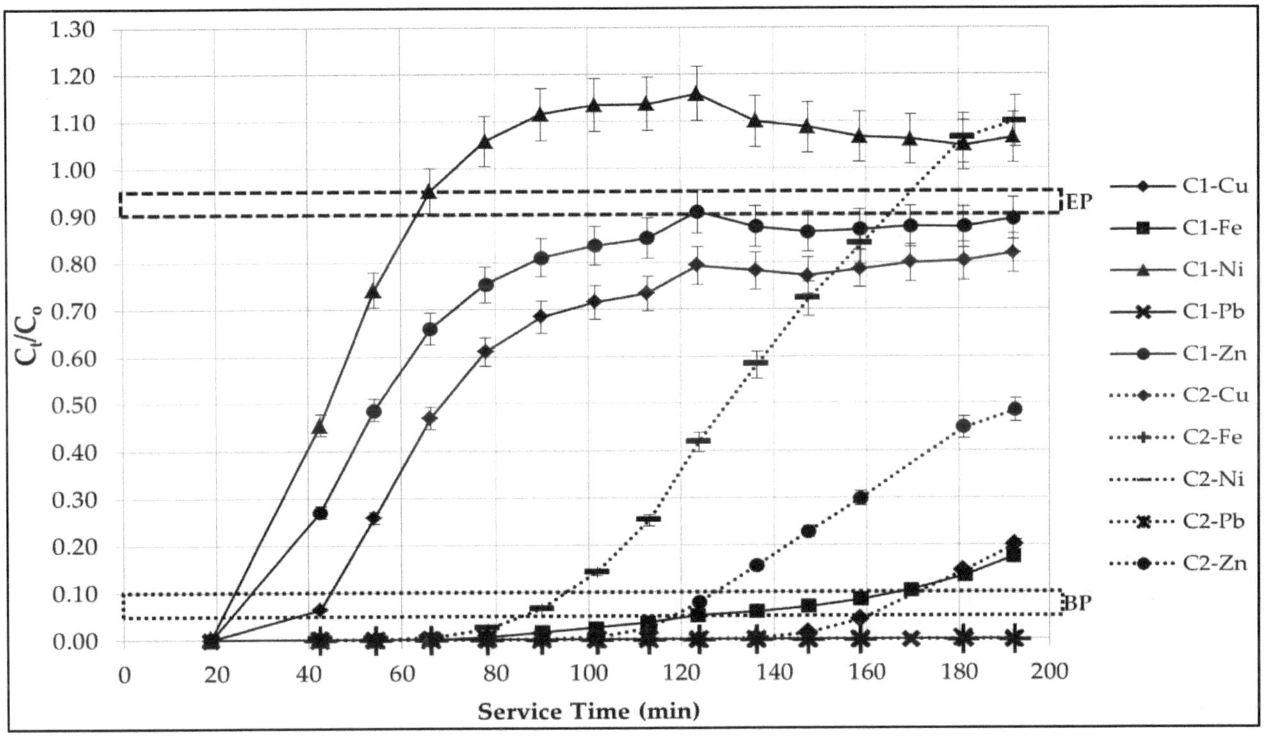


**Figure 3.** The multi-component system breakthrough curve.

The influent stock concentration of the $Ni^{2+}$ ion is exceeded in the effluent solution, where the normalized $C_t/C_o$ ratio surpasses 1 to reach an approximate maximum of 1.16 at 125 min and 1.10 at 190 min of service time, in sorption columns C1 and C2, respectively. The final ratio readings plateau at the end of service to approximately 5%–10% of the value of 1, given the nature of this experimental investigation. The effluent concentration that overshoots the influent concentration translates to concentration wave extremes inside the column [50]. Nuic et al. [12] investigates the breakthrough curves of $Pb^{2+}$ and $Zn^{2+}$ ions in dual-component solutions by natural zeolite; a similar trend is observed compared to this present study, where the $C_t/C_o$ ratio even reaches a value of 2 for one set of operation conditions. This is attributed to the displacement of the bound $Zn^{2+}$ by the $Pb^{2+}$ from the influent, which is supported by a lower breakthrough capacity and higher exhaustion capacity in favour of the $Pb^{2+}$ ion, specifically [12]. It is important to note that the ion-exchange mechanism that attributes to the sorption process of HMIs transpires through the zeolite's framework of pores and channels. The presence of stronger binding HMIs, such as $Pb^{2+}$, weaken the chemical bonds between the functional group on the surface of zeolite and the weaker HMIs, such as $Ni^{2+}$ ions [9]. Given that zeolite demonstrates its highest preference towards $Pb^{2+}$, sorption site availability has reached its threshold, which may cause the leaching out of ions that zeolite holds a lower preference towards during this process. Therefore, careful screening on the selectivity of HMIs by zeolite should be conducted prior to adaptation.

In summary, the major trends observed from the breakthrough curves are as follows:

1. The zeolite holds the greatest preference towards to $Pb^{2+}$ ion, based on its complete removal throughout the analysis period;
2. The zeolite demonstrates the least preference towards the $Ni^{2+}$ ion;
   a. A more sudden breakpoint occurring after just 25 min and 90 min of service time in columns C1 and C2, respectively;
   b. An approximate exhaustion point after just 65 min and 165 min of service time in columns C1 and C2, respectively;
3. The $Fe^{3+}$ ion is removed entirely and sustained throughout the analysis period in C2, and;
4. The removal of both the $Cu^{2+}$ and $Zn^{2+}$ ions begin to plateau at 120 min of service time in C1, acting in parallel and do not reach the lower threshold of the exhaustion point in both columns throughout the analysis period.

The breakthrough curves provide significant information from a perspective of sorption process performance, feasibility and optimization, which are vital for scaling-up the sorption system for industrial applications [28,50].

The overall column performance efficiency and its relationship between breakthrough capacity ($C_{BP}$) and exhaustion capacity ($C_{EP}$) of each sorption column are unique to the individual HMIs selected. As observed in the data displayed Table 6 and the trends visualized in Figure 3, the EP is only attained by $Ni^{2+}$; the remaining four HMIs have yet to reach this point due to the constraints of the 3-h analysis period. Evidently, the optimization of future works would be to prolong the service time in order to quantify the overall columns' performance efficiency. Qualitatively speaking, for both columns' effluent and total waste (TW), the $Pb^{2+}$ ion is completely removed throughout the analysis period, demonstrating the utmost efficiency; neither BP nor EP are attained. On the opposite end of the spectrum, the $Ni^{2+}$ ion reaches exhaustion quite rapidly.

The volume of the effluent treated at breakthrough ($V_{BP}$) is determined with the use of the mean flow rates of 8.18 mL/min ($Q_{C1}$) and 5.39 mL/min ($Q_{C2}$) for the sorption columns C1 and C2, respectively. These flow rates are applied to the time data of Table 4, at which BP (approximately 95% removal or 5% of the 2.0 meq/L influent concentration per HMI) is observed; as indicated by the superscript in Table 6. The 2:26 min:s time required for inlet priming to the base of C1 as well as the 24:08 min:s time observed for solute contact to the base of C2 are deducted from these BP times. As summarized in Table 7, the approximate effluent volumes treated at BP ($V_{BP}$) are provided for both sorption columns and each HMI selected; based on the zeolite mass in each bed (m) of 152.10 g, Equations (2) and (5) are employed to determine the corresponding breakthrough capacity ($C_{BP}$) and usage rate ($v_U$), respectively.

**Table 7.** System breakthrough point performance.

| HMI | Sorption Column 1 | | | Sorption Column 2 | | |
|---|---|---|---|---|---|---|
| | $V_{BP,1}$ (L) | $C_{BP,1}$ (meq/g) | $v_{U,1}$ (g/L) | $V_{BP,2}$ (L) | $C_{BP,2}$ (meq/g) | $v_{U,2}$ (g/L) |
| $Cu^{2+}$ | 0.3295 | 0.00433 | 461.67 | 0.7257 | 0.00954 | 209.59 |
| $Fe^{3+}$ | 0.9926 | 0.01305 | 153.24 | - | - | - |
| $Ni^{2+}$ | 0.1342 | 0.00176 | 1133.54 | 0.3547 | 0.00466 | 428.80 |
| $Pb^{2+}$ | - | - | - | - | - | - |
| $Zn^{2+}$ | 0.1342 | 0.00176 | 1133.54 | 0.5380 | 0.00707 | 282.71 |

The $Pb^{2+}$ and $Fe^{3+}$ ions are completely removed in sorption column C2, which essentially transforms the influent stock of a five HMI multi-component solution to a triple-component solution containing $Cu^{2+}$, $Ni^{2+}$, and $Zn^{2+}$ ions within the 3-h analysis period. The zeolite does not have to address the competition of the two preferred HMIs, which provides greater sorption site availability for the three HMIs remaining in solution. It is important to develop a relationship between the EBCT and usage rate. The time that the fluid element is in contact with the zeolite bed in sorption column C2 is ten minutes greater than the detention time in C1. This thereby demonstrates a stronger overall treatment availability in C2; reflected in the average 195.2% increase of bed capacity at breakthrough and the average 63.94% decrease in usage rate of the zeolite between the columns; based on the trends observed for the $Cu^{2+}$, $Ni^{2+}$ and $Zn^{2+}$ ions detected in solution.

Evidently, the columns' usage rate provides significant insight into the operation and management required for this unique sorption system. It has a direct impact on the financial viability of performing either replacement (disposal) or regeneration (on- or off-site), and is affected by factors that include HMI influent concentration, zeolite bed depth, and flow rate. Research into other sorbent materials demonstrates that the order of usage rate is consistent with the sorption capacity [51]. Due to the unique automated variable influent feeding rate and sampling technique proposed in this study, the usage rate and performance efficiency are very complex [28]. However, the major removal trend of Pb2+ >> Fe3+ > Cu2+ > Zn2+ >> Ni2+ is well-established and supports previous results [22,27,32,33,38,42,43], providing significant validation of this design.

The use of natural zeolites as sorbents in industrial wastewater treatment and environmental management is motivated by the non-toxicity of these minerals, their abundant global availability, and economic feasibility. The removal and recovery processes of HMIs from aqueous solutions by natural zeolites take into consideration the regeneration potential of the zeolite bed to be reused in multiple cycles, as well as the use of the recovered metals [52] in applicable industrial applications. Metal processing effluents contain high concentrations of recoverable metals, triggering a movement towards technologies to recover these metals from AMD waste [53]. The removal-regeneration-recovery process has the potential to generate additional revenue streams with the use of metals of value [54]; such as the HMIs investigated in this innovative study.

## 4. Conclusions

This research has demonstrated the performance of natural zeolite (clinoptilolite) to remove multiple heavy metallic ions (HMIs) commonly found in acid mine drainage. With the design and development of a novel dual-column sorption system, the lead ($Pb^{2+}$) ion is removed completely and sustained throughout the analysis period. The relationships between empty bed contact time (EBCT), breakthrough capacity, and usage rate are evident. The additional ten minutes of EBCT in the second sorption column contributes to an enhancement in overall removal for $Cu^{2+}$, $Fe^{3+}$, and $Zn^{2+}$ by 75.67%, 99.90%, and 45.72%, respectively, from the first sorption column. This improvement is also apparent in the greater breakthrough capacity and lower usage rate in the second column, and visualized in an improved S-shape to the characteristic breakthrough curve. Based on the multi-component influent stock of 10 meq/L total concentration, the second column demonstrates a removal of 99.98%, 99.98%, 80.07%, 51.53%, and 0.00%, for $Pb^{2+}$, $Fe^{3+}$, $Cu^{2+}$, $Zn^{2+}$, and $Ni^{2+}$, respectively; and the final cumulative collection of effluent reports a removal of 100.00%, 99.72%, 91.71%, 79.56% and 48.97%, for $Pb^{2+}$, $Fe^{3+}$, $Cu^{2+}$, $Zn^{2+}$, and $Ni^{2+}$, respectively at the completion of the analysis period. These HMI sorption removal trends confirm the consistency between batch and continuous mode operations.

The modular design inventively incorporated a 'circulation-pulse' method to distribute the flow, rather than operating on a more commonly implemented fixed flow rate. With the consideration of this unique stock feed method, the findings of the service time and flow rate with respect to the removal trends are both interesting and significant. Forthcoming works in this research project include the advancement of service life and regeneration cycles, with further design development and optimization. The potential for variable flow rate operation and automatic adjustable sampling in a packed fixed-bed dual-column sorption design reveals practicality for treatment applications. This study has provided greater insight into the immense potential that the natural mineral zeolite holds for the future of industrial wastewater treatment.

**Acknowledgments:** This research was conducted with the financial support of a Natural Sciences and Engineering Research Council of Canada (NSERC) Discovery Grant to Grace K. Luk.

**Author Contributions:** Amanda L. Ciosek and Grace K. Luk conceived and designed the experiments; Amanda L. Ciosek constructed the prototype, performed the experiments and analytical simulations, and analyzed the data; Amanda L. Ciosek and Grace K. Luk wrote the paper.

## References

1. Vaca-Mier, M.; Lopez-Callejas, R.; Gehr, R.; Jimenez-Cisneros, B.E.; Alvarez, P.J.J. Heavy metal removal with mexican clinoptilolite: multi-component ionic exchange. *Wat. Res.* **2001**, *35*, 373–378. [CrossRef]
2. Johnson, D.B.; Hallberg, K.B. Acid mine drainage remediation options—A review. *Sci. Total Environ.* **2005**, *338*, 3–14. [CrossRef] [PubMed]
3. Akcil, A.; Koldas, S. Acid Mine Drainage (AMD): Causes, treatment and case studies. *J. Clean. Prod.* **2006**, *14*, 1139–1145. [CrossRef]
4. Mohan, D.; Chander, S. Removal and recovery of metal ions from acid mine drainage using lignite—A low cost sorbent. *J. Hazard. Mater.* **2006**, *B137*, 1545–1553. [CrossRef] [PubMed]
5. Motsi, T.; Rowson, N.A.; Simmons, M.J.H. Kinetic studies of the removal of heavy metals from acid mine drainage by natural zeolite. *Int. J. Miner. Process.* **2011**, *101*, 42–49. [CrossRef]
6. Baraket, M.A. New trends in removing heavy metals from industrial wastewater. *Arab. J. Chem.* **2011**, *4*, 276–282. [CrossRef]
7. Vukojevic Medvidovic, N.; Peric, J.; Trgo, M.; Nuic, I.; Ugrina, M. Design of fixed bed column for lead removal on natural zeolite based on batch Studies. *Chem. Biochem. Eng. Q.* **2013**, *27*, 21–28.
8. Nuic, I.; Trgo, M.; Vukojevic Medvidovic, N. The application of the packed bed reactor theory to Pb and Zn uptake from the binary solution onto the fixed bed of natural zeolite. *Chem. Eng. J.* **2016**, *295*, 347–357. [CrossRef]
9. Han, R.; Zou, W.; Li, H.; Li, Y.; Shi, J. Copper(II) and lead(II) removal from aqueous solution in fixed-bed columns by manganese oxide coated zeolite. *J. Hazard. Mater.* **2006**, *137*, 934–942. [CrossRef] [PubMed]
10. Stylianou, M.A.; Inglezakis, V.J.; Moustakas, K.G.; Malamis, S.P.; Loizidou, M.D. Removal of Cu(II) in fixed bed and batch reactors using natural zeolite and exfoliated vermiculite as adsorbents. *Desalination* **2007**, *215*, 133–142. [CrossRef]
11. Stylianou, M.A.; Hadjiconstantinou, M.P.; Inglezakis, V.J.; Moustakas, K.G.; Loizidou, M.D. Use of natural clinoptilolite for the removal of lead, copper and zinc in fixed bed column. *J. Hazard. Mater.* **2007**, *143*, 575–581. [CrossRef] [PubMed]
12. Nuic, M.; Trgo, J.; Peric, N.; Vukojevic Medvidovic, N. Analysis of breakthrough curves of Pb and Zn sorption from binary solutions on natural clinoptilolite. *Micropor. Mesopor. Mater.* **2013**, *167*, 55–61. [CrossRef]
13. Peric, J.; Trgo, M.; Vukojevic Medvidovic, N. Removal of zinc, copper and lead by natural zeolite—A comparison of adsorption isotherms. *Water Res.* **2004**, *38*, 1893–1899. [CrossRef] [PubMed]
14. Wang, C.; Li, J.; Sun, X.; Wang, L.; Sun, X. Evaluation of zeolites synthesized from fly ash as potential adsorbents for wastewater containing heavy metals. *J. Environ. Sci.* **2009**, *21*, 127–136. [CrossRef]
15. Anari-Anaraki, M.; Nezamzadeh-Ejhieh, A. Modification of an Iranian clinoptilolite nano-particles by hexadecyltrimethyl ammonium cationic surfactant and dithizone for removal of Pb(II) from aqueous solution. *J. Colloid Interf. Sci.* **2015**, *440*, 272–281. [CrossRef] [PubMed]
16. Borandegi, M.; Nezamzadeh-Ejhieh, A. Enhanced removal efficiency of clinoptilolite nano-particles toward Co(II) from aqueous solution by modification with glutamic acid. *Colloids Surf. A Physicochem. Eng. Asp.* **2015**, *479*, 35–45. [CrossRef]
17. Curkovic, L.; Cerjan-Stefanovic, S.; Filipan, T. Metal ion exchange by natural and modified zeolites. *Water. Res.* **1997**, *31*, 1379–1382. [CrossRef]

18. Helfferich, F. Equilibria; Kinetics; Ion-Exchange in Columns. In *Ion Exchange*; Series in Advanced Chemistry; McGraw-Hill Book Company: New York, NY, USA, 1962; pp. 95–322, 421–506.
19. Inglezakis, V.J.; Poulopoulos, S.G. Chapter 4—Adsorption and Ion-Exchange (Kinetics). In *Adsorption, Ion Exchange and Catalysis—Design of Operations and Environmental Applications*, 1st ed.; Elsevier Science: Amsterdam, The Netherlands, 2006; pp. 262–266. ISBN 13 978-0-444-52783-7.
20. Trgo, M.; Peric, J.; Vukojevic Medvidovic, N. A comparative study of ion exchange kinetics in zinc/lead—Modified zeolite-clinoptilolite systems. *J. Hazard. Mater.* **2006**, *136*, 938–945. [CrossRef] [PubMed]
21. Tsitsishvili, G.V. Perspectives of Natural Zeolite Applications. Occurrence. In *Properties and Utilization of Natural Zeolites—2nd International Conference 1985*; Akademiai Kiado: Budapest, Hungary, 1988; pp. 367–393.
22. Wang, S.; Peng, Y. Natural zeolites as effective adsorbents in water and wastewater treatment. *Chem. Eng. J.* **2010**, *156*, 11–24. [CrossRef]
23. Nezamzadeh-Ejhieh, A.; Shirzadi, A. Enhancement of the photocatalytic activity of Ferrous Oxide by doping onto the nano-clinoptilolite particles towards photodegradation of tetracycline. *Chemosphere* **2014**, *107*, 136–144. [CrossRef] [PubMed]
24. Inglezakis, V.J.; Loizidou, M.D.; Grigoropoulou, H.P. Equilibrium and kinetic ion exchange studies of $Pb^{2+}$, $Cr^{3+}$, $Fe^{3+}$ and $Cu^{2+}$ on natural clinoptilolite. *Water Res.* **2002**, *36*, 2784–2792. [CrossRef]
25. Nuic, I.; Trgo, M.; Peric, J.; Vukojevic Medvidovic, N. Uptake of Pb and Zn from a binary solution onto different fixed bed depths of natural zeolite—The BDST model approach. *Clay Miner.* **2015**, *50*, 91–101. [CrossRef]
26. Ersoy, B.; Celik, M.S. Electrokinetic properties of clinoptilolite with mono- and multivalent electrolytes. *Micropor. Mesopor. Mater.* **2002**, *55*, 305–312. [CrossRef]
27. Inglezakis, V.J.; Grigoropoulou, H. Effects of operating conditions on the removal of heavy metals by zeolite in fixed bed reactors. *J. Hazard. Mater.* **2004**, *112*, 37–43. [CrossRef] [PubMed]
28. Inglezakis, V.J. Ion exchange and adsorption fixed bed operations for wastewater treatment—Part I: Modelling fundamentals and hydraulics analysis. *J. Eng. Stud. Res.* **2010**, *16*, 29–41.
29. Erdol Aydin, N.; Nasun Saygili, G. Column experiments to remove copper from wastewaters using natural zeolite. *Int. J. Environ. Waste Manag.* **2009**, *3*, 319–326. [CrossRef]
30. Inglezakis, V.J.; Papadeas, C.D.; Loizidou, M.D.; Grigoropoulou, H.P. Effects of pretreatment on physical and ion exchange properties of natural clinoptilolite. *Environ. Technol.* **2001**, *22*, 75–82. [CrossRef] [PubMed]
31. Inglezakis, V.J. Ion exchange and adsorption fixed bed operations for wastewater treatment—Part II: scale-up and approximate design methods. *J. Eng. Stud. Res.* **2010**, *16*, 42–50.
32. Ciosek, A.L.; Luk, G.K. Lead Removal from mine tailings with multiple metallic ions. *Int. J. Water Wastewater Treat.* **2017**, *3*, 1–9. [CrossRef]
33. Ciosek, A.L.; Luk, G.K. Kinetic modelling of the removal of multiple heavy metallic ions in mine waste by natural zeolite sorption. *Water* **2017**, *9*, 482. [CrossRef]
34. Vukojevic Medvidovic, N.; Peric, J.; Trgo, M. Column performance in lead removal from aqueous solutions by fixed bed of natural zeolite–clinoptilolite. *Sep. Purif. Technol.* **2006**, *49*, 237–244. [CrossRef]
35. Reed, B.E.; Jamil, M.; Thomas, B. Effect of pH, empty bed contact time and hydraulic loading rate on lead removal by granular activated carbon columns. *Water Environ. Res.* **1996**, *68*, 877–882. [CrossRef]
36. Peric, J.; Trgo, M.; Vukojevic Medvidovic, N.; Nuic, I. The Effect of Zeolite Fixed Bed Depth on Lead Removal from Aqueous Solutions. *Sep. Sci. Technol.* **2009**, *44*, 3113–3127. [CrossRef]
37. Bear River Zeolite Co. Inc. Zeolite—Specifications and MSDS. Available online: http://www.bearriverzeolite.com (accessed on 1 September 2012 and 1 April 2017).
38. Inglezakis, V.J.; Hadjiandreou, K.J.; Loizidou, M.D.; Grigoropoulou, H.P. Pretreatment of natural clinoptilolite in a laboratory-scale ion exchange packed bed. *Water Res.* **2001**, *35*, 2161–2166. [CrossRef]
39. Mullin, J. Physical and thermal properties. In *Crystallization*, 4th ed.; Read Educational and Professional Publishing Ltd: Woburn, MA, USA, 2001; pp. 76–77, IBSN 0-7506-4833-3.
40. Wilson, L.J. Canada-Wide Survey of Acid Mine Drainage Characteristics. Project Report 3.22.1—Job No. 50788. Mineral Sciences Laboratories Division Report MSL 94–32 (CR). Ontario Ministry of Northern Development and Mines. Mine Environment Neutral Drainage (MEND) Program. Canada, 1994. Available online: http://mend-nedem.org/wp-content/uploads/2013/01/3.22.1.pdf (accessed on 30 October 2014).
41. Canadian Minister of Justice—Metal Mining Effluent Regulations. Consolidation SOR/2002-222. Justice Laws—Government of Canada. Available online: http://laws-lois.justice.gc.ca (accessed on 1 September 2014).

42. Wingenfelder, U.; Hansen, C.; Furrer, G.; Schulin, R. Removal of Heavy Metals from Mine Waters from Natural Zeolites. *Environ. Sci. Technol.* **2005**, *39*, 4606–4613. [CrossRef] [PubMed]
43. Inglezakis, V.J.; Loizidou, M.D.; Grigoropoulou, H.P. Ion exchange of $Pb^{2+}$, $Cu^{2+}$, $Fe^{3+}$, and $Cr^{3+}$ on natural clinoptilolite: Selectivity determination and influence of acidity on metal uptake. *J. Colloid Interface Sci.* **2003**, *261*, 49–54. [CrossRef]
44. Minceva, M.; Fajgar, R.; Markovska, L.; Meshko, V. Comparative Study of $Zn^{2+}$, $Cd^{2+}$, and $Pb^{2+}$ Removal From Water Solution Using Natural Clinoptilolitic Zeolite and Commercial Granulated Activated Carbon: Equilibrium of Adsorption. *Sep. Sci. Technol.* **2008**, *43*, 2117–2143. [CrossRef]
45. Ouki, S.K.; Kavannagh, M. Treatment of metals-contaminated wastewaters by use of natural zeolites. *Water. Sci. Tech.* **1999**, *39*, 115–122. [CrossRef]
46. Rice, E.W.; Baird, R.B.; Eaton, A.D.; Clesceri, L.S. Part 1000-Introduction, Part 3000-METALS. In *Standard Methods for the Examination of Water and Wastewater*, 22nd ed.; The American Public Health Association (APHA): Washington, DC, USA; The American Water Works Association (AWWA): Denver, CO, USA; The Water Environment Federation (WEF): Alexandria, VA, USA, 2012; pp. 1.1–68, 3.1–112, ISSN 978-087553-013-0.
47. Perkin Elmer Inc. *Atomic Spectroscopy—A Guide to Selecting the Appropriate Technique and System: World Leader in AA, ICP-OES, and ICP-MS*; Perkin Elmer Inc.: Waltham, MA, USA, 2011.
48. Perkin Elmer Inc. *WinLab32 for ICP—Instrument Control Software, Version 5.0*; Perkin Elmer Inc.: Waltham, MA, USA, 2010.
49. ASTM D2434–68. Standard Test Method for Permeability of Granular Soils (Constant Head). ASTM International, West Conshohocken, PA. 2000. Available online: www.astm.org (accessed on 1 March 2016).
50. Naja, G.; Volesky, B. Multi-metal biosorption in a fixed-bed flow-through column. *Colloid. Surf.* **2006**, *281*, 194–201. [CrossRef]
51. Othman, M.Z.; Roddick, F.A.; Snow, R. Removal of Dissolved Organic Compounds in Fixed-Bed Columns: Evaluation of Low-Rank Coal Adsorbents. *Water Res.* **2001**, *35*, 2943–2949. [CrossRef]
52. Sprynskyy, M.; Buszewski, B.; Terzyk, A.P.; Namiesnik, J. Study of the selection mechanism of heavy metal ($Pb^{2+}$, $Cu^{2+}$, $Ni^{2+}$, and $Cd^{2+}$) adsorption on clinoptilolite. *J. Colloid Interface Sci.* **2006**, *304*, 21–28. [CrossRef] [PubMed]
53. Zinck, J. Review of Disposal, Reprocessing and Reuse Options for Acidic Drainage Treatment Sludge. Report 3.42.3. Mine Environment Neutral Drainage (MEND) Program. Mining Association of Canada. CANMET Mining and Mineral Sciences Laboratories. Canada, 2005. Available online: http://mend-nedem.org/wp-content/uploads/2013/01/3.42.3.pdf (accessed on 30 October 2014).
54. Dinardo, O.; Kondos, P.D.; MacKinnon, D.J.; McCready, R.G.L.; Riveros, P.A.; Skaff, M. Study on Metals Recovery/Recycling from Acid Mine Drainage Phase IA: Literature Survey. Report 3.21.1a. Mine Environment Neutral Drainage (MEND) Program. CANMET, Energy, Mines and Resources Canada and WTC, Environment Canada. 1991. Available online: http://mend-nedem.org/wp-content/uploads/2013/01/3.21.1a.pdf (accessed on 30 October 2014).

# 10

# Textile Wastewater Purification using an Elaborated Biosorbent Hybrid Material (*Luffa–Cylindrica*–Zinc Oxide) Assisted by Alternating Current

**Amina Othmani [1,2], Aida Kesraoui [1], Roberto Boada [3], Mongi Seffen [1,*] and Manuel Valiente [3,*]**

1 Laboratory of Energy and Materials (LabEM): LR11ES34, Higher School of Science and Technology of Hammam Sousse, University of Sousse, 4011 Hammam Sousse, Tunisia
2 Faculty of Sciences of Monastir (Monastir University), 5000 Monastir, Tunisia
3 GTS-UAB Research Group, Department of Chemistry, Faculty of Science, Universitat Autònoma Barcelona, 08193 Bellaterra, Spain
* Correspondence: mongiseffen@yahoo.fr (M.S.); Manuel.Valiente@uab.cat (M.V.).

**Abstract:** This paper aims to synthesize hybrid materials with high pollutant-uptake capacity and low costbased based on *Luffa cylindrica* (*L.C*) and different percentage of $Zn^{2+}$ in the presence and absence of alternating current (AC). Physico-chemical, morphological and structural characterizations of the hybrid materials were performed by Boehm method, point zero charge ($pH_{pzc}$), infrared characterizations (IR), scanning electron microscopy (SEM), energy–dispersive spectroscopyand and X-ray photoelectron spectroscopy. The efficiency of the designed hybrid materials was optimized based on their performance in water depollution. Methylene blue (MB) and industrial textile wastewater were the investigated pollutants models. IR characterizations confirmed the fixation of $Zn^{2+}$ onto the *L.C* by the creation of Zn-OH, Zn-O and Zn-O-C bonds. Boehm titration showed that the fixation of $Zn^{2+}$ onto *L.C* is accompanied by an increase of the basic functions of its surface and subsequently an increase in the $pH_{pzc}$. SEM results confirmed the fixation of $Zn^{2+}$ onto the *L.C* coupling AC with biosorption showed an increase in the adsorbed amount of MB and speed when adding the 4% of $Zn^{2+}$ compared to the pure *L.C* the $Q_m$ shifted from 3.22 to 9.84 and 8.81 mg/g, respectively, for hybrid materials synthesized under AC, in absence of AC and pure *L.C*.

**Keywords:** alternating current; coupling; hybrid material; biosorption; wastewater reuse

---

## 1. Introduction

Textile dying wastewaters are classified among the most highly toxic effluents [1,2]. Numerous processes have been suggested for treatment and purification such as biological treatment [3], coagulation-flocculation [4], adsorption [5], ultrafiltration [6], electrocoagulation [7], reverse osmosis [8] and anodic oxidation [9]. The current research was devoted to water remediation through the valorization of the abundant renewable resource of cellulosic fibers [10]. Brown algae [11], *Bacillus maceran* [12], *Posidonia oceanica* [13], corn stigmas [14], *Agave Americana* [15], *Luffa cylindrica* [16], fly ash and red mud [17], raw date pits [18] and *Phragmites australis* [19] are widely used as a natural and cheap biosorbent for a pollutant removal. However, achieving a quick adsorption kinetic with a high possibility of water reuse after filtration absolutely depends on the biosorbent efficiency.

Hybrid materials are praised for not only intermediate properties between mineral and organic matter but also for new interesting behaviors that allow them to be used in several fields of applications. Such domains include catalytic applications, medical and pharmaceutical applications, optoelectronics, the environment, and biomaterials [20,21]. Hybrid materials can be prepared by several methods such as chemical vapor deposition (CVD) [22], physical vapor deposition (PVD) [23], laser ablation and sputtering [24]. The synthesis of hybrid materials based on cellulose and zinc oxide is the objective of recent studies. Perelshtein et al. (2009) developed a composite based on cotton fibers and zinc oxide. They synthesized the ZnO particles and then deposited them on the cotton surface by ultrasonic irradiation (sonochemical method) [25]. Weili et al. (2010) have succeeded in synthesizing a hybrid matrix based on wurtzite-cellulose ZnO bacteria by the thermal decomposition method. This new hybrid material obtained plays a very important role in improving the photo catalytic activity of anionic dyes such as, orange methyl [26].

However, the techniques used for the synthesis of hybrid materials are expensive. Other researchers have developed less expensive techniques such as the electrochemical method [27], spray pyrolysis [28] and sol-gel process [29]. Recently, Kesraoui et al. (2018) reported that the precipitation method can be a simple and cheap method used for the synthesis of hybrid materials. They have successfully synthesized hybrid material base on *L.C* and metal oxides (ZnO, $Al_2O_3$). These last showed a good ability for dyes removals compared to the pure *Luffa cylidnrica* (*L.C*) [30]. According to literature, the synthesis of hybrid materials based on compounds containing cellulose and ZnO can present interesting properties allows it to be used in several applications.

However, the removal process often requires a lot of time to reach the balance. Therefore, coupling environmental security (there is no transformation of the initial molecule to toxic compounds as for the oxidation process), low cost, final efficiency, and quick process remains a very important stake.

Taking into account these considerations, a set of aims are proposed in this paper, the synthesis of a high performing hybrid material based on *Luffa cylinrdica* (*L.C*) and metal oxide with high pollutant-uptake capacity and low cost by an easy precipitation method in absence and presence of alternating current (AC). Since their good affinity for the removal of several pollutants and their interesting physical and chemical properties, *L.C* and ZnO were considered for the synthesis of the hybrid materials [31,32]. Two main pollutants were taken as models in this study; a cationic dye the methylene blue (MB) taken as a model of dye and an industrial textile wastewater. A detailed investigation of the physicochemical, morphological and structural characterizations was carried out using Boehm titration. Also, Drift method, scanning electron microscopy (SEM), Energy-dispersive spectroscopy (EDS), infrared spectroscopy (IR) and X-ray photoelectron spectroscopy (XPS) analysis were performed. The possible pathway of the $Zn^{2+}$ fixation into the *L.C* fibers and the mechanism of possible interaction of MB onto the lignocellulosic surface are suggested. The mathematic modeling was also supplied using the stochastic model of Brouers-Sotolongo.

## 2. Material and Methods

### *2.1. Materials*

Methylene blue (MB) (purity: 95%) from Fluka™ (manufacturer, Streinheim, Germany), with chemical formula $C_{16}H_{18}N_3SCl$ and molar mass 319.85 g/mol, was chosen as commercial dye model for this study.

Zinc nitrate ($Zn(NO_3)_2.6H_2O$; Hexahydrate Purified), sodium chloride (NaCl), sodium bicarbonate ($NaHCO_3$), sodium carbonate ($Na_2CO_3$) and sodium hydroxide (NaOH) (purity 99%) were obtained from LOBA CHEMIE (Wodehouse road, Mumbai, India).

The biosorbent used for the synthesis of the hybrid material was fruits of Tunisian *Luffa cylindrica* fibers. This biosorbent was purchased at the local market Sousse. The chemical composition of these fibers was determined by Kesraoui et al. (2016). These fibers are composed of 54% of cellulose, 11% of lignin, 5% of pectin, 7% of fats and waxes and 23% hemicelluloses [33].

### 2.2. Methods

#### 2.2.1. Biosorbent and Adsorbate Preparation

Tunisian *Luffa cylidnrica* (*L.C*) was chosen as a natural biosorbent in this paper. The preparation of this biosorbent consisted of cutting the fibers finely, washing them several times to remove all impurities and drying them at 70 °C until the material was completely dried. As for the adsorbate preparation, it consisted of dissolving 10 mg of MB in 1 L of distilled water to obtain the desired concentration (10 mg/L).

#### 2.2.2. Preparation of Hybrid Materials *L.C*+ (1%, 2%, and 4% $Zn^{2+}$) in Presence and Absence of AC

The $Zn^{2+}$ precursors of solution were obtained from ($Zn(NO_3)_2{\cdot}6H_2O$ (Hexahydrate Purified) Sigma-Altrich (99%) (Wodehouse road, Mumbai, India).

The preparation of the hybrid material with different percentages of $Zn^{2+}$ (1%, 2%, and 4%) consists inmixing 5 g of *L.C* (size 250 µm) with 0.05, 0.1 and 0.2 g of $Zn(NO_3)_2{\cdot}6H_2O$ respectively. Each composite was dissolved with the biomass in 100 mL of distilled water at 298 K at pH = 10, under stirring for two hours. The size of the fibers used was chosen after sieving using an electric sifter and after studying the effect of fiber grain size (40, 80, 125 and 250 µm on the biosorption efficiency). Then, the product was washed several times using distilled water to remove the excess of $Zn(NO_3)_2{\cdot}6H_2O$ which has not been fixed onto the surface of *L.C*. Thereafter, the product obtained was transferred to a sand bath to dry it at 393 K for two hours. A similar experiment was done for the synthesis of the hybrid material under AC using two zinc electrodes (1.3 × 2.5 $cm^2$; purity 99%) immersed into the solution while stirring. The electrical mounting comprising an AC source, a voltmeter to fix the current density at 0.5 $A/m^2$ and the voltage at 15 volts.

### 2.3. Morphological and Crystallographic Characterizations of Hybrid Materials

Morphological and structural characterizations were done in order to characterize the hybrid material synthesized in the presence and absence of AC.

Infrared characterizations (IR) were performed using a Perkin Elmer Spectrum using KBr pellet technique in the frequency range of 4000 to 500 $cm^{-1}$. Scanning electron microscopy (SEM) was performed using a JEOL JSM 5400 scanning microscope (USA) after coating them with gold using a JEOL JFC-1199E ion sputtering device (USA). Energy dispersive spectroscopy (EDS) was planned to assess the surface elemental compositions of raw and the hybrid materials *L.C* +4% $Zn^{2+}$ using a JEOL JSM 5400 scanning microscope. EDS was performed after coating them with gold using a JEOL JFC-1199E ion sputtering device. X-photoelectron spectroscocopy (XPS) measurements were performed at room temperature with a SPECS PHOIBOS 150 hemispherical analyzer (SPECS GmbH, Berlin, Germany) in a base pressure of 5 × $10^{-10}$ mbar using monochromatic Al Kalpha radiation (1486.74 eV) as an excitation source.

### 2.4. Quantitative and Qualitative Characterization of the Pure L.C and the Hybrid Material L.C-$Zn^{2+}$

The physicochemical characterization of the pure *L.C* and the synthesized hybrid materials was determined by the Drift method and the Boehm titration. The determination of the zero charge point pH ($pH_{pzc}$) consisted of placing 0.1 g of *L.C* in 30 mL of NaCl solution (0.01 M) at different pH ranging from 2 to 12. The initial pH is obtained by adding a certain amount of NaOH or HCl (1 M). For the first 24 h, the pH must be measured and then this operation must be repeated during the second 24 h in order to register the difference in pH. By plotting the $pH_f$ = f ($pH_i$) curve, the $pH_{pzc}$ corresponds to the intersection of this curve with the straight line $pH_f$ = $pH_i$ [34].

The determination of acid-basic properties by Boehm method consisted in bringing onto contact 0.5 g of each lignocellulosic material with 25 mL of one of these bases: $NaHCO_3$, $Na_2CO_3$ and NaOH

(0.1 M) for 48 h with stirring. 10 mL of each solution was back-dosed with NaOH solution (0.1 M) after acidification with an excess of 0.1 M of hydrochloric acid [35].

*2.5. Biosorption of MB onto the Synthesized Hybrid Material*

The experiments used for the MB removal by biosorption have been performed in batch reactor by adding 0.1 g of adsorbent (Pure *L.C*, hybrid material elaborated by precipitation) in 100 mL of MB solution (pH = 10, $C_i$ = 10 mg/L, J = 0.5A/m$^2$, voltage = 15 volts, T = 298K). The electrical mounting comprising an AC source and a voltmeter. All the analyses were carried out in triplicate.

After biosorption, residual concentrations were determined by UV visible spectrophotometer biochrome Libra S.22 at λ = 663 nm based on the following equation [36]. Plastic and quartz cuvettes were used for absorbance tests.

$$\text{Dye removal } (\%) = \frac{(C_i - C_t)}{C_i} \times 100 \tag{1}$$

where $C_i$ is the initial dye concentration (mg/L) and $C_t$ is the dye concentration at any time (mg/L).

*2.6. Kinetics Studies*

The evaluation of the MB biosorption in terms of adsorbed quantity was done following the Equation (2) [37].

$$Q_t = \frac{C_i - C_e}{m} \times V \text{ (mg/g)} \tag{2}$$

where $Q_t$ is the adsorbed quantity at equilibrium time, $C_i$ is the initial MB concentration (mg/L), $C_e$ is the residual MB concentration at any time (mg/L), V is the volume of solution (L) and m is the mass of the adsorbent (g). At equilibrium, $C_i$ is equal to $C_e$ and Q is equal to $Q_e$ [38].

Brouers-Sotolongo (B.S) model was used to fit the experimental data of MB biosorption. The generalized kinetic equation of the Brouers-Sotolongo model has been developed to provide a universal function for the kinetics of complex systems characterized by exponential law and/or power exponent behaviors. This kinetic model unifies and generalizes previous theoretical attempts to describe what is called "fractal kinetics". The mathematical development of this model and its application are presented in Brouers and Sotolongo [39–41].

The pseudo BSf (n, α) sorption kinetics is given by Equation (3):

$$Q_t = Q_e[1 - (1 + (n-1)\left(\frac{t}{\tau_c}\right)^{\alpha})^{\frac{-1}{n-1}}] \tag{3}$$

where, n is order of the fractional reaction, α is "fractal time" exponent.

$\tau_c$: characteristic time, $Q_e$: the adsorbed quantity at saturation, $Q_t$: the adsorbed quantity at any time.

*2.7. Evaluation of the Treated Water Qualities*

2.7.1. Determination of Chemical Oxygen Demand (COD) and Total Organic Carbon (TOC)

The global mineralization was determined by measuring the chemical oxygen demand (COD) by a photometric method. The % COD removal can be estimated by (Equation (4)) [37].

$$\text{COD removal } (\%) = \frac{COD_i - COD_t}{COD_i} \times 100 \tag{4}$$

where $COD_i$ is the initial COD (g/L $O_2$) and $COD_t$ is the residual COD at any time (g/L $O_2$).

The total organic carbon (TOC) in the solutions of MB and industrial wastewater was measured by a TOC analyzer (HACHIL, 550-TOC-TN model, Schwerte, Germany) and the TOC removal can be estimated as follows:

$$\text{TOC removal } (\%) = \frac{\text{TOC}_i - \text{TOC}_t}{\text{TOC}_i} \times 100 \tag{5}$$

where, $\text{TOC}_i$ and $\text{TOC}_t$ are respectively the initial total organic carbon and the total organic carbon obtained after fixed time t of electrolysis treatment (ppm C.O).

#### 2.7.2. Germination Tests

The phytotoxicity test consists in the determination of the inhibitory effect of the treated water on the germination and growth potential of lettuce based on the GI. All sample experiments, including reference (pure water), were tripled. The results obtained are finally expressed according to the following relation [42]:

$$\text{GI}(\%) = \frac{\text{number of seeds sprouted in the sample}}{\text{number of seeds sprouted in the reference}} \times \frac{\text{averagelength of root in the sample}}{\text{average root length in the reference}} \times 100 \tag{6}$$

## 3. Results and Discussion

### 3.1. Surface Characterizations

#### 3.1.1. Energy-Dispersive Spectroscopy (EDS) Analysis

The zinc deposit was analyzed with EDS (Table 1). Carbon and oxygen were the major compounds of *L.C* and hybrid materials due to their chemical composition (cellulose, hemicellulose, and lignin). In addition, the appearance of the Zn atom confirmed its fixation onto the surface of *L.C*. The AC enhanced the distribution of $Zn^{2+}$ into the *L.C* fibers where the rates according to the mass of Zn reached about 1.19% and 0.95% respectively for the hybrid material synthesized in presence and absence of AC.

**Table 1.** Percentages by mass of the chemical elements present on the fiber surface of the raw *Luffa cylindrica* (L.C) and the hybrid materials *L.C* + 4% $Zn^{2+}$ and *L.C* + 4% $Zn^{2+}$ + AC.

| | C | O | Na | Mg | Al | Si | P | S | Cl | Ca | Mo | Zn |
|---|---|---|---|---|---|---|---|---|---|---|---|---|
| L.C | 54.46 | 40.60 | 0.45 | 0.38 | 0.80 | 1.10 | 0.29 | 0.00 | 0.35 | 1.40 | 0.17 | 0.00 |
| L.C + 4% $Zn^{2+}$ + AC | 57.01 | 35.64 | 0.00 | 0.38 | 4.78 | 0.00 | 0.00 | 0.00 | 1.00 | 0.00 | 0.00 | 1.19 |
| L.C + 4% $Zn^{2+}$ | 50.43 | 40.18 | 0.00 | 0.79 | 4.58 | 2.40 | 0.00 | 0.08 | 0.59 | 0.00 | 0.00 | 0.95 |

#### 3.1.2. Infrared Spectroscopy (IR)

Figure 1 shows that the spectrum of pure *L.C* has a vibration band at 3438 $cm^{-1}$ which corresponds to the O-H bond present in cellulose, hemicellulose, and lignin. In the presence of different percentages of $Zn^{2+}$ (1%, 2%, and 4%), obtained by precipitation (Figure 1a), the band of hydroxyl groups is observed at 3435, 3430 and 3414 $cm^{-1}$. The decrease in the position of these bands can be attributed to the presence of elongation vibration of the Zn-OH bond, between $Zn^{2+}$ and cellulose [43]. These results indicate that the hydroxyl groups have a strong interaction with $Zn^{2+}$.

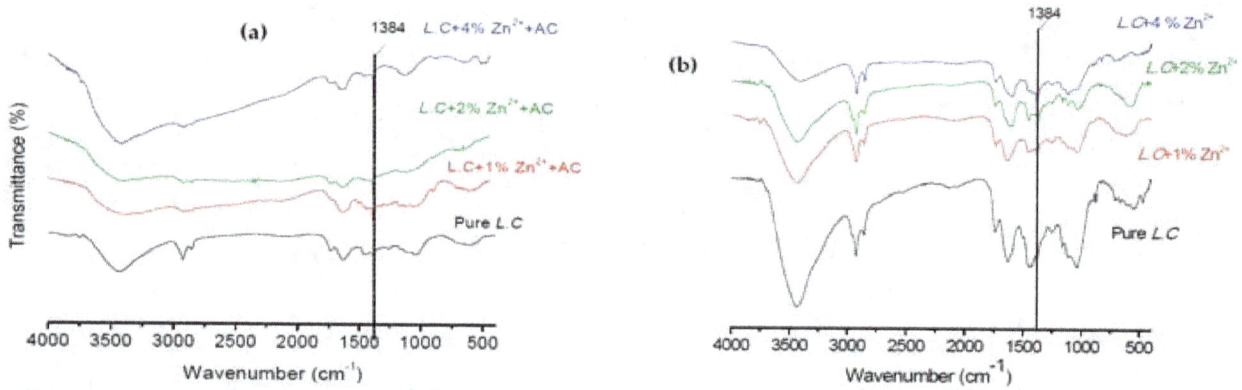


**Figure 1.** Spectrum of pure *L.C* and *L.C* + (1%, 2% and 4% $Zn^{2+}$) synthesized (**a**) in presence of alternating current (AC) and (**b**) in the absence of AC.

The appearance of 1383 peaks at 4% *L.C*-$Zn^{2+}$ is probably due to symmetric Zn-O-C vibration [44]. The bands at 1032.56, 1016.77 and 1016 $cm^{-1}$, which correspond to the vibration of the C-O bond, increase in intensity, which can be attributed to the appearance of the Zn-O-C bond.

Based on Figure 1b, the rising of the percentages of $Zn^{2+}$ added during the synthesis of hybrid materials by precipitation under AC caused the decrease of the position of the band of hydroxyl groups. This last passed from 3438 $cm^{-1}$ to 3430, 3427and 3420 $cm^{-1}$. This decrease in position can be attributed to the formation of Zn-OH stretching vibration [45].

Furthermore, we note the appearance of the Zn-O bond is proven by the appearance of peaks at 1383, 517 and 485 $cm^{-1}$ [46].

The bonds located at 1032.56, 1016.77 and 1016 $cm^{-1}$ correspond to the vibration of the C-O bond. An increase in the intensity of these bonds was shown when adding the $Zn^{2+}$. This increase can be explained by the appearance of the Zn-O-C bond.

The band located at 1383 $cm^{-1}$ presents the main difference between the IR spectrum obtained the hybrid materials synthesized in the presence and absence of AC. For the hybrid material synthesized under AC, the band at 1383 $cm^{-1}$ appears from 1% of $Zn^{2+}$. While the hybrid material synthesized by precipitation this band appears only for 4 % of $Zn^{2+}$. This behavior is probably due to the effect of AC on the fixation of $Zn^{2+}$ ions into the *L.C* fibers by the creation of Zn-O-C bonds.

### 3.1.3. X-ray Photoelectron Spectroscopy (XPS) Analysis

XPS measurements were performed in order to clarify the main difference between the syntheses of the hybrid material with and without AC. Based on IR characterizations; Zn is strongly bound to the lignocellulosic surface by OH hydroxyl ions which are confirmed by the presence of O-Zn-O and O-Zn bonds. Figure 2 displays the main XPS results for the hybrid material *L.C* + 4% $Zn^{2+}$ synthesized in the presence and absence of AC. The high-resolution XPS spectra of C(1s), O(1s), Zn(2p), Zn $LMM^{+}$ show noticeable differences between the material synthesis by the two methods. However, neither of them showed any impurities. The C(1s) peak, depicts a chemical shift by +/−0.5 eV in comparison with the C(1s) spectra observed for the hybrid material synthesized in absence of AC (not AC) (282.5 eV) towards a relatively high binding energy positioned at 283 eV specific to C=O and O–C–O [47]. Similar behavior was observed for the O(1s) peak which is positioned at the binding energies of 529.9 and 531.2 eV for the hybrid material *L.C* + 4% Zn synthesized in absence and presence of AC, respectively. These contributions can be assigned to $O^{2-}$ ions in the Zn-O bonding and to O–H groups of adsorbed water molecules [48]. The four peaks observed at the binding energies of 1017.5, 1020 and 1040 and 1044.5 eV, correspond respectively to the zinc oxide (ZnO) and the spin orbit of Zn ($2p_{3/2}$) and Zn ($2p_{1/2}$) for the material synthesized in absence of AC [48]. In presence of AC, two main peaks located at 1021 and 1043.5 eV are assigned to the Zn($2p_{3/2}$) and Zn($2p_{1/2}$) energy levels.

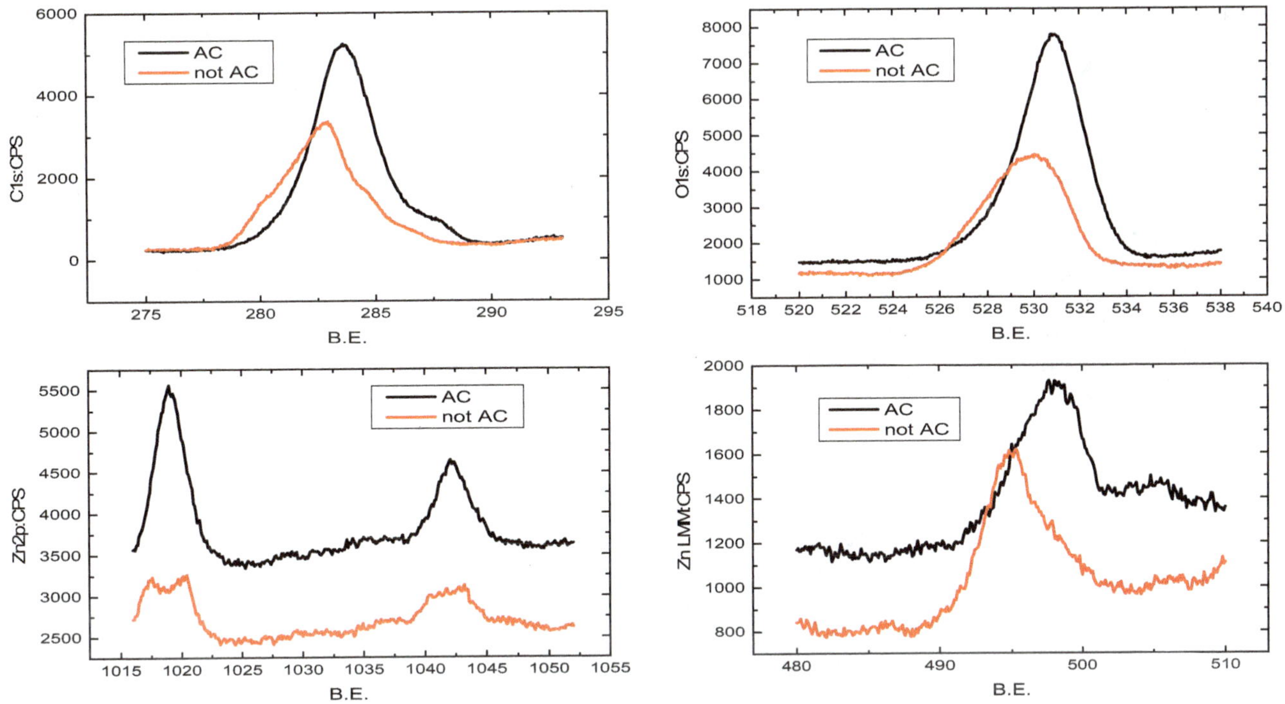


**Figure 2.** Spectra of C1S, O 1s, Zn 2p and Zn LMM of various morphologies of *L.C* + 4% $Zn^{2+}$ synthesized in presence of AC (-) and absence of AC (not AC -).

Both of these were symmetrical and narrow, indicating the absence of the zinc oxide. Despite this difference, these peaks of the Zn(2p) detected in both hybrid materials are related to Zn-O bonding as confirmed by previous studies [49]. Furthermore, we determined a small offset followed by a big discrepancy in terms of the intensities of peaks which were higher in case of the use of AC. These findings indicated that the obtained Zn does not occur in the same oxidation state and even if they are similar in number. It can also be attributed to the depth distribution of the atoms. This increase in intensity is explained by the diffusion of atoms towards the surface which was more pronounced compared to the hybrid material synthesized without AC. This behavior might stem from the difference in the local chemical environment. It can also be directly related to the number of atoms in the corresponding chemical state, which is confirmed by the signal Zn LMM Auger region. Another possibility is the sharing of the element with its neighbor atoms which applied a crystal field on it. This can also be interpreted in terms of the asymmetry in the bond arrangement which can create extra pressure on the element—so, the binding energy required for the ejection of the electron is slightly higher than its previous case. In the light of these results, we confirm the fixation of $Zn^{2+}$ onto the *L.C* by the creation of Zn-O bond as proven by IR characterizations for both methods used for the synthesis of the hybrid material. However, the environment of Zn bonding is slightly different.

### 3.1.4. Scanning Electron Microscopy (SEM) Analysis

SEM characterizations were performed to gain further insight into the distribution of $Zn^{2+}$ obtained from $(Zn(NO_3)_2{\cdot}6H_2O$ on the *L.C* surface by precipitation in presence and absence of AC. Figure 3 shows the obtained micrograph of the pure *L.C* (Figure 3a), hybrid material (*L.C* + 4% $Zn^{2+}$) synthesized by precipitation (Figure 3b), and hybrid material (*L.C* + 4% $Zn^{2+}$) synthesized under AC (Figure 3c). *L.C* presents a rough structure and a homogeneous appearance which is formed by bonded multicellular fibers (Figure 3a). This indicates that the existence of a large number of hydroxyl groups could provide effective interaction between *L.C* and $Zn^{2+}$ [26,50]. Figure 3b,c refer to the hybrid material (*L.C* + 4% $Zn^{2+}$) synthesized in the absence and presence of AC. Here the $Zn^{2+}$ deposit is well visible and well-fixed onto the *L.C* and exhibits good dispersion without significant aggregation. The use of AC has enhanced the distribution of $Zn^{2+}$ into the *L.C* fibers.

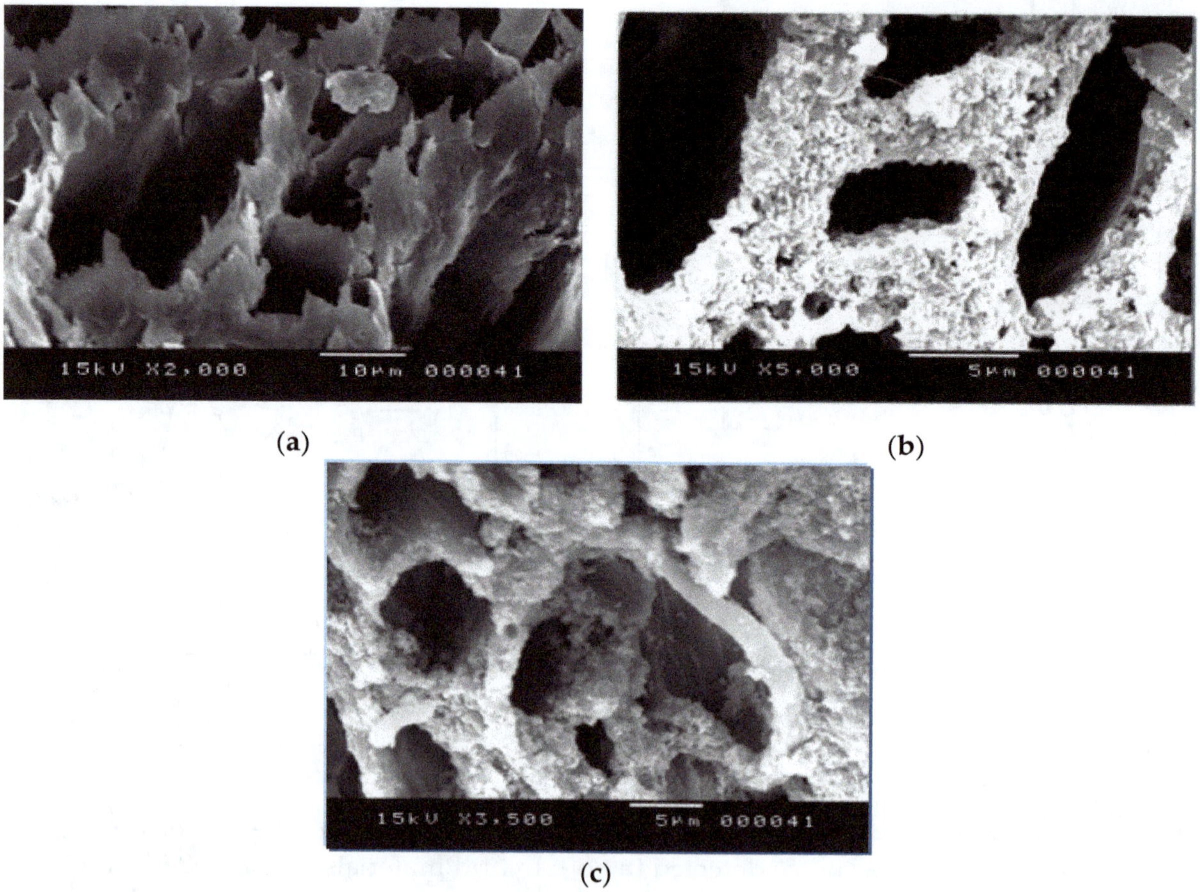


**Figure 3.** Images of (**a**) pure *L.C*, (**b**) *L.C* + 4% $Zn^{2+}$ synthesized in absence of AC, (**c**) *L.C* + 4% $Zn^{2+}$ synthesized in presence of AC.

### 3.1.5. Surface Chemical Analysis of Pure *L.C* and Hybrid Materials

Table 2 shows the main $pH_{pzc}$ values obtained for the pure *L.C* and the synthesized hybrid materials. The presence of zinc increased the $pH_{pzc}$ which made the fibers slightly basic. In addition, $pH_{pzc}$ increases with increasing the percentage of zinc added.

**Table 2.** $pH_{pzc}$ of hybrid material synthesized in absence and presence of AC.

| Samples | $pH_{pzc}$ under AC | $pH_{pzc}$ without AC |
|---|---|---|
| Pure *L.C* | 7.38 ± 0.021 | 7.38 ± 0.021 |
| *L.C* + 1% $Zn^{2+}$ | 7.57 ± 0.008 | 7.41 ± 0.016 |
| *L.C* + 2%$Zn^{2+}$ | 7.77 ± 0.005 | 7.69 ± 0.012 |
| *L.C* + 4% $Zn^{2+}$ | 7.96 ± 0.005 | 7.86 ± 0.012 |

The $pH_{PZC}$ of *L.C* is equal to 7.38. This result shows that the surface of *L.C* has a positive charge at pH values below $pH_{PZC}$ and should, therefore, be able to adsorb anions and a negative charge at pH values above $pH_{PZC}$ and should, therefore, be able to adsorb cations [51].

On the other hand, according to Table 3, for the pure *L.C*, the amount of base and the total amount of acid groups are approximately equal. These results suggest that *L.C* fibers have an amphoteric character. This amorphous character confirms the result found by the $pH_{PZC}$ (7.38).

**Table 3.** Different functional groups obtained on the adsorbent surfaces.

| Adsorbent | Acidic Function ($\times10^{-4}$ mmol/g) | | | | Basic Function ($\times10^{-4}$ mmol/g) |
|---|---|---|---|---|---|
| | Carboxylic | Phenolic | Lactonic | Total | |
| Pure *L.C* | 0.397 ± 0.002 | 0.935 ± 0.003 | 0.178 ± 0.008 | 1.510 ± 0.005 | 1.560 ± 0.0008 |
| *L.C* + 1%$Zn^{2+}$ | 0.280 ± 0.002 | 0.790 ± 0.004 | 0.200 ± 0.001 | 1.280 ± 0.001 | 1.610 ± 0.0005 |
| *L.C* + 2% $Zn^{2+}$ | 0.201 ± 0.002 | 0.544 ± 0.003 | 0.310 ± 0.002 | 1.055 ± 0.002 | 1.700 ± 0.0004 |
| *L.C* + 4% $Zn^{2+}$ | 0.170 ± 0.002 | 0.400 ± 0.002 | 0.430 ± 0.002 | 1.000 ± 0.001 | 1.810 ± 0.0007 |
| *L.C* + 1% $Zn^{2+}$ + AC | 0.220 ± 0.001 | 0.503 ± 0.001 | 0.270 ± 0.002 | 0.993 ± 0.0009 | 1.560 ± 0.0006 |
| *L.C* + 2%$Zn^{2+}$ + AC | 0.198 ± 0.0008 | 0.399 ± 0.002 | 0.390 ± 0.0008 | 0.987 ± 0.0006 | 1.790 ± 0.0003 |
| *L.C* + 4%$Zn^{2+}$ + AC | 0.155 ± 0.001 | 0.236 ± 0.001 | 0.511 ± 0.0005 | 0.902 ± 0.0005 | 1.920 ± 0.0001 |

The carboxylic and phenolic functions of hybrid materials synthesized under AC (*L.C* + $Zn^{2+}$ + AC) are lower than those synthesized in absence of AC (*L.C* + $Zn^{2+}$) (0.155 ± 0.001 and 0.170 ± 0.002 for the carboxylic functions, 0.400 and 0.230 for the phenolic functions, respectively). These findings confirm the increase of $pH_{pzc}$ obtained during the synthesis of the hybrid material in presence of AC (7.96 ± 0.005) which was slightly higher than that obtained during the synthesis in its absence (7.86 ± 0.012). It is apparent that the addition of $Zn^{2+}$ and the use of AC slightly influenced the functional groups of hybrid materials obtained. Such behavior can be explained by the increase in the number of available sites on the surface of the materials, which enhances the uptake capacity of such materials and accelerates the biosorption kinetics.

Furthermore, the basic and lactonic functions of *L.C*-$Zn^{2+}$ + AC are higher than those of *L.C*-$Zn^{2+}$(1.920 ± 0.0001 and 1.810 ± 0.0007 for basic functions and 0.511 ± 0.0005 and 0.430 ± 0.002 for lactonic functions, respectively). According to Mahdoudi et al. (2015), the increase in basic functions is probably due to partial or total deprotonation of the active sites of the hybrid material. Therefore, the fixation of zinc onto *L.C* is accompanied by an increase of the basic functions of its surface and subsequently an increase in the $pH_{pzc}$. The dissociation of the surface oxygen groups of the acidic groups (carboxylic, lactone and phenol) confers a negative charge on the surface of Luffa cylindrica. Thus, the surface acid sites are of Brönsted type [52]. With regard to the positive charge, this may result from the existence of basic oxygen groups such as pyrones or benzopyran.

### *3.2. Kinetics of MB Bbiosorption Assisted by AC*

The MB biosorption tests were performed at an initial dye concentration of 10 mg /L, pH = 10, a mass of the adsorbent of 0.1 g and a temperature of 298 K, V = 15 volts, J = 0.5 A/$m^2$.

Figure 1 shows that the method of elaboration of the hybrid material has a great influence on the biosorption capacity. Indeed, the adsorbed quantity of the highest MB is obtained by the hybrid material (*L.C* + 4%$Zn^{2+}$ + AC). As a result, the adsorbed amount of MB increased from 3.22 mg/g for pure *L.C* to 8.81 mg/g for *L.C* + 4% $Zn^{2+}$ and 9.84 mg/g for *L.C* + 4% $Zn^{2+}$ + AC. Similarly, the MB removal rates reached about 98.4, 88.1 and 31.4% respectively for the hybrid material *L.C* + 4%$Zn^{2+}$ synthesized in the presence and absence of AC and the pure *L.C* (Figure 4).

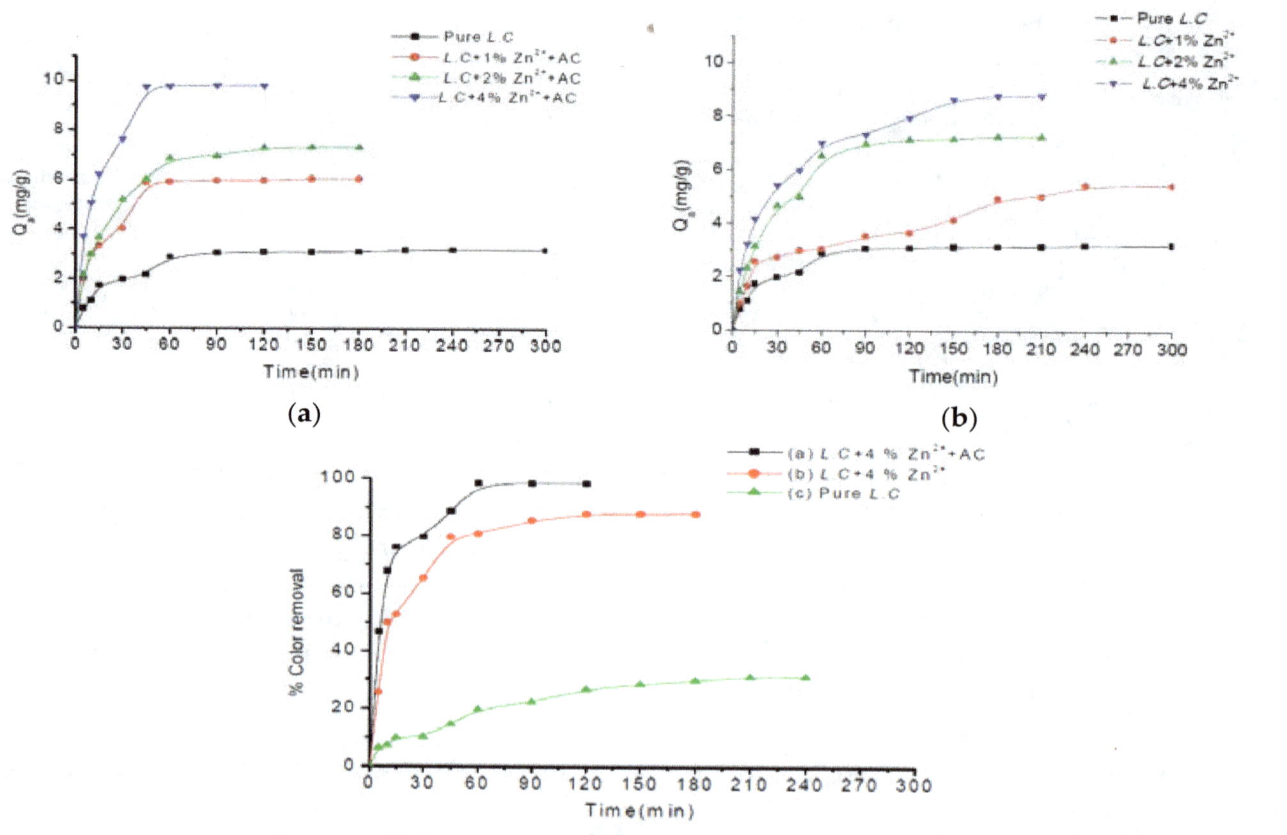


**Figure 4.** Kinetics of retention of methylene blue (MB) onto different materials used *L.C*; pH = 10, $C_i$ = 10 mg/L, m = 0.1 g, T = 298 K. (**a**) hybrid material *L.C* + 4% $Zn^{2+}$ synthesized in presence of AC, (**b**) hybrid material *L.C* + 4% $Zn^{2+}$ synthesized in absence of AC, (**c**) % color removal using different materials.

The use of AC in the synthesis of the hybrid material for different percentages of $Zn^{2+}$ improved the quantities adsorbed as well as the time required for the process. This behavior can be explained by the facility of the access of pollutants into the biosorbent pores and the availability of more active sites for biosorption offered by the synthesized hybrid materials.

Contrariwise, the best results were obtained for hybrid materials *L.C* + 4% $Zn^{2+}$ synthesized in the presence and absence of AC as the $Q_{max}$ reached, respectively, 9.83 mg/g and 8.81 mg/g. Interestingly, the high amount of added $Zn^{2+}$ assures the increase of superficial negatively charged groups obtained by lactonic function (as confirmed by Boehm titration). Therefore, this increase of negatively charged groups increased the biosorption capacity.

When comparing biosorption when using hybrid material synthesized in the presence of AC and those synthesized in its absence as well as the pure *L.C*, definitely, the speed of biosorption and the capacity uptake were greatly enhanced. While the time required for the biosorption of MB is almost lower than that required when using the pure *L.C* and hybrid materials synthesized in the absence of AC. This is probably due to the existence of the electric field which increased the speed of movement of the molecules of MB and facilitated their access into the active sites.

### *3.3. UV Characterizations*

To forecast the main phenomena that may occur during the MB biosorption, an UV-visible characterization was performed as shown in Figure 5. The characterization was done in order to compare the initially existing bands and the new ones that may appear during the biosorption of MB assisted by AC. The initial MB spectra display three main peaks located at 293,602 and

663 nm corresponding to the chromophore (dimethylamino group) and the aromatic rings in the MB molecule [53]. After MB biosorption, a decrease in the intensity of peaks recorded in the visible range at $\lambda_{max}$ = 663 nm relative to the naphthalene group [54] was observed for all biosorbent used. No new band appeared, thus indicating that no intermediates appeared and confirming that the color removal is probably due to a biosorption and not to an oxidation process which was probably due to the azo groups. It is interesting to note that the biosorption of MB was particularly important by using the hybrid material *L.C* + 4% $Zn^{2+}$ synthesized by precipitation under AC (Figure 5a) compared to the pure *L.C* (Figure 5c) and hybrid material synthesized in absence of AC (Figure 5b). In addition, the biosorption was rapidly obtained and within 120 min and a steady state concentration was achieved indicating a great ability of hybrid material for dye removal. On the contrary, the MB biosorption was tardier and the removal efficiency was lower when using the pure *L.C* (Figure 5c).

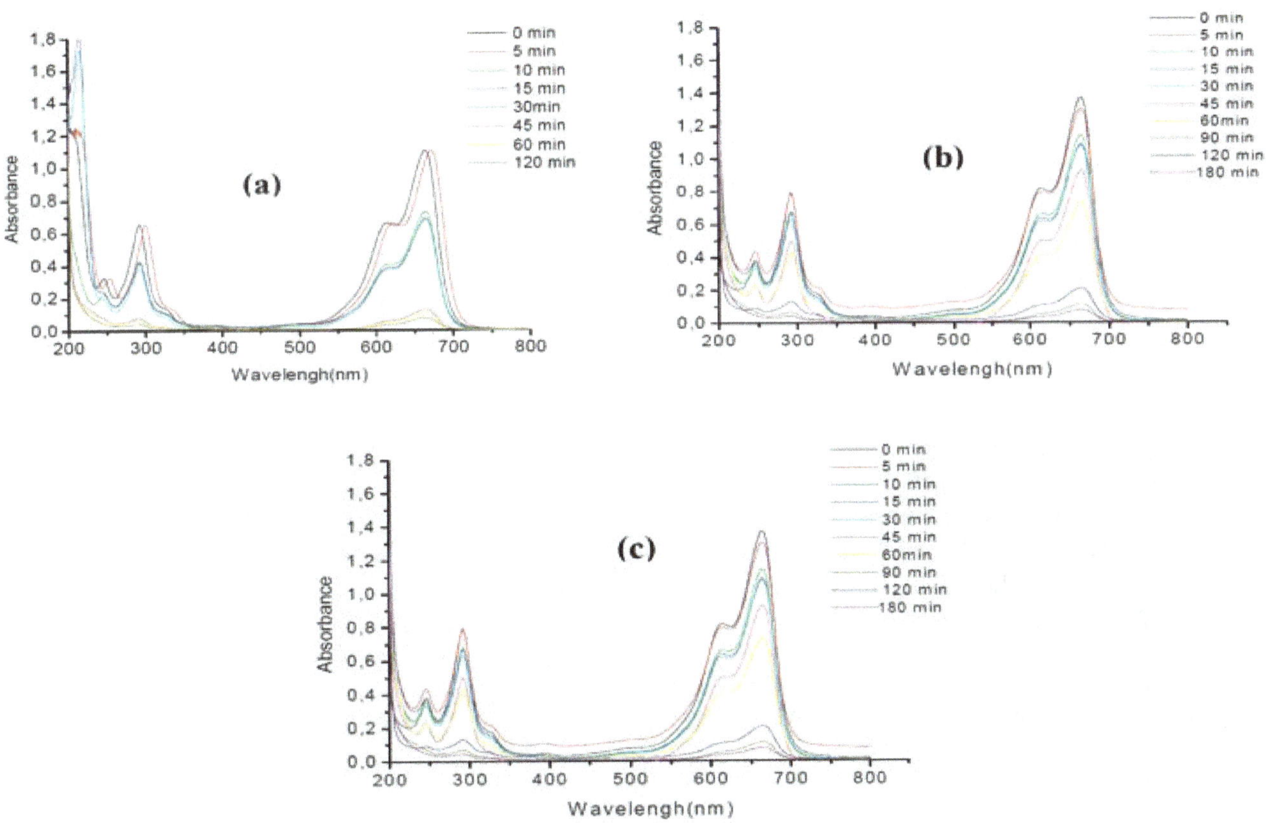


**Figure 5.** Spectra of MB biosorpion assisted by AC using (**a**) hybrid material *L.C* + 4% $Zn^{2+}$ synthesized in presence of AC, (**b**) hybrid material *L.C* + 4% $Zn^{2+}$ synthesized in absence of AC and (**c**) pure *L.C*, $C_i$ = 10 mg/L, pH = 10, $m_{L.C}$ = 0.1g, V = 0.1 L, T = 298 K.

### *3.4. Modeling of the Biosorption of MB Assisted by AC Using Brouers Sotolongo (B.S) Model*

The biosorption mechanism of MB onto pure *L.C* and hybrid materials elaborated by precipitation in the presence and absence of AC was studied by B.S model. The surface of the biosorbent is heterogeneous. The functions are distributed randomly hence the need to use a stochastic model (B.S). This model was chosen based on the coefficient of determination ($R^2$), the non-linear chi-square test ($\chi^2$) and the calculated $Q_e$ (Table 4 and Table S1 in the complementary information).

**Table 4.** Comparative results for the B.S model.

| Biosorbent Used | $Q_{exp}$ (mg/g) | $Q_{the}$ (mg/g) | $R^2$ | $\chi^2$ | $\alpha$ | $\tau_c$ |
|---|---|---|---|---|---|---|
| | | | **Brouers Sotolongo n = 1** | | | |
| Pure *L.C* | 3.22 | 4.33 | 0.9999 | 0.0001 | 0.99 | 91.52 |
| *L.C* + 1% $Zn^{2+}$ + AC | 6.06 | 5.97 | 0.9995 | 0.0005 | 1.11 | 56.14 |
| *L.C* + 2% $Zn^{2+}$ + AC | 7.33 | 7.41 | 0.9999 | 0.0005 | 1.27 | 47.98 |
| *L.C* + 4% $Zn^{2+}$ + AC | 9.84 | 9.31 | 0.9999 | 0.0001 | 5.41 | 36.83 |
| *L.C* + 1% $Zn^{2+}$ | 5.47 | 5.40 | 0.9999 | 0.0001 | 0.99 | 59.01 |
| *L.C* + 2% $Zn^{2+}$ | 7.27 | 7.22 | 0.9999 | 0.0004 | 1.06 | 49.13 |
| *L.C* + 4% $Zn^{2+}$ | 8.81 | 8.64 | 0.9999 | 0.0001 | 3.64 | 39.55 |

Table 4 shows that B.S model BSf (1, $\alpha$) presents the best fit to experimental data related to MB biosorption by *L.C* and hybrid materials. Indeed, the B.S model BSf (1, $\alpha$) has the highest coefficient of determination and the lowest nonlinear chi-square test. In addition, the Qe calculated by this model gives acceptable values.

The comparison of the values of $\tau_c$ calculated from BSf (1, $\alpha$) shows that *L.C* + 4% $Zn^{2+}$ + AC has the lowest value of $\tau_c$ (36.83). These results indicate that the AC synthesized hybrid materials remarkably improve both the adsorption capacity and the reaction rate of pure *L.C*.

On the other hand, the fractal constant $\alpha$ of *L.C* is less than one—which shows that the surface of *L.C* is heterogeneous [55]. Although the fractal constant $\alpha$ of the hybrid materials exceeds one, the kinetic is not clearly fractal [56].

For the hybrid materials, the fractal constant $\alpha$ increases with the increase of the percentage of $Zn^{2+}$ added. The hybrid material elaborated by AC (*L.C* + 4% $Zn^{2+}$) presents a higher value of $\alpha$. In fact, Selmi et al. (2018) showed that the fractal constant $\alpha$ increases with the number of functional surface groups [57]. Therefore, the hybrid material elaborated by AC (*L.C* + 4% $Zn^{2+}$) has the highest number of functional groups.

### *3.5. Evaluation of the Elaborated Hybrid Material for MB Biosorption: A Comparative Study*

To estimate the efficiency of the hybrid material synthesized by the precipitation method in presence and absence of AC and other low-cost material widely used for dye removal, a comparative study in terms of biosorption capacity ($Q_m$) and efficiency was done as illustrated in Table 5. Results obtained when using the hybrid material *L.C* + (1%, 2%, and 4% $Zn^{2+}$) synthesized by precipitation under AC are largely higher than those obtained when using the hybrid material synthesized in absence of AC and the other low-cost materials. The $Q_m$ shifted from 3.22 to 8.81 and 9.84 mg/g, respectively, for the hybrid material *L.C* + 4% $Zn^{2+}$ synthesized under AC, the hybrid material *L.C* + 4% $Zn^{2+}$ synthesized in absence of AC and the pure *L.C*. The $Q_m$ increased greatly when adding the 4% of $Zn^{2+}$ compared to the pure *L.C*. These findings can be explained by the high dispersion of active sites due to the availability of a large number of sites. The modification of pure *L.C* by different zinc rates in the presence and absence of AC increased the active sites available for MB retention. Furthermore, the rapid change in current direction has enhanced the access of pollutants into the biosorbent pores—which explained the main difference between rates achieved in both cases.

**Table 5.** Comparison of biosorption characteristics between hybrid material synthesized and other studied biosorbents used for MB biosorption.

| Biosorbents/MB | $Q_m$ (mg/g) | References |
|---|---|---|
| *Luffa cylindrica* | 3.9 | This study |
| L.C + 1% $Zn^{2+}$ | 5.47 | This study |
| L.C + 2% $Zn^{2+}$ | 6.55 | This study |
| L.C + 4% $Zn^{2+}$ | 8.77 | This study |
| L.C + 1% $Zn^{2+}$ + AC | 6.06 | This study |
| L.C + 2% $Zn^{2+}$ + AC | 7.33 | This study |
| L.C + 4% $Zn^{2+}$ + AC | 9.01 | This study |
| Algae Caolina | 6 | [57] |
| Langsat peel | 8 | [58] |
| Agave Americana | 6 | [59] |
| Posidonia Oceanica | 5.51 | [60] |
| Phragmites Australis | 6.31 | [61] |

### 3.6. Evaluation of the Synthesized Hybrid Material on the Purification of Industrial Textile Wastewater

Some experiments were performed in order to foresee the variation of pH, COD, turbidity, TOC and processing time during the purification of industrial wastewater (Table 6). In fact, the treated wastewater is composed of sodium carbonate, caustic soda, sodium sulfate, sea salt, wetting agent, reactive dyes, chlorine, and enzymes. *L.C* + 4% $Zn^{2+}$ synthesized in the presence and absence of AC and the pure *L.C* were evaluated for the purification.

**Table 6.** Comparative results obtained for the purification of industrial textile wastewater.

| Parameters | Industrial Textile Wastewater | | |
|---|---|---|---|
| **Biosorbent Used** | **Pure *L.C*** | ***L.C* + 4% $Zn^{2+}$** | ***L.C* + 4% $Zn^{2+}$ + AC** |
| $pH_i$ | 11.22 | 11.22 | 11.22 |
| $pH_f$ | 8.99 | 8.08 | 8 |
| $COD_i$ (mg/L)$O_2$ | 1265 | 1265 | 1265 |
| $COD_f$ (mg/L)$O_2$ | 597 | 297 | 159 |
| COD removal (%) | 52.80 | 76.5 | 87.43 |
| $Turbidity_i$ (NTU) | 11.70 | 11.70 | 11.70 |
| $Turbidity_f$ (NTU) | 4.58 | 2.23 | 1.99 |
| Turbidity removal (%) | 60.85 | 80.94 | 82.99 |
| $TOC_i$ (ppm(C.O)) | 28.9 | 28.9 | 28.9 |
| $TOC_f$ (ppm(C.O)) | 11.9 | 3.2 | 2.29 |
| TOC removal (%) | 58.82 | 88.92 | 92.07 |
| Processing time (min) | 300 | 240 | 180 |

Results show that the hybrid material (*L.C* + 4% $Zn^{2+}$) synthesized under AC is more efficient when synthesized in the absence of AC and pure *L.C* for purification of industrial textile wastewater.

The % of TOC, % of COD and the % of turbidity removals obtained for the hybrid material synthesized in presence of AC were higher than those obtained by hybrid material *L.C* + 4% $Zn^{2+}$ synthesized in absence of AC and pure *L.C*.

Both 87.43% of COD and 92.07% of TOC were removed from the industrial textile wastewater when using the hybrid material *L.C* + 4% $Zn^{2+}$ synthesized in presence of AC. 52.80% of COD and 58.82% of TOC were removed from the tested industrial wastewater when using the pure *L.C*. 76.5% of COD and 80.94% of TOC were removed when using the hybrid material *L.C* + 4% $Zn^{2+}$ synthesized in absence of AC, for the purification of the industrial wastewater.

The same behavior was observed for the % of turbidity removal. Values reached about 60.85 for pure *L.C*, 80.94% for the hybrid material (*L.C* + 4% $Zn^{2+}$) synthesized in absence of AC and 82.99%, respectively, for hybrid material (*L.C* + 4% $Zn^{2+}$) synthesized in presence of AC.

While for pH, both *L.C* and hybrid materials (*L.C* + 4% $Zn^{2+}$) had a significantly poor effect. This considerable fluctuation in the evaluated parameters may be attributed to the increase of the sites available for the color retention offered by the elaborated hybrid material as indicated by the Brouers-Sotolongo modeling. This analysis revealed that almost all of the parameters satisfied the Tunisian standards TN-106-02 indicating that the treated wastewater can be discharged to any receiving water body.

### *3.7. Phytotoxicity Test*

In order to evaluate the suitability of the raw and treated industrial textile wastewater for agricultural use, a phytotoxicity test was done. The evolution of the germination index (GI), the length of root, stem and leaf of lettuce using distilled water, raw and treated industrial textile wastewater was evaluated. The most used model for weeds is lettuce "*Lactuca Sativa*". It has been used extensively thanks to its fast germination and high sensitivity for a large variety of pollutants. Furthermore, it is often used for the examination of plants interaction in an aquatic environment [62,63].

Figure 6 (Figure S1 in the complementary information) presents the main results obtained. The germination index (GI) of the initial concentrations of the tested industrial wastewater was equal to 29.87%. After biosorption assisted by AC, the percentage of the germination index increased indicating a decrease in the toxicity of the wastewater tested. It reached respectively 72.33, 59.79, and 52.14%, respectively, when using the hybrid material (*L.C* + 4% $Zn^{2+}$) synthesized in presence of AC, the hybrid material *L.C* + 4% $Zn^{2+}$ synthesized in presence of AC and the pure *L.C*. Results showed that all GI were higher than 50%, indicating the purity of the treated wastewater and its suitability to agricultural use.

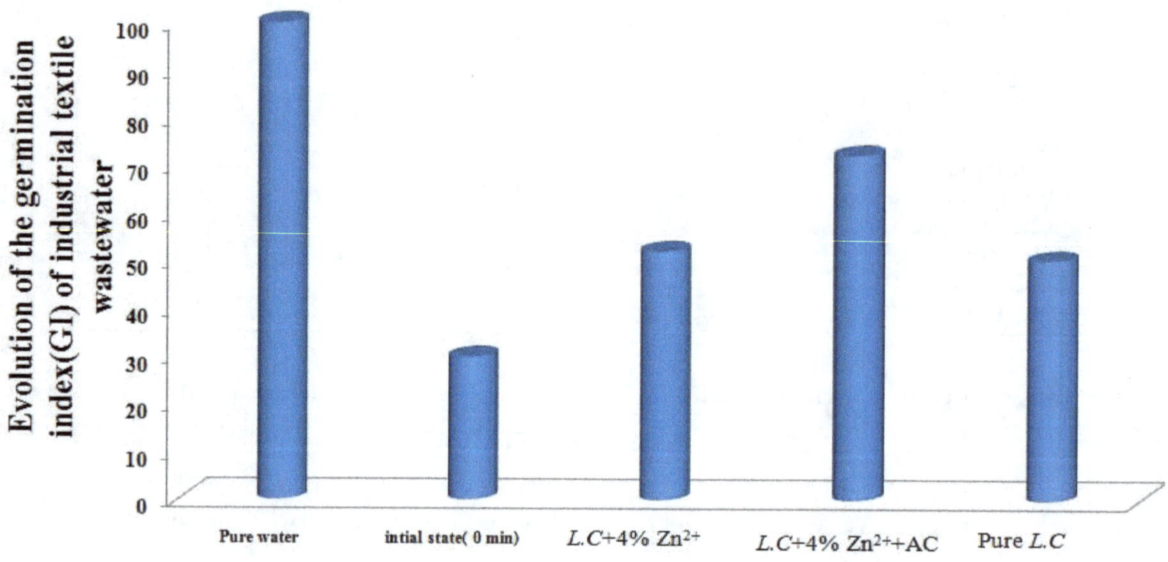


**Figure 6.** Evolution of the % GI after coupling biosorption with AC using pure *L.C* and hybrid material *L.C* + 4% $Zn^{2+}$ synthesized in presence and absence of AC.

### *3.8. Mechanism of Biosorption of MB*

An assumption of the pathway of the fixation of $Zn^{2+}$ ions onto the *L.C* fibers and the possible interactions between MB and the synthesized hybrid material were predicted as shown in Figures 7 and 8.

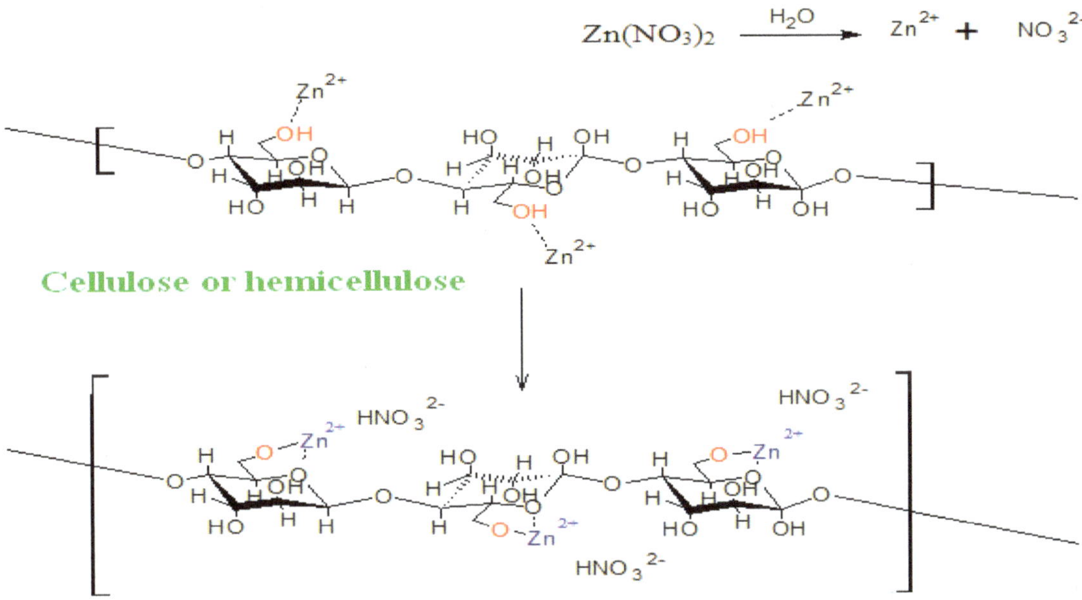


**Figure 7.** Possible pathway for the fixation of $Zn^{2+}$ onto the *L.C* surface.

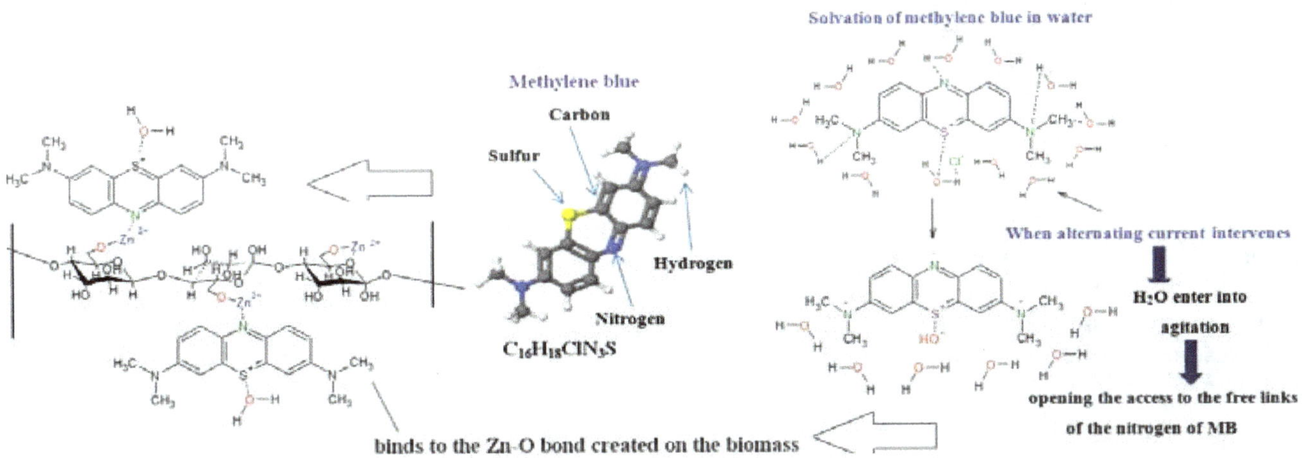


**Figure 8.** Possible pathway for the interactions between methylene blue and the synthesized hybrid material.

Figure 7 (Figure S2 in the complementary information) presents the possible pathway of the fixation of $Zn^{2+}$ ions onto the *L.C* fibers in the presence and absence of AC.

For both methods used the $Zn^{2+}$ ions generated from the dissolution of $Zn(NO_3)_2$ in water were fixed on the surface of the *L.C* through cellulose or hemicellulose, where oxygen is found, ensuring the creation of Zn-O bond on the biomass as confirmed by IR characterizations. The use of AC has enhanced the access of the $Zn^{2+}$ ions onto the fibers. Therefore, favoring the participation of all $Zn^{2+}$ ions presents in solution in the creation of Zn-O and Zn-O-C bonds. These findings can be confirmed by IR characterizations. The band located at 1383 $cm^{-1}$ appears from 1% of $Zn^{2+}$ for the hybrid material synthesized under AC, while it appears only for 4 % of $Zn^{2+}$ for the hybrid materials synthesized in absence of AC ensuring an increase in the number of sites available onto the hybrid material surface.

Figure 8 (Figure S3 in the complementary information) presents the proposal of a possible pathway of the interactions between MB and the synthesized hybrid material and the effect of alternating current on the proposal process.

Water solvates MB by making hydrogen bridges hydrogen obtained from and nitrogen obtained from MB and a polar bond between oxygen and sulfur ($\pi$–$\pi$ Bond).

When alternating current intervenes, the molecules of $H_2O$ enter into agitation opening the access to the free links of the nitrogen of MB which subsequently binds to the Zn-O bond created on the biomass.

## 4. Conclusions

We have successfully synthesized a performing hybrid material based on *Luffa cylindrica* fibers and different percentage of zinc oxide (1%, 2%, and 4%) by an easy precipitation method under AC. The fast and efficient pollutant removals are the basic benefits of this research. Results confirmed clearly the effect of AC on the modification of acidic-basic properties of the *L.C* compared to the pure one. The addition of the different percentage of $Zn^{2+}$ (1%, 2%, and 4%) has increased the basicity of the surface functionality by increasing the $pH_{pzc}$. Boehm titration indicated that the use of AC in the preparation of the hybrid materials has decreased the carboxylic and phenolic groups and increased the lactonic group. The presence of more lactonic groups has increased the density of the negative charge on the hybrid material surface which will play a crucial effect on the positively charged dye (MB) biosorption capacity. AC has increased the fixation and the distribution of $Zn^{2+}$ into the *L.C* fibers by increasing the sites' number as confirmed by B.S (1, $\alpha$). Physicochemical analyses revealed a prominent decrease in the COD, the TOC and the turbidity of the treated industrial textile wastewater, complied with the Tunisian Standards TN-106-02. The treated textile wastewater can be discharged to any receiving water. Furthermore, the germination indexes are higher than 50% confirmed their suitability to agricultural use.

**Supplementary Materials:**
Figure S1: Evolution of the % GI after coupling biosorption with AC using pure *L.C* and hybrid material *L.C* + 4% $Zn^{2+}$ synthesized in presence and absence of AC, Figure S2: Possible pathway for the fixation of $Zn^{2+}$ onto the *L.C* surface, Figure S3: Possible pathway for the interactions between methylene blue and the synthesized hybrid material, Table S1: Comparative results for the B.S model.

**Author Contributions:** Conceptualization, A.O., A.K. and M.S.; Formal analysis, A.O., A.K. and M.S.; Methodology, A.O.; resources, A.O., A.K., M.S. and R.B.; Writing—Original Draft preparation, A.O.; Writing—Review and Editing, A.O., A.K., M.S., R.B. and M.V.

**Acknowledgments:** The authors express their sincere thanks to the Laboratory of Energy and Materials (High School of Sciences and technology of Hammam Sousse)for the financial support and the Unitat de QuímicaAnalítica, Departament de Química, Centre Grup de Técniques de Separacio'en Química (GTS), Universitat Autónoma de Barcelona, Bellaterra, Spain.

## References

1. Adegoke, K.A.; Bello, O.S. Dye sequestration using agricultural wastes as adsorbents. *Water Resour. Ind.* **2009**, *12*, 8–24. [CrossRef]
2. Laasri, L.; Elamrani, M.K.; Cherkaoui, O. Removal of two cationic dyes from a textile effluent by filtration-adsorption on woodsawdust. *Environ. Sci. Pollut. Res.-Int.* **2007**, *14*, 237–240. [CrossRef] [PubMed]
3. Abidin, F.C.Z.A.; Rahmat, N.R. Multi-stage ozonation and biological treatment for removal of azo dye industrial effluent. *Int. J. Environ. Sci. Dev.* **2010**, *1*, 193–198.
4. Othmani, A.; Kesraoui, A.; Seffen, M. The alternating and direct current effect on the elimination of cationic and anionic dye from aqueous solutions by electrocoagulation and coagulation flocculation. *Euro-Mediterr. J. Environ. Integr.* **2017**, *2*, 2–6. [CrossRef]
5. Selmi, T.; Sanchez-Sanchez, A.; Gadonneix, P.; Jagiello, J.; Seffen, M.; Sammouda, H.; Celzard, A.; Fierro, V. Tetracycline removal with activated carbons produced by hydrothermal carbonisation of Agave americana fibres and mimosa tannin. *Ind. Crop. Prod.* **2018**, *115*, 146–157. [CrossRef]

6. Căilean, D.; Barjoveanu, G.; Musteret, C.P.; Sulitanu, N.; Manea, L.R.; Teodosiu, C. Reactive dyes removal from wastewater by combined advanced treatment. *Environ. Eng. Manag. J.* **2009**, *8*, 503–511. [CrossRef]
7. Vasudevan, S.; Lakshmi, J.; Sozhan, G. Effects of alternating and direct current in electrocoagulation process on the removal of cadmium from water. *J. Hazard. Mater.* **2011**, *192*, 26–34. [CrossRef] [PubMed]
8. Vijayageetha, V.A.; Rajan, A.P.; Arockiaraj, S.P.; Annamalai, V.; Janakarajan, V.N.; Balaji, M.D.S.; Dheenadhayalan, M.S. Treatment study of dyeing industry effluents using reverse osmosis technology. *Res. J. Recent Sci.* **2014**, *3*, 58–61.
9. Elaissaoui, I.; Akrout, H.; Grassini, S.; Fulginiti, D.; Bousselmi, L. Role of SiOx interlayer in the electrochemical degradation of Amaranth dye using SS/$PbO_2$anodes. *Mater. Des.* **2016**, *10*, 633–643. [CrossRef]
10. Vendruscolo, F.; Ferreira, G.L.R.; Filho, N.R.A. Biosorption of hexavalent chromium by microorganisms. *Int. Biodeterior. Biodegrad* **2017**, *119*, 87–95. [CrossRef]
11. Davis, T.A.; Volesky, B.; Mucci, A. A review of the biochemistry of heavy metal biosorption by brown algae. *Water Res.* **2003**, *37*, 4311–4330. [CrossRef]
12. Atar, N.; Olgun, A.; Colk, F. Thermodynamic, Equilibrium and Kinetic Study of the Biosorption of Basic Blue 41 using *Bacillus maceran*. *Eng. Life Sci.* **2008**, *8*, 499–506. [CrossRef]
13. Champenois, W.; Borges, A.V. Determination of dimethyl sulfonio propionate and dimethyls ulfoxide in Posidonia oceanic leaf tissue. *MethodsX* **2019**, *6*, 56–62.
14. Mbarki, F.; Kesraoui, A.; Seffen, M.; Ayrault, P. Kinetic, thermodynamic and adsorption behavior of cationic and anionic dyes onto Corn stigmata: Non-linear and stochastic analyses. *Water Air Soil* **2018**, *229*, 95. [CrossRef]
15. Kesraoui, A.; Mabrouk, A.; Seffen, M. Valuation of Biomaterial: *Phragmites australis* in the Retention of Metal Complexed Dyes. *Am. J. Environ. Sci.* **2017**, *3*, 266–276. [CrossRef]
16. Demir, H.; Top, A.; Balkose, D.; Ulk, S. Dye adsorption behavior of Luffa cylindrica fibers. *J. Hazard. Mater.* **2008**, *153*, 389–394. [CrossRef] [PubMed]
17. Wang, S.; Boyjoo, Y.; Choueib, Z.H.A. Removal of dyes from aqueous solution using fly ash and red mud. *Water Res.* **2015**, *39*, 129–138. [CrossRef]
18. Hachani, R.; Sabir, N.S.; Zohra, K.F.; Nesrine, M.N. Performance Study of a Low-cost Adsorbent-Raw Date Pits-for Removal of Azo Dye in Aqueous Solution. *Water Environ. Res.* **2017**, *98*, 827–839. [CrossRef]
19. El Shahawy, A.; Ghada Heikal, G. Organic pollutants removal from oily wastewater using clean technology economically, friendly biosorbent (Phragmites australis). *Ecol. Eng.* **2018**, *122*, 207–218. [CrossRef]
20. Osman, C. Synthèse De Nouveaux Matériaux Hybrides Pour Les Catalyses En Atrp Supporté Du Méthacrylate De Méthyle. Ph.D. Thesis, University of Pierre and Marie Curie, Paris, France, 2014.
21. Shivaramakrishnan, B.; Gurumurthy, B.; Balasubramanian, A. Potential biomedical applications of metallic nanibiomaterials: A review. *Int. J. Pharm. Sci. Res.* **2017**, *8*, 98500.
22. Ye, J.D.; Gu, S.L.; Qin, F.; Zhu, S.M.; Liu, S.M.; Zhou, X.; Liu, W.; Hu, L.Q.; Zhang, R.; Shi, Y.; et al. MOCVD growth and properties of ZnO films using dimethylzinc and oxygen. *Appl. Phys. A* **2005**, *81*, 809–812. [CrossRef]
23. Tului, M.A.; Bellucci, A.; Migliozzi, G. Zinc oxide targets for magnetron sputtering PVD prepared by plasma spray. *Surf. Coat. Technol.* **2005**, *205*, 1070–1073. [CrossRef]
24. Zamiri, R.; Zakaria, H.A.; Ahangar, M.; Zak, A.K.; Drummen, G.P.C. Aqueous starch as a stabilizer in zinc oxide nanoparticle synthesis via laser ablation. *J. Alloys Compd.* **2012**, *516*, 41–48. [CrossRef]
25. Perelshtein, L.; Applerot, G.; Perkas, N.; Wehrschetz-Sigl, E.; Hasmann, A.; Guebitz, G.M.; Gedanken, A. Antibacterial Properties of an In Situ Generated and Simultaneously Deposited Nanocrystalline ZnO on Fabrics. *ACS Appl. Mater. Interfaces* **2008**, *1*, 361–366. [CrossRef] [PubMed]
26. Hu, W.; Chen, S.; Zhou, B.; Wang, H. Facile synthesis of ZnO nanoparticles based on bacterial cellulose. *Mater. Sci. Eng. B* **2010**, *170*, 88–92. [CrossRef]
27. Ching, C.G.; Lee, S.C.; Ooi, P.K.; Ng, S.S.; Hassan, Z.; Hassan, H.A.; Abdullah, M.J. Optical and structural properties of porous zinc oxide fabricated via electrochemical etching method. *Mater. Sci. Eng. B* **2013**, *178*, 956–959. [CrossRef]
28. Chen, C.Y.; Chang, H.W.; Shih, S.J.; Tsay, C.Y.; Chang, C.J.; Lin, C.K. High super capacitive stability of spray pyrolyzed ZnO-added manganese oxide coatings. *Ceram. Int.* **2013**, *39*, 1885–1892. [CrossRef]
29. Brinker, C.J.; Scherer, G.W. *Sol-Gel Science: The Physics and Shemistry of Sol-Gel Processing Academic*; ET Press: Frankfurt, Germany, 1990.

30. Kesraoui, A.; Bouzaabia, S.; Seffen, M. The combination of Luffa cylindrical fibers and metal oxides offers a highly performing hybrid fiber material in water decontamination. *Environ. Sci. Pollut. Res.* **2018**, *26*, 11524–11534. [CrossRef]
31. Kesraoui, A.; Moussa, A.; Ali, G.B.; Seffen, M. Biosorption of alpacide blue from aqueous solution, by lignocellulosicbiomass: *Luffa cylindrical* fibers. *Environ. Sci. Pollut. Res.* **2015**, *23*, 15832–15840. [CrossRef]
32. Shafei, A.; Nikzar, M.; Arami, M. Photocatalytic degradation of terephlatic acid using titania and zinc oxide ZnO photocalyts comparative study. *Desalination* **2013**, *252*, 8–16. [CrossRef]
33. Kesraoui, A.; Selmi, T.; Seffen, M.; Brouer, F. Influence of alternating current on the adsorption of indigo carmine. *Environ. Sci. Pollut. Res.* **2016**, *24*, 9940–9950. [CrossRef] [PubMed]
34. Boumediene, M.; Benaïssa, H.; George, B.; Molina, S.; Merlin, A. Effects of pH and ionic strength on methylene blue removal from synthetic aqueous solutions by sorption onto orange peel and desorption study. *J. Mater. Environ. Sci.* **2018**, *9*, 1700–1711.
35. Goertzen, L.; ThériaultKim, S.D.; Oickle, M.A.; Tarasuk, C.A.; Andreas, A.H. Standardization of the Boehm titration. Part I. $CO_2$ expulsion and endpoint determination. *Carbon* **2010**, *48*, 1252–1261. [CrossRef]
36. Nava, J.L.; Quiroz, M.A.; Martinez-Huitle, C.A. Electrochemical treatment of synthetic wastewaters containing Alphazurine A dye: Role of electrode material in the color and COD removal. *J. Mex. Chem. Soc.* **2008**, *52*, 249–255.
37. Sharma, A.; Bhattacharya, S.; Sen, R.; Reddy, B.S.B.; Fecht, H.J.; Das, K.; Das, S. Influence of current density on microstructure of pulse electrodeposited tin coatings. *Mater. Charact.* **2012**, *68*, 22–32. [CrossRef]
38. Amrhar, O.; Nassali, H.; Elyoubi, M.S. Adsorption of a cationic dye, Methylene Blue, ontoMoroccan Illitic Clay. *J. Mater. Environ. Sci.* **2015**, *6*, 3054–3065.
39. Brouers, F. Statistical foundation of empirical isotherms. *Open J. Stat.* **2014**, *4*, 687–701. [CrossRef]
40. Brouers, F. The fractal (BSf) kinetics equation and its approximations. *J. Mod. Phys.* **2014**, *5*, 1594–1601. [CrossRef]
41. Brouers, F.; Al-Musawi, T.J. On the optimal use of of isotherm models for the characterization of biosorption of lead onto algae. *J. Mol. Liq.* **2015**, *212*, 46–51. [CrossRef]
42. Khot, L.R.; Sankaran, S.; Maja, J.M.; Ehsani, R.; Schuster, E.W. Applications of nanomaterials in agricultural production and crop protection: A review. *Crop Prot.* **2012**, *35*, 64–70. [CrossRef]
43. Antonio, R.; Tapia-Benavides, A.R.T.; Tlahuextl, M.; Tlahuext, H.; Galán-Vidala, C. Synthesis of Zn compounds derived from *1H*-benzimidazol-2-ylmethanamine. *Arkivoc* **2008**, *5*, 172–186.
44. Xu, Q.; Chena, C.; Rosswurma, K.; Yaoa, T.; Janaswamy, S. A facile route to prepare cellulose-based films. *Carbohydr. Carbohydr. Polym.* **2016**, *149*, 274–281. [CrossRef]
45. Vergnat, V. Matériaux Hybrides Organique-Inorganique Par Greffage Covalent De Polymères Sur Des Oxydes Métalliques. Ph.D. Thesis, University of Strasbourg, Strasbourg, France, 2011.
46. Zeleňák, V.; Vargová, Z.; Györyová, K. Correlation of infrared spectra of zinc (II) carboxylates with their structures. *Spectrochim. Acta A Mol. Biomol. Spectrosc.* **2007**, *66*, 262–272. [CrossRef]
47. Li, L.; Deng, J.; Deng, H.; Liu, Z.; Li, X. Preparation, characterization and antimicrobial activities of chitosan/Ag/ZnO blend films. *Chem. Eng. J.* **2010**, *160*, 378–382. [CrossRef]
48. Liu, X.F.; Sun, J.; Ren, J.; Curling, S.; Sun, R. Physicochemical characterization of cellulose from perennial leaves (Loliumperenne). *Carbohydr. Res.* **2006**, *341*, 2677–2687. [CrossRef] [PubMed]
49. Awala, H.; El-Roz, M.; Toufaily, J.; Mintova, S. Template-Free Nanosized EMT Zeolite Grown on Natural Luffa Fibers for Water Purification. *Adv. Porous Mater.* **2014**, *4*, 141–182. [CrossRef]
50. Arefi, M.R.; Zarchi, S.R. Synthesis of Zinc Oxide Nanoparticles and Their Effect on the Compressive Strength and Setting Time of Self-Compacted Concrete Paste as Cementitious Composites. *Int. J. Mol. Sci.* **2012**, *13*, 4340–4350. [CrossRef]
51. Wan, C.; Jian, L. Embedding ZnO nanorods into porous cellulose aerogels via a facile one-step low-temperature hydrothermal method. *Mater. Des.* **2015**, *83*, 620–625. [CrossRef]
52. Mahmoudi, K.H.; Hamdi, N.; Srasra, E. Kinetics and equilibrium studies on removal of methylene blue and methyl orange by adsorption onto activated carbon prepared from date pits—A comparative study. *Water Air Soil Pollut.* **2018**, *229*, 95. [CrossRef]
53. Cenens, J.; Schoonheydt, R.A. Visible spectroscopy of methylene blue on hectorite, laponite and barasym in aqueous suspension. *Clays Clay Min.* **1988**, *36*, 214–224. [CrossRef]

54. Kalmár, J.; Lente, G.; Fabian, I. Kinetics and mechanism of the adsorption of methylene blue from aqueous solution on the surface of a quartz cuvette by on-line UV-Vis spectrophotometry. *J. Dyes Pigments.* **2016**, *127*, 170–178. [CrossRef]
55. Selmi, T.; Seffen, M.; Sammouda, H.; Mathieu, S.; Jagiello, J.; Celzard, A.; Fierro, V. Physical meaning of the parameters used in fractal kinetic and generalised adsorption models of Brouers–Sotolongo. *Adsorption* **2017**, *24*, 11–27. [CrossRef]
56. Hammud, H.H.; Fayoumi, L.M.A.; Holail, H.; Mostafa, E.S.M.E. Biosorption studies of methylene blue by Mediterranean algae Carolina and its chemically modified forms. Linear and nonlinear models' prediction based on statistical error calculation. *Int. J. Chem.* **2011**, *3*, 147–163. [CrossRef]
57. Salleh, M.A.M.; Mahmoud, D.K.; Binti Awang Abu, N.A.; Abdul Karim, W.A.W.; Idris, A.B. Methylene blue adsorption from aqueous solution by langsat (lansium domesticum) peel. *J. Purity Util. React. Environ.* **2012**, *1*, 442–465.
58. Hamissa, A.M.B.; Brouers, F.; Ncibi, M.C.; Seffen, M. Kinetic Modeling Study on Methylene Blue Sorption onto Agave americana fibers: Fractal Kinetics and Regeneration Studies. *Sep. Sci. Technol.* **2013**, *48*, 2834–2842. [CrossRef]
59. Ncibi, M.C.; Mahjoub, B.; Seffen, M. Studies on the biosorption of textile dyes from aqueous solutions using Posidonia Oceanic (L.) leaf sheaths fibres. *Adsorpt. Sci. Technol.* **2006**, *24*, 461–473. [CrossRef]
60. García-Ávila, F.; Patiño-Chávez, J.; Zhinín-Chimbo, F.; Donoso-Moscoso, S.; del Pino, L.F.; Avilés-Añazco, L. Performance of Phragmites Australis and Cyperus Papyrus in the treatment of municipal wastewater by vertical flow subsurface constructed wetlands. *Int. Soil Water Conserv. Res.* 2019. [CrossRef]
61. Dallel, D.; Kesraoui, A.; Seffen, M. Biosorption of cationic dye onto "Phragmites australis" fibers: Characterization and mechanism. *J. Environ. Chem. Eng.* **2018**, *6*, 6. [CrossRef]
62. Louhichi, G.; Bousselmı, L.; Ghrabi, A.; Khouni, I. Process optimization via response surface methodologyin the physico-chemical treatment of vegetable oil refinery wastewater. *Environ. Sci. Pollut. Res.* **2018**, 1–19. [CrossRef]
63. Mendes, P.M.; Becker, R.; Corrêa, L.B.; Bianchi, I.; Dai Prá, M.A.; Lucia, T., Jr.; Corrêa, E.K. Phytotoxicity as an indicator of stability of broiler production residues. *J. Environ. Manag.* **2016**, *167*, 156–159. [CrossRef]

11

# Wastewater Reclamation in Major Jordanian Industries: A Viable Component of a Circular Economy

**Motasem N. Saidan [1,*], Mohammad Al-Addous [2], Radwan A. Al-Weshah [3], Ibrahim Obada [4], Malek Alkasrawi [5] and Nesrine Barbana [6]**

1 Department of Chemical Engineering, School of Engineering, The University of Jordan, Amman 11942, Jordan
2 Department of Energy Engineering, School of Natural Resources Engineering and Management, German Jordanian University, Amman 11180, Jordan; mohammad.addous@gju.edu.jo
3 Civil Engineering Department, School of Engineering, The University of Jordan, Amman 11942, Jordan; weshah11@yahoo.com
4 Ministry of Water and Irrigation, Amman 11181, Jordan; ibrahim.obadah@gmail.com
5 Department of PS & Chemical Engineering, University of Wisconsin, Stevens Point, WI 54481, USA; malek.alkasrawi@uwsp.edu
6 Department of Environmental Technology, Technische Universität Berlin, 10623 Berlin, Germany; nesrine@campus.tu-berlin.de
* Correspondence: m.saidan@gmail.com or m.saidan@ju.edu.jo

**Abstract:** Water scarcity remains the major looming challenge that is facing Jordan. Wastewater reclamation is considered as an alternative source of fresh water in semi-arid areas with water shortage or increased consumption. In the present study, the current status of wastewater reclamation and reuse in Jordan was analyzed considering 30 wastewater treatment plants (WWTPs). The assessment was based on the WWWTPs' treatment processes in Jordan, the flowrates scale, and the effluents' average total dissolved solid (TDS) contents. Accordingly, 60% of the WWTPs in Jordan used activated sludge as a treatment technology; 30 WWTPs were small scale ($<1 \times 10^4$ $m^3$/day); and a total of 17.932 million $m^3$ treated wastewater had low TDS (<1000 ppm) that generally can be used in industries with relatively minimal cost of treatment. Moreover, the analysis classified the 26 million $m^3$ groundwater abstraction by major industries in Jordanian governorates. The results showed that the reclaimed wastewater can fully offset the industrial demand of fresh water in Amman, Zarqa, and Aqaba governorates. Hence, the environmental assessment showed positive impacts of reclaimed wastewater reuse scenario in terms of water depletion (saving of 72.55 million $m^3$ groundwater per year) and climate change (17.683 million kg $CO_{2Eq}$ reduction). The energy recovery assessment in the small- and medium-scale WWTPs ($<10 \times 10^4$ $m^3$/day) revealed that generation of electricity by anaerobic sludge digestion equates potentially to an offset of 0.11–0.53 kWh/$m^3$. Finally, several barriers and prospects were put forth to help the stakeholders when considering entering into an agreement to supply and/or reuse reclaimed water.

**Keywords:** reclaimed water; circular economy; anaerobic digestion; biogas; reuse; water pricing; water depletion; industrial sector

---

## 1. Introduction

Water is becoming a limited resource in terms of quantity and quality due to the growing global economy and population, accelerating urbanization, and climate change effects [1–3]. Water reuse has been employed as an alternative water supply in arid and semi-arid regions [4,5]. In this context, wastewater and water reclamation plays a vital role in sustainable water resource management and mainly in various application such as agricultural irrigation, industrial processes, aquaculture, and for any non-human contact utilization, etc. [5–8]. Moreover, reclaimed wastewater is a resource that can be continuously produced unaffected by climatic conditions [9,10], especially in the Mediterranean region, one of the most vulnerable areas to climate change and with limited water resources [11–13].

However, the potential of recycling and reusing treated wastewater in a transition to a circular economy should be exploited thoroughly in arid and semi-arid areas, since it could synergize the wide adoption of water reuse as an alternate water supply [14–16].

*Wastewater Reclamation Overview in Jordan*

Jordan is classified as a semi-arid to arid country, with scarce water resources compared with other countries in the Middle East, and is ranked among the poorest countries in the world in terms of water availability [17–25]. Figure 1 shows the water resources in Jordan including locations of wastewater treatment plants (WWTPs). The Syria crisis is still adding strain on Jordan's economy and infrastructure and has put pressure on all sectors including water, municipal services, and electricity supply [26–34]. This problem is even more intense in areas with high population due to refugee influx that caused unsustainable over-exploitation of groundwater, and consequently led to increasing groundwater salinity and depleting resources (i.e., the water table was reduced by 5 m in areas like Dhuleil-Hallabat, area of the Amman-Zaraq basin, and tripled in salinity) [35].

Most of the published literature on water reuse in the Middle East focused on reclaimed wastewater uses in agricultural fields [36–40]. For instance, Hussain et al. (2019) reviewed 124 recent publications on the multiple aspects of safe use of treated wastewater for agriculture, landscape, and forestry and for non-conventional water resources management [40]. Moreover, it is also reported that approximately 20 million hectares of arable land worldwide is irrigated with wastewater [41].

Considering the water scarcity situation, Jordan has given top priority to the use of reclaimed wastewater in agriculture and industrial sectors [42–45], hence, the reuse of wastewater in agriculture has replaced freshwater resources, which were previously used for irrigation, allowing freshwater to be reallocated to the municipal sector where there is higher demand and quality water is needed for potable use. Despite of that, the agricultural sector accounts for 75% of all water consumption in Jordan and produces only 2% of the Gross Domestic Product (GDP) [46]. On the other hand, to reuse the reclaimed wastewater eco-efficiently in the industrial sector, most of the industrial facilities need to improve their wastewater management practices and upgrade their on-site treatment units to treat the wastewater before use [47].

The WWTPs play a vital role to decrease the environmental impacts of municipal and industrial discharges [48], while having an advanced (tertiary) treatment, the wastewater recycling and reuse can be promoted [49], as well as, enhancing the recovery of materials or energy [50]. Wastewater reclamation is one of the recommended solutions for the problem of water scarcity although the process may be complex, costly in terms of resources, and energy demanding depending on the quality of treated wastewater and the adopted technology for tertiary treatment [51]. However, shifting of the WWTP effluents from their application in agricultural irrigation to the industrial sector will require recognition of the fact that some agricultural activity would no longer have access to water for irrigation. Despite this, such a shift is recommended since most of the existing conventional treatment of WWTPs (mechanical chemical and biological treatment) does not eliminate emerging pollutants (i.e., pharmaceuticals and personal care products, hormones and steroids, persistent organic pollutants, etc.) from the wastewater, which can be induced into the food chain, subsequently causing adverse ecological and human health effects [52].

So far, wastewater reclamation and reuse in the context of water shortage in Jordan is not high, overall, whereas the potentiality of wastewater reuse is huge. The objectives of this paper are to compressively analyze the current status of wastewater reclamation and its reuse in major industries in Jordan, and to summarize the opportunities and the challenges of expanding wastewater reuse, and then to put forth prospects for future wastewater reclamation and reuse in Jordan.

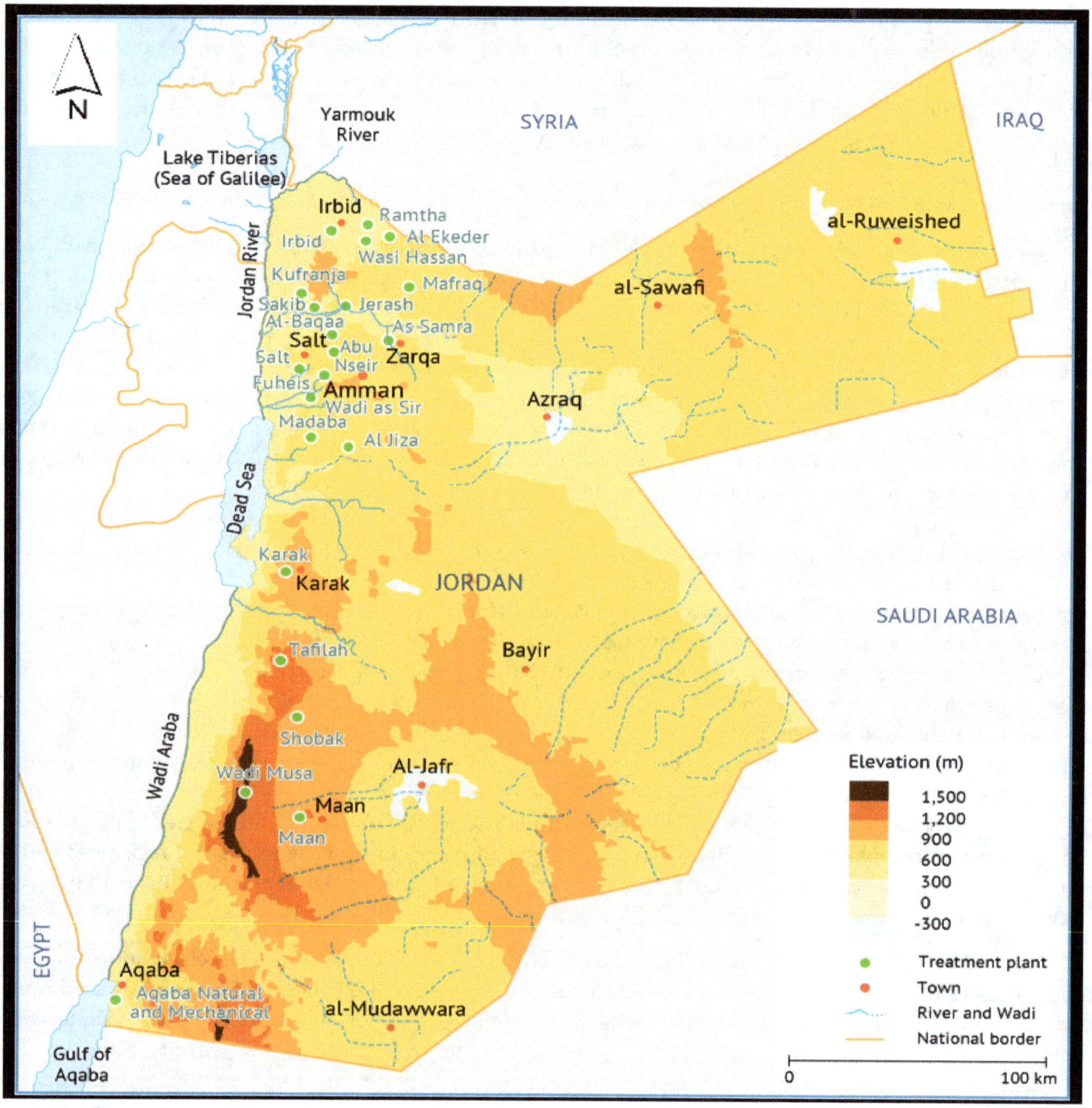


**Figure 1.** Water resources in Jordan.

## 2. Materials and Methods

### 2.1. Wastewater Treatment Plants in Jordan

Jordan has a fair operational capacity in wastewater treatment, although it is highly cost-intensive. The 34 central WWTPs are expected to treat 240 million $m^3$ per year (MCM/year) by 2025 [18]. Increasing sanitation coverage is expensive, and the proposed shift in water sector expenditures from water supply to sanitation in 2011–2013 is a significant step toward increasing coverage. In 2013, collection costs amounted to JOD 47 million (1$ is 0.71 Jordan Dinar (JOD)) and treatment costs to JOD 43.1 million [53]. Moreover, water and sanitation service costs are subsidized. Combined water and sewer bills amount to less than 0.92% of the total household annual expenditures. With Jordan's population expected to almost double by 2050, water demand will exceed the available water resources by more than 26% [18].

Figure 2 shows the variety and distribution of 34 different processes in WWTPs in Jordan. The most widely used technologies are the activated sludge (AS) process with a share of 60%. Followed by the wastewater stabilization pond (WSP) process with a share of 19%. While the trickling filter (TF) and AS process, Membrane Bioreactor (MBR) and TF process, and oxidation sludge (OS) process were evenly having the same use share of 6%, respectively. The TF process was the least used technology with a share of 3%. Moreover, one of these WWTPs is of super-large scale (>30 × $10^4$ $m^3$/day), 4 WWTPs are of medium scale (1 × $10^4$–10 × $10^4$ $m^3$/day), and 30 WWTPs are small scale (<1 × $10^4$ $m^3$/day), which are generally built in medium and small size cities and refugees camps.

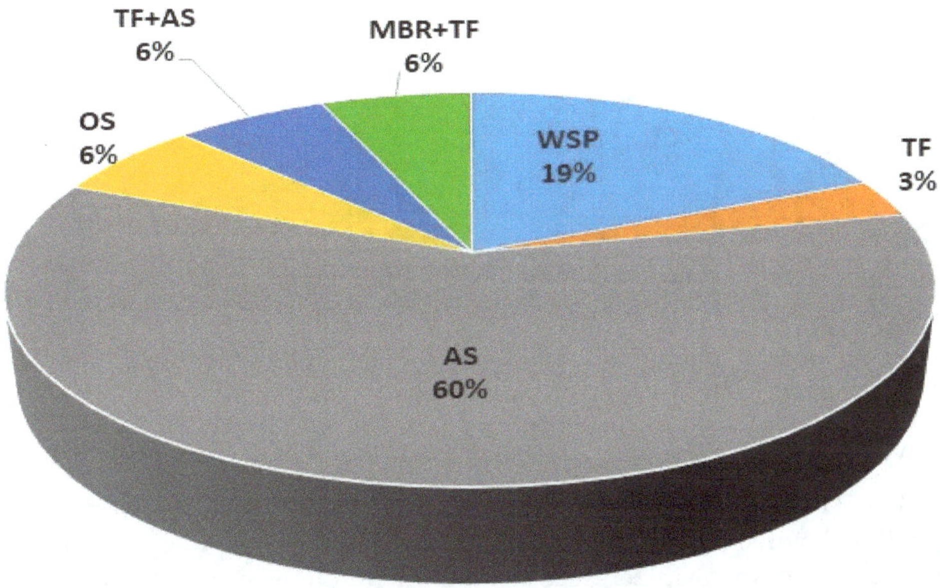


**Figure 2.** The variety and distribution of different processes in wastewater treatment plants (WWTPs) in Jordan. AS stands for activation sludge; OS is oxidation sludge; TF is trickling filter; WSP is wastewater stabilization pond; MBR + TF is Membrane Bioreactor and TF process; and TS + AS is trickling filter and activation sludge process.

### *2.2. Data Gathering and Analysis*

The analysis carried out in the present study is divided into four main steps as illustrated in Figure 3, which shows the methodological approach to addressing the specific objectives of this study.

A desk study was carried out for the available baseline documents (i.e., unpublished, monthly progress reports, internal memos, and minutes of meetings) and other references for collecting the technical data. The data and information used in the present study were gathered via semi-structured interviews with key stakeholders in the water (Ministry of Water and Irrigation, Ministry of Agriculture, Ministry of Environment, etc.) and industrial sectors (Ministry of Trade and Industry, Chambers of Industry, etc.), and with international funding agencies (i.e., USAID, GIZ, etc.) involved in the ongoing projects targeting integrated water resource management in Jordan. In addition, qualitative and quantitative data and information have been derived from unpublished government reports.

Moreover, before the interviews, a brief session was hosted to probe respondents for greater clarity in answers and consistency in relation to the objectives of the questions.

Information obtained through the interviews was crosschecked with the objective to reassess gaps and divergences of information.

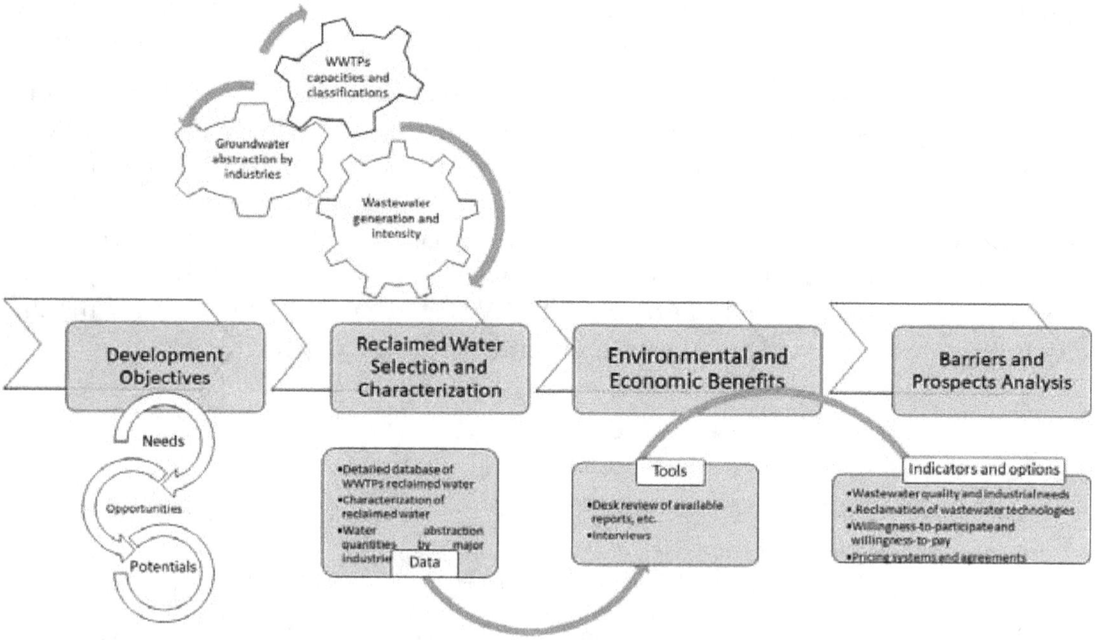


**Figure 3.** Diagram of the framework.

## 3. Results and Discussion

### 3.1. Wastewater Reclamation: Current Capacity and Potential Reuse

#### 3.1.1. Reclaimed Wastewater Production: Overview and Potentials

Most of the WWTPs in Jordan provide secondary treatment with a variety of activated sludge processes followed by disinfection with chlorine. The exception is the Aqaba treatment facility, which provides tertiary filtration of the oxidized secondary effluent followed by ultraviolet disinfection and a chlorine residual. The total effluent of the wastewater flow from the WWTPs is around 166 million m$^3$ based on data obtained from the Ministry of Water for the year 2018, as shown in Table 1.

The industrial sector mostly relies on fresh water, which could be used for domestic purposes. For instance, the industry uses 32.2 million m$^3$ groundwater, 4.8 million m$^3$ surface water, and 1.7 million m$^3$ of treated wastewater [18]. Thus, this provides a great opportunity for groundwater-to-recycled water substitution.

In the present study, the WWTP effluents were classified according to their average total dissolved solids contents (TDSs) as follows: <1000 ppm; 1000 < TDS <1500; and >1500, based on wastewater analysis data (average data 2010–2016). Figure 4 shows the classification of WWTPs according to their effluents' TDS.

Table 1 shows the annual WWTP effluents' flow rate according to the TDS classifications. The first class (TDS < 1000 ppm), which relatively has the lowest TDS, can be reused several times in most industrial applications, especially in thermal units, cooling towers, etc. For instance, Aqaba recycled water, which has the lowest salinity among the WWTPs in Jordan (TDS = 587 ppm), is most readily usable in industrial applications. So potentially, this class represents 9 WWTPs distributed in different locations in Jordan, as shown in Table 1, and, in total, 17.932 million m$^3$ of treated wastewater of this class can be used directly with no or low cost of on-site treatment in the industrial sector depending on the fit-for-purpose water criteria.

However, the second class (1000 < TDS < 1500), which has medium TDS, has the highest annual effluent flow rate of 147.323 million m$^3$ in total out of 18 WWTPs distributed in widely different locations in Jordan, as shown in Table 1. The most effluent wastewater flowrate in this class is generated from Al Samra WWTP with 117.1 million m$^3$ per year by offering sanitation services to about two million in Amman and Zarqa governorates; the first and third most populated cities in

Jordan, respectively [44,54]. With such large capacity and modern technology to ensure the highest purifications, Al Samra is considered as one of the largest plants in the region [40], which treats about 70.54% of total reclaimed wastewater in Jordan. This class represents 18 WWTPs distributed in different locations in Jordan, and, in total, 147.33 million $m^3$ of treated wastewater of this class can be used with medium cost of some necessary modification in the plant process in the industrial sector depending on the fit-for-purpose water criteria.

The third class has a TDS > 1500, the WWTP effluents in this class cannot be used without further intensive treatment such as: demineralization; blending with low-salinity water; and some change in the industrial process. This class represents three WWTPs with 0.788 million $m^3$ of treated wastewater, as shown in Table 1. Therefore, due to the high capital cost of investment and relatively expensive operating cost, this class is excluded from the present study analysis.

**Table 1.** Annual WWTP effluent flow rate according to the total dissolved solids (TDSs) classifications.

| TDS Classification | WWTP | Effluent Flow Rate (Million $m^3$/year) | Total Flowrate of Grouped WWTPs (Million $m^3$/year) |
|---|---|---|---|
| <1000 ppm | Aqaba-Tertiary | 3.90 | 17.93 |
| | Aqaba-Lagoon | 2.22 | |
| | Wadi Essir | 1.71 | |
| | Wadi Musa | 1.02 | |
| | Salt | 3.19 | |
| | Fuhis | 1.15 | |
| | Abu Nusseir | 1.31 | |
| | Madaba | 2.53 | |
| | Ma'an | 0.92 | |
| 1000 < TDS < 1500 | Karak | 0.54 | 147.32 |
| | Mafraq | 1.29 | |
| | Mu'taa | 0.55 | |
| | Wadi Hassan | 0.38 | |
| | Al Samra | 117.10 | |
| | Irbid | 3.10 | |
| | Wadi Shalallah | 3.43 | |
| | Kufranja | 1.25 | |
| | Jeza | 0.29 | |
| | South Amman | 4.72 | |
| | Tafileh | 0.80 | |
| | Wadi Arab | 4.98 | |
| | Ain Albasha | 5.12 | |
| | Al-Me'rad | 1.16 | |
| | North Shouneh | 0.15 | |
| | Tal-Almanttah | 0.15 | |
| | Akeeder | 0.82 | |
| | Ramtha | 1.50 | |
| >1500 | Jerash | 0.42 | 0.78 |
| | Shobak | 0.05 | |
| | Al Lujjon | 0.30 | |

Excluding food and pharmaceutical industries, the total groundwater abstraction for industrial purposes was approximately 26 million $m^3$ in Jordan in 2015 [55]. The major industries considered in the present study as the major groundwater abstracting industries are clarified in Table 2. Considering this, Figure 5 shows the total groundwater abstraction by major industries in Jordanian governorates, where the industries in Karak governorate were the most groundwater abstracting, with approximately 11.5 million $m^3$ per year. Followed by the industries in Ma'an (4.38 million $m^3$ per year). While the industries in Zarqa and Aqaba governorates were close to each other in terms of groundwater abstraction with 2.74 and 1.825 million $m^3$ per year, respectively. The industries in the Capital Amman were the least groundwater abstracting with 0.611 million $m^3$ per year.

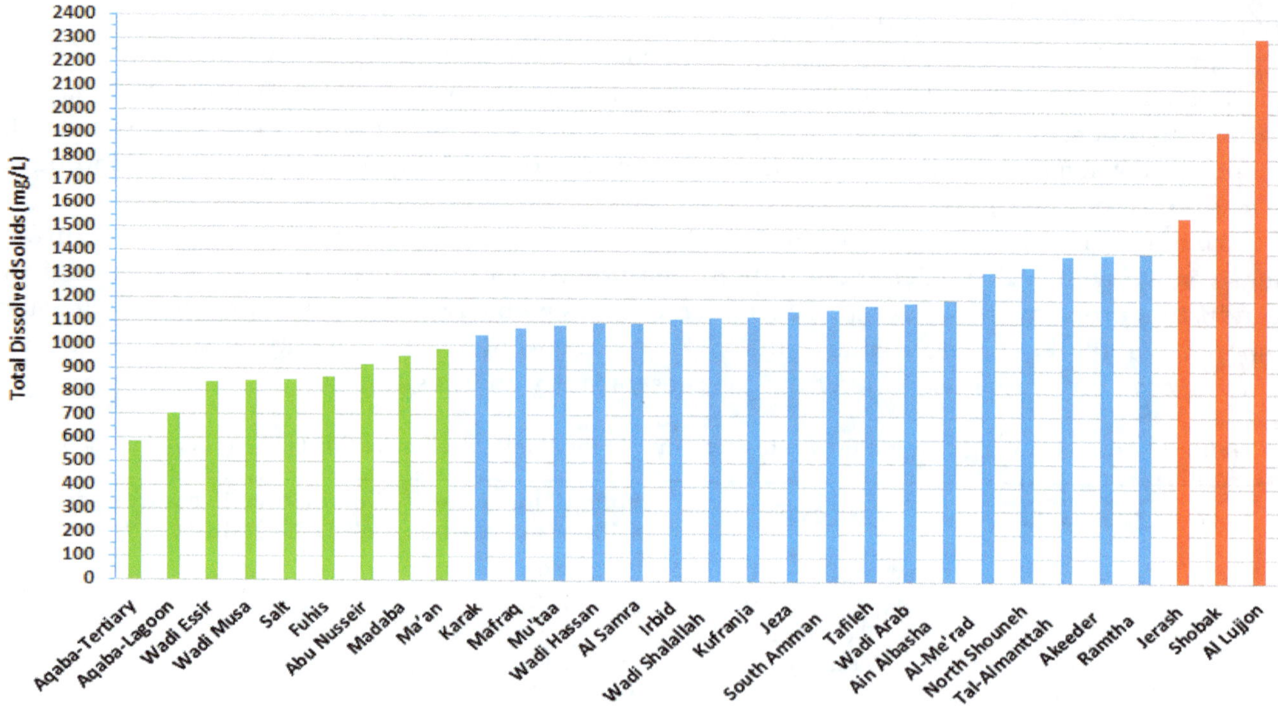


**Figure 4.** Classification of WWTPs according to their effluents' TDS.

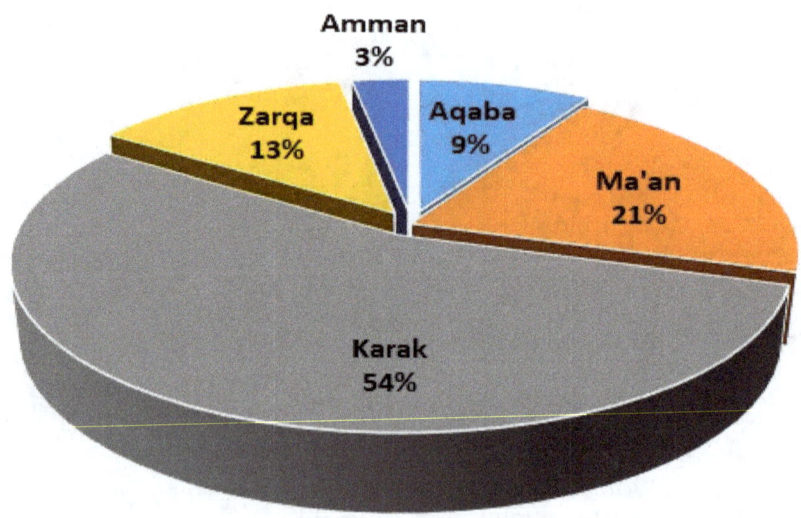


**Figure 5.** Percentages of total groundwater abstraction by major industries in Jordanian governorates.

**Table 2.** Major groundwater abstraction industries in Jordan according to governorates.

| Governorate | Major Groundwater Abstraction Industries |
|---|---|
| Amman | Cement, metals, concrete, paper and carton, etc. |
| Zarqa | Petroleum refinery, metals and pipes, paper and cardboard, thermal power and electricity plants, cement, etc. |
| Karak | Phosphate mines, potash, chemical fertilizers, cement, mining, etc. |
| Ma'an | Phosphate, cement, etc. |
| Aqaba | Phosphate, fertilizers, etc. |

Hence, based on the data of first class and second class in Table 1, the potential reclaimed wastewater substitution in major industries in Jordanian governorates is shown in Figure 6. It is obvious that the reclaimed wastewater in Zarqa governorate can fully substitute the industrial demand of fresh water (Figure 6a) and the needs for irrigation of 3000 donums for 20–30 farmers adjacent to Al Samra WWTP as reported by Hussein (2018) [54] and Maldonado (2017) [44]. The full substitution of industrial demand is also noticed in both Amman and Aqaba governorates with 13.13- and 3.36-fold, respectively. However, the shortage of industrial demand substitution is significantly clear in both of Ma'an and Karak governorates with substitution amounts of 2.45 and 10.4 million $m^3$ per year, respectively, as clearly shown in Figure 6b. Therefore, for the WWTPs in the governorates with a substitution factor less than one (mainly Ma'an and Karak governorates) it is preferable to prioritize their effluents (reclaimed wastewater) for irrigation use where applicable.

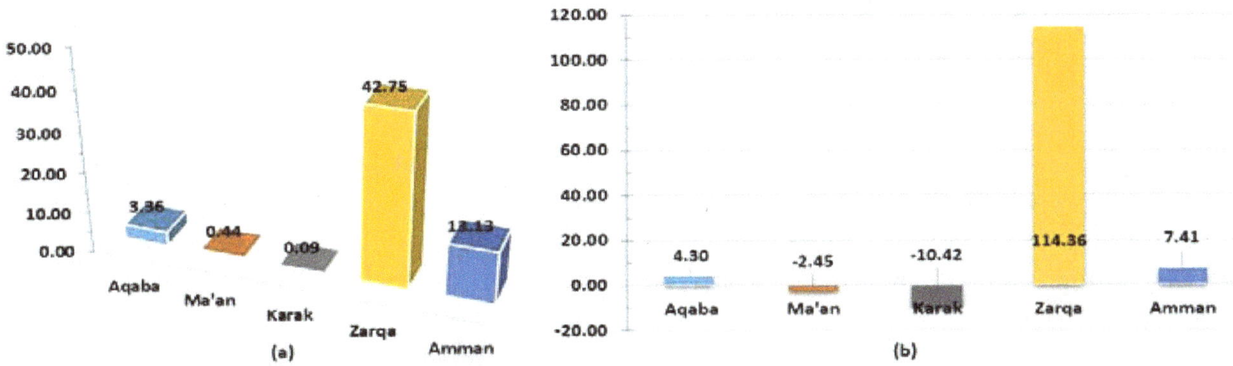


**Figure 6.** Reclaimed wastewater substitution in major industries in Jordanian governorates: (**a**) substitution factors, and (**b**) substitution amounts (million $m^3$ per year).

Figure 7 shows the responses of the interviewed industries (17 samples from those shown in Table 2). It is drastically indicated that low TDS (water salinity) is the major requirement that was requested by 35% of the responses. Interestingly, the sample responses showed willingness to accept to replace the groundwater with reclaimed wastewater.

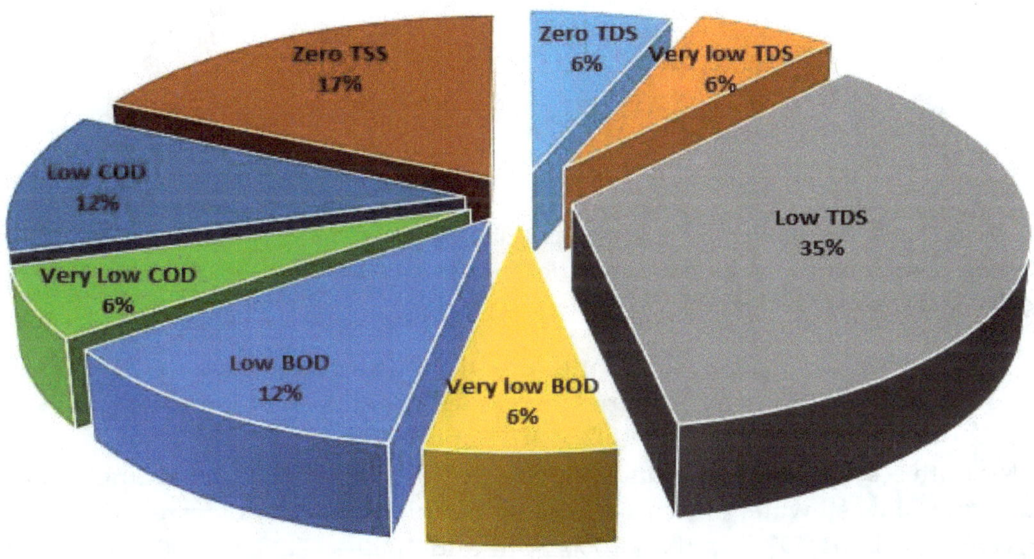


**Figure 7.** Responses of reclaimed wastewater quality requirements by the major industries.

However, 6% of the responses requested advanced treatment to receive very low values of TDS, biological oxygen demand (BOD), and chemical oxygen demand (COD). Zero total suspended solids (TSS) was requested by 17% of the responses, and this was mainly required for the cooling of power generators.

### 3.1.2. Environmental and Economic Benefits

Pintilie et al. (2016) studied the life cycle assessment (LCA) of substituting fresh water with treated wastewater obtained from tertiary treatment and concluded that it does not lead to a substantial improvement of environmental impact for most of the indicators [48]. However, only water depletion (WD) and climate change (CC) were considered in the present study to compare the environmental impact between reclaimed wastewater reuse and no reuse scenarios. WD is recommended for water-stressed situations because a net saving of water from nature represents the most important effect of water reuse. The WD indicator values proposed by Pintilie et al. (2016) were considered in the present assessment as the following: $5.74 \times 10^{-4}$ $m^3$ per $m^3$ entering the whole system for the no reuse scenario, and $-4.39 \times 10^{-1}$ $m^3$ per $m^3$ entering the whole system for the reclaimed wastewater reuse scenario [48]. Negative values mean benefits to the environment, and positive values mean damages. Accordingly, using the data in Table 1, the annual wastewater effluent amounts (mainly the total flowrates of grouped WWTPs (million $m^3$/year)) of both TDS less than 1000 ppm and 1000 < TDS < 1500 ppm were 17.93 and 147.33 million $m^3$ per year, respectively. The sum of them is 165.26 million $m^3$ per year, and using the aforementioned WD indicators, the analysis revealed that 94,860 $m^3$ of fresh water are depleted for the scenario of no-reuse of reclaimed wastewater; however, 72.55 million $m^3$ of water can be saved in reclaimed wastewater reuse in major industries in Jordan, as shown in Figure 8. Results of a similar tendency were founded in literature [48,56].

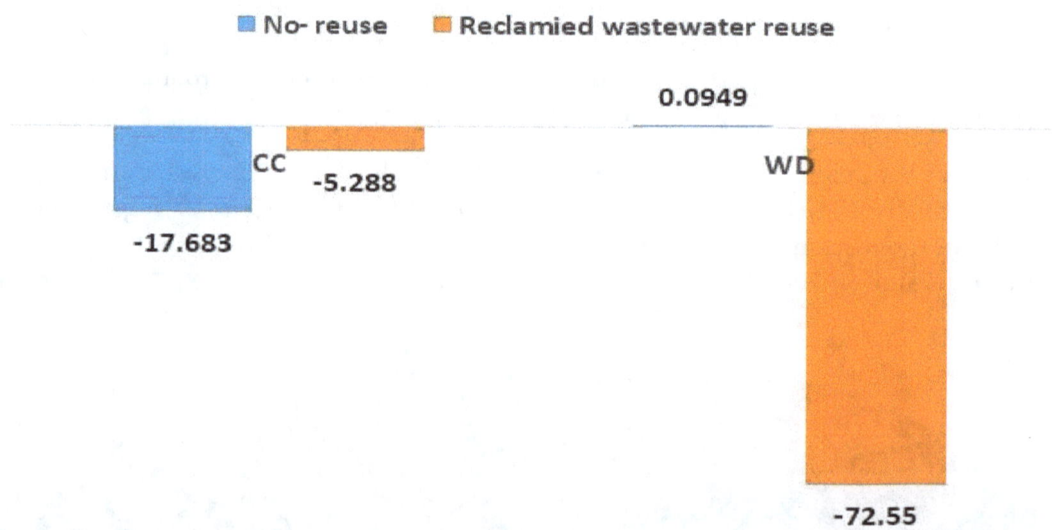


**Figure 8.** Water depletion (million $m^3$ of water) and climate change (million kg $CO_{2Eq}$) analysis for both reclaimed wastewater reuse and no reuse scenarios (negative values mean benefits to the environment, and positive values mean damages).

The CC indicator values proposed by Pintilie et al. (2016) were considered as stated above [48]. The CC indicators were with negative values (indicates benefits to the environment) according to Pintilie et al. (2016) are the following: $-1.07 \times 10^{-1}$ and $-3.20 \times 10^{-2}$ kg $CO_{2Eq}$ per $m^3$ reclaimed wastewater for both scenarios of reuse and no reuse, respectively [48]. Accordingly, using the data in Table 1, the annual wastewater effluent amounts (mainly the total flowrates of grouped WWTPs (million $m^3$/year)) of both TDS less than 1000 ppm and 1000 < TDS < 1500 ppm were 17.93 and 147.33 million $m^3$ per year, respectively. The sum of them is 165.26 million $m^3$ per year, and using the aforementioned CC indicators, as shown in Figure 8, both scenarios showed beneficial impacts (negative values) to the environment in terms of climate change impacts. The no reuse scenario has relatively higher benefits with 17.683 million kg $CO_{2Eq}$ reduction compared to a 5.288 million kg $CO_{2Eq}$ reduction for the reuse scenario.

Normally, several factors influence the reclaimed wastewater provision and exploitation as a substitute [57,58]. According to the economic analysis of wastewater reclamation in Jordan, the difference between water price and reclaimed wastewater price plays a vital role in the willingness of the industries to accept the reclaimed wastewater as substitute. Therefore, for the low TDS (<1000) reclaimed wastewater (Table 1), the average cost of one $m^3$ of reclaimed wastewater is estimated at

0.55 JOD (including the pipeline installation, pumping electricity, and operation and naintenance (O&M) costs), while the cost of fresh water is 1 JOD/$m^3$. In this case, the reclaimed wastewater is competitive to some extent with regard to its price advantage. Moreover, based on experts' estimation, the environmental value of groundwater saved in the groundwater aquifer is 1.5 JOD/$m^3$. Hence, the cost–benefit analysis of this case (water of TDS < 1000) is attractive for the consumer and the government.

While for reclaimed wastewater with TDS higher than 1000 ppm, a treatment is needed based on the application. Therefore, excluding the reuse of reclaimed in cement and concrete industries, the average cost of one $m^3$ of reclaimed wastewater is estimated at 2 JOD (including treatment, pipeline installation, pumping electricity, and O&M costs). It is worth mentioning that the long-distance pipelines from WWTPs to industrial zones and clusters, were the major cause for such costly per $m^3$ water cost, especially in southern Jordan clusters. In order to overcome the hesitance of industries to reuse reclaimed wastewater when advanced treatment is required, subsidies by way of discounted cost of water should be provided in addition to fund allocation for capital cost coverage when on-site treatment is needed, as well as policy reforms to enhance the financial sustainability of the water sector.

### *3.2. Energy Recovery from Wastewater Reclamation*

Wastewater treatment in WWTPs (mainly AS treatment process) requires around 0.38–2.74 kWh/$m^3$ in Jordan, as shown in Figure 9. Additionally, 0.95–1.25 kWh/$m^3$ is needed for wastewater as reported in literature [59,60]. The difference in energy use needed for wastewater reclamation and supply can be reduced by recovering organic energy during the wastewater treatment process [59]. Currently, only in the Al Samra WWTP, biogas production from sludge treatment is undertaken in Jordan. As shown in Figure 10, the two types of thickened sludge are mixed in two covered tanks of 98 $m^3$ volume before being pumped and introduced in seven anaerobic digesters of a capacity of 15,900 $m^3$ each. In the digesters, the sludge is mixed thoroughly by Cannon®mixers (Trevose, PA USA) using the recycled compressed biogas. The sludge stays for three weeks at 35 °C in the digesters. Heating is done by hot water recovered from the cooling of the engines in a shell-and-tube heat exchanger. Through hydro energy and biogas production, the Al Samra WWTP has a potential energy recovery of 95% of its needs, only 5% is drawn from the national grid. Moreover, 300,000 tons of $CO_2$ is saved per year through energy recovery and renewable energy utilization [61].

The introduction of anaerobic sludge digestion is generally expected to offset 25–50% of an aerobic wastewater treatment plant's energy needs [59,63,64], however, based on WWTP data gathered in Jordan, having anaerobic sludge digestion in the small- and medium-scale WWTPs ($<10 \times 10^4$ $m^3$/day) can potentially produce electricity that would equate to an offset of 0.11–0.53 kWh/$m^3$. Consequently, this may help in reducing the costs of reclaimed wastewater reuse with further treatment requirements mainly for reclaimed wastewater with TDS higher than 1000 ppm as stated before.

However, energy produced from anaerobic sludge digestion can be feasibly increased by co-digestion with kitchen or other organic wastes [65–69]. Currently, the co-digestion is only applied at a laboratory scale in Jordan. Al-Addous et. al. (2019) evaluated the potential biogas production from the co-digestion of municipal food waste and wastewater sludge at a refugee camp. Accordingly, a possible ratio to start with is 60–80% organic waste, which can produce 21–65 $m^3$ biogas $ton^{-1}$ of fresh matter [70].

Notwithstanding that co-digestion does not exist in Jordan yet, the anaerobic digestion systems tend to be well operated in Jordan (i.e., Al Samra WWTP). Hence, when co-digestion is utilized in Jordan, this will be a vital opportunity to make cost-effective use of existing facilities and improve sludge biogas potential [71–74].

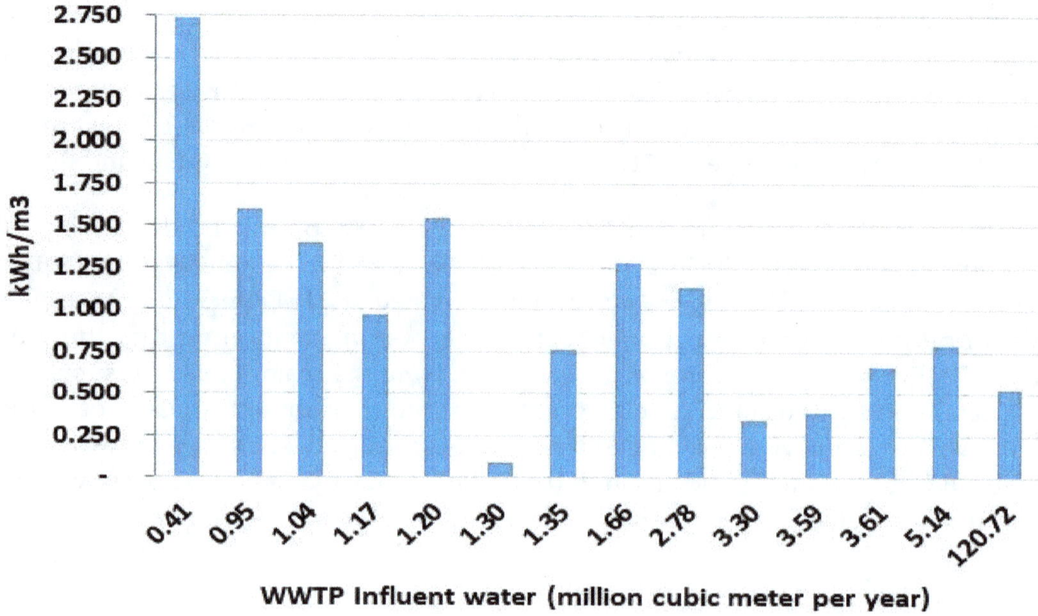


**Figure 9.** WWTPs' electricity consumption data for the activated sludge (AS) treatment process.

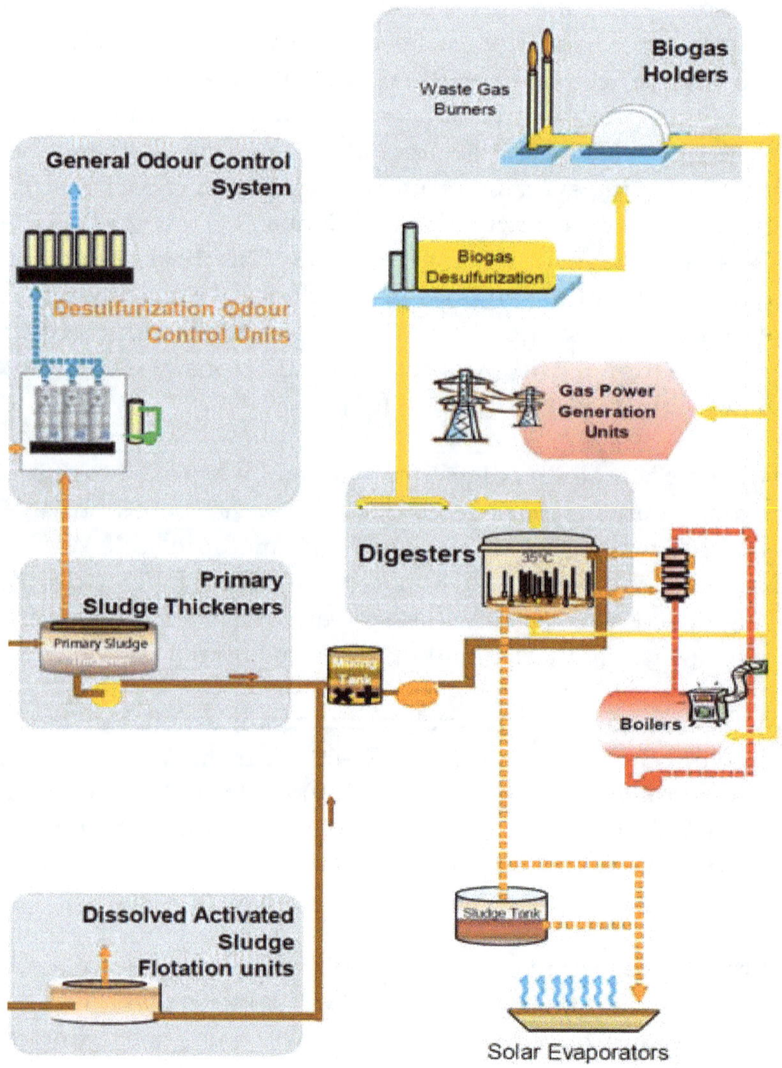


**Figure 10.** Biogas production in anaerobic digesters at the Al Samra WWTP [62].

### 3.3. *Reclaimed Wastewater Reuse: Barriers and Prospects*

#### 3.3.1. Reclaimed Wastewater Quality and Industrial Needs

Industrial uses of reclaimed wastewater come in many different ways such as cooling-water, processing, and boiler feed water. Therefore, the process water requirements for water quality vary depending on the industry. Some of the concerns for industrial use of reclaimed wastewater are corrosion, scaling, and biological growth; however, these concerns are applicable to potable water as well. Most of cooling water is treated already to address these concerns. For instance, corrosion is a concern in cooling water no matter whether the facility uses potable water or reclaimed wastewater. Scaling from dissolved minerals such as calcium, magnesium, and phosphates can be controlled by monitoring and chemically treating the water to prevent scaling. Magnesium–phosphorus precipitation from sludge and the recovery of struvite after anaerobic sludge treatment process is conducted in order to prevent clogging in pumps and pipes in any further reuse applications [75–78]. Biological concerns can be addressed by adding chlorine to levels of 2.0 mg/L that will kill most microorganisms that causes corrosion or deposits in cooling systems [79].

To facilitate the use of recycled water in industrial applications, the information on the quality of the municipal recycled water should be provided and available to the industrial users. Moreover, opportunities to improve water quality for specific purposes, either by the supplier through additional treatment and/or source control, or the industrial user can improve treatment and control processes to levels specific to its process needs.

#### 3.3.2. Reclamation of Wastewater Technologies

Lyu et al. (2016) discussed all advances in technology by which wastewater may be treated to meet the most stringent quality requirements and be used for any purposes desired [49]. For instance, the technologies applied in wastewater reuse include: (1) oxidants for disinfection purposes using sodium hypochlorite; ultraviolet radiation [80]; and ozone for high bactericidal disinfection [81] and the removal of 90–99% for antibiotics and estrogens [82]; (2) biological treatments such as anaerobic, maturation ponds and constructed wetlands [83–85]; (3) physical separations such as membrane filtration for 81% removal of electroconductivity, 83% for $Na^+$, and 80% for $Cl^-$ [86]; the removal of 95% of heavy metals [87]; the removal of >89% of pharmaceuticals [88]; (4) electrochemical treatments to completely remove *Escherichia coli* [89–91]; as well as, (5) solar photocatalysis with $TiO_2$ for >90% removal of emerging pollutants (i.e., pharmaceuticals and personal care products) [92], and removals of 33% for Cd and 75% for Co [93].

#### 3.3.3. Reclaimed Wastewater Supply Continuity

The industry demands a constant non-interrupted flow of reclaimed wastewater throughout the day [94]. In Jordan, although the reclaimed wastewater supply volumes vary diurnally and seasonally, its continuity is not so critical since WWTP effluents have relatively uninterruptable higher flows than the demand flows needed by the nearby main industries. It is noteworthy that flow equalization and water conveying capacities should be investigated to match the supplies with the demands and vice versa [95].

#### 3.3.4. Willingness to Participate and Willingness to Pay

As deduced based on the results of these interviews, most of main industries considered in the present study expressed a positive stance toward reclaimed wastewater reuse, while they are willing to pay a significantly less amount of money than they already pay, for freshwater. Therefore, a comprehensive survey about the willingness of the industrial sector to switch to the use of reclaimed wastewater instead of groundwater is of high significance. Such surveys will help in providing more accurate data for the financial evaluation of the recycled water service and a basis for negotiation with the industries.

Factors that influence industrial user's 'willingness to pay' for reclaimed wastewater include: (1) price of alternative water sources (i.e., potable, surface water, and groundwater supplies); (2) perception of the scarcity of alternative sources; (3) capital and operating costs of switching to reclaimed wastewater supply; (4) reclaimed wastewater quality, quantity, and levels of service and reliability of supply.

#### 3.3.5. Pricing Systems

A range of pricing systems for reclaimed wastewater can be proposed in Jordan and assessed on a win–win situation. The pricing systems can be employed alone or in combination [96], which are, but not limited to: (1) A usage fee scheme in which the industries finance the infrastructure installation, and then the usage charge offsets the supply cost of the reclaimed wastewater. For instance, such type of pricing was adopted in 2003 by the Australian government under the national water reform process [97]. (2) A connection fee which is a once-off contribution toward the cost of infrastructure needed to deliver reclaimed wastewater to the industry's delivery point. This fee may be subject to negotiation between the supplier and the industries to agree on a financial arrangement where both parties may fully or partially cover the fee of the actual work to deliver the reclaimed water to the delivery point. (3) A flat fee regardless of use ("take or pay" arrangement). For instance, regardless of actual use, the industries are obliged to pay for 75–100% of the contracted recycled water volume, and for all water consumed by the industries above the contracted level. Although this pricing scheme provides the WWTPs with guaranteed income that sustains the financials of running the scheme, it may encourage overuse of reclaimed wastewater by the industry and improper discharges to the environment.

#### 3.3.6. Reclaimed Wastewater Agreements

Specific reclaimed wastewater guidelines are important in managing the supply and use of reclaimed wastewater particularly in relation to quantity and quality [98]. Through the agreement negotiations between the supplier of reclaimed wastewater and the customers (i.e., industries). Wherein, the parties agree to a set of obligations and responsibilities under which the reclaimed wastewater reuse scheme will operate [99]. Key issues that reclaimed wastewater agreements should cover include: (a) price, quantity, and quality of reclaimed wastewater; (b) security of the reclaimed wastewater supply; (c) measures to identify, allocate, and manage risks and ensure safe use of reclaimed wastewater; (d) liabilities and insurance for potential damages caused by supply and use; and (e) compliance with legislative and common law requirements.

## 4. Conclusions

The following findings can be concluded in the present study:

- Jordan is classified as a semi-arid to arid country and is ranked among the poorest countries in the world in terms of water availability. Therefore, reclaimed wastewater reuse has been driven as an alternative water supply in such looming challenges of water scarcity. For instance, a total of 26 million $m^3$ of groundwater abstraction is exploited annually for industrial purposes.
- In the present study, the 34 processes in WWTPs in Jordan were assessed in terms of their treatment processes, scale, and effluent TDS. The most widely used technologies are AS (60%) and WSP (19%), while the TF and AS process, MBR and TF process, and OS processes were had an even use share of 6% each. Moreover, 30 WWTPs were classified as small scale ($<1 \times 10^4$ $m^3$/day), which were generally built in medium- and small-size cities and refugee camps. Moreover, the analysis showed that 17.932 million $m^3$ of treated wastewater has low TDS $< 1000$ ppm and can be reused several times in most industrial applications, especially in thermal units, cooling towers, etc. However, highest annual effluents flow rate of 147.323 million $m^3$ in total out of 18 WWTPs distributed in widely different locations in Jordan have $1000 < \text{TDS} < 1500$, which can be used with medium cost depending on the fit-for-purpose water criteria.

- Full substitution of industrial demand by reclaimed wastewater reuse can be achieved in both Amman and Aqaba governorates with 13.13- and 3.36-fold, respectively. However, the shortage of industrial demand substitution by reclaimed wastewater is significantly clear in both of Ma'an and Karak governorates with substitution amounts of 2.45 and 10.4 million $m^3$ per year, respectively.
- The environmental assessment showed positive impacts of reclaimed wastewater reuse scenario in terms of water depletion (saving of 72.55 million $m^3$ of groundwater per year) and climate change (17.683 million kg $CO_{2Eq}$ reduction).
- From circular economic perspective, and based on WWTP data gathered in Jordan, having anaerobic sludge digestion in the small- and medium-scale WWTPs ($<10 \times 10^4$ $m^3$/day) can potentially produce electricity that would equate to an offset of 0.11–0.53 kWh/$m^3$. Consequently, this may help in reducing the costs of reclaimed wastewater reuse with further treatment requirements mainly for reclaimed wastewater with TDS higher than 1000 ppm as stated before.
- It is recommended in the present study that reclaimed wastewater agreement negotiations should be promoted between the supplier of reclaimed wastewater and the customers (i.e., industries). Moreover, indicators such as willingness to participate and willingness to pay need to be significantly determined in order to reach a win–win scheme of reclaimed water pricing model.

**Author Contributions:** Conceptualization, M.N.S. and I.O.; Methodology, M.N.S., I.O., and R.A.A.-W.; Formal analysis, M.N.S., M.A., M.A.-A., N.B., and I.O.; Investigation, M.N.S., R.A.A.-W., and I.O.; Resources, M.N.S. and I.O.; Writing—original draft preparation, M.N.S., M.A., M.A.-A., and I.O.; Writing—review and editing, M.N.S. and R.A.A.-W.; Project administration, M.N.S., N.B. and I.O. All authors have read and agreed to the published version of the manuscript.

**Acknowledgments:** The corresponding author does acknowledge with gratitude the University of Jordan for granting him sabbatical assistance to undertake this research study. The authors also thank the anonymous reviewers and the editor of the journal for their constructive feedback.

## References

1. Forslund, A.; Malm Renöfält, B.; Barchiesi, S.; Cross, K.; Davidson, S.; Ferrel, T.; Korsgaard, L.; Krchnak, K.; McClain, M.; Meijer, K.; et al. *Securing Water for Ecosystems and Human Well-being: The Importance of Environmental Flows (Swedish Water House Report 24)*; Swedish International Water Institute: Stockholm, Sweden, 2009.
2. Roccaro, P.; Verlicchi, P. Wastewater and reuse. *Curr. Opin. Environ. Sci. Health* **2018**, *2*, 61–63. [CrossRef]
3. Sun, Y.; Chen, Z.; Wu, G.; Wu, Q.; Zhang, F.; Niu, Z.; Hu, H.Y. Characteristics of water quality of municipal wastewater treatment plants in China: Implications for resources utilization and management. *J. Clean. Prod.* **2016**, *131*, 1–9. [CrossRef]
4. Shannon, M.; Bohn, P.W.; Elimelech, M.; Georgiadis, J.G.; Mariñas, B.J.; Mayes, A.M. Science and technology for water purification in the coming decades. *Nature* **2008**, *452*, 301–310. [CrossRef] [PubMed]
5. Akhoundi, A.; Nazif, S. Sustainability assessment of wastewater reuse alternatives using the evidential reasoning approach. *J. Clean. Prod.* **2018**, *195*, 1350–1376. [CrossRef]
6. Chen, W.; Geng, Y.; Dong, H.; Tian, X.; Zhong, S.; Wu, Q.; Li, S. An emergy accounting based regional sustainability evaluation: A case of Qinghai in China. *Ecol. Indic.* **2018**, *88*, 152–160. [CrossRef]
7. Yan, H.; Wu, F.; Dong, L. Influence of a large urban park on the local urban thermal environment. *Sci. Total Environ.* **2018**, *622*, 882–891. [CrossRef]
8. Zhou, Z.; Zhang, X.; Dong, W. Fuzzy comprehensive evaluation for safety guarantee system of reclaimed water quality. *Procedia Environ. Sci.* **2013**, *18*, 227–235. [CrossRef]
9. Ait-Mouheb, N.; Bahri, A.; Thayer, B.B.; Benyahia, B.; Bourrié, G.; Cherki, B.; Condom, N.; Declercq, R.; Gunes, A.; Héran, M.; et al. The reuse of reclaimed water for irrigation around the Mediterranean Rim: A step towards a more virtuous cycle? *Reg. Environ. Change* **2018**, *8*, 693–705. [CrossRef]
10. Licciardello, F.; Milani, M.; Consoli, S.; Pappalardo, N.; Barbagallo, S.; Cirelli, G. Wastewater tertiary treatment options to match reuse standards in agriculture. *Agric. Water Manag.* **2018**, *210*, 232–242. [CrossRef]

11. Collet, L.; Denis, R.; Borrell Estupina, V.; Dezetter, A.; Servat, E. Water supply sustainability and adaptation strategies under anthropogenic and climatic changes of a meso-scale Mediterranean catchment. *Sci. Total Environ.* **2015**, *536*, 589–602. [CrossRef]
12. La Jeunesse, I.; Cirelli, C.; Aubin, D.; Larrue, C.; Sellami, H.; Afifi, S.; Bellin, A.; Benabdallah, S.; Bird, D.N.; Deidda, R.; et al. Is climate change a threat for water uses in the Mediterranean region? Results from a survey at local scale. *Sci. Total Environ.* **2016**, *543*, 981–996. [CrossRef] [PubMed]
13. European Environmental Agency. *Climate Change, Impacts and Vulnerably in Europe—An Indicator-Based Report*; European Environmental Agency (2012) EEA report no 12/2012; European Environmental Agency: København, Denmark, 2012.
14. Voulvoulis, K. Water reuse from a circular economy perspective and potential risks from an unregulated approach. *Curr. Opin. Environ. Sci. Health* **2018**, *2*, 32–45. [CrossRef]
15. Coats, E.R.; Wilson, P.I. Toward nucleating the concept of the water resource recovery facility (WRRF): Perspective from the principal actors. *Environ. Sci. Technol.* **2017**, *51*, 4158–4164. [CrossRef]
16. Sgroi, M.; Vagliasindi, F.G.A.; Roccaro, P. Feasibility, sustainability and circular economy concepts in water reuse. *Curr. Opin. Environ. Sci. Health* **2018**, *2*, 20–25. [CrossRef]
17. Matouq, M.; El Hasan, T.; Al Bilbisi, H.; Abdelhadi, M.; Hindiyeh, M.; Eslaiman, S.; Duheisat, S. The climate change implication on Jordan: A case study using GIS and artificial neural networks for weather casting. *J. Sci. Taibah Univ.* **2013**, *7*, 44–55. [CrossRef]
18. Ministry of Water and Irrigation. *Jordan's Water Strategy 2016–2025*; Ministry of Water and Irrigation: Amman, Jordan, 2016.
19. Aboelnga, H.; Saidan, M.; Al-Weshah, R.; Sturm, M.; Ribbe, L.; Frechen, F.B. Component analysis for optimal leakage management in Madaba, Jordan. *J. Water Supply Res. Technol.* **2018**, *67*, 384–396. [CrossRef]
20. Saidan, M.N.; Al-Weshah, R.A.; Obada, I. Potential Rainwater Harvesting: Adaptation Measure for Urban Areas in Jordan. *Am. Water Works Assoc.* **2015**, *107*, 594–602. [CrossRef]
21. Al-Weshah, R.; Saidan, M.; Al-Omari, A. Environmental ethics as a tool for sustainable water resource management. *Am. Water Works Assoc.* **2016**, *108*, 175–181. [CrossRef]
22. Al-Awad, T.K.; Saidan, M.N.; Gareau, B.J. Halon management and ozone-depleting substances control in Jordan. *Int. Environ. Agreem.* **2018**, *18*, 391–408. [CrossRef]
23. Aldayyat, E.; Saidan, M.N.; Abu Saleh, M.A.; Hamdan, S.; Linton, C. Solid Waste Management in Jordan: Impacts and Analysis. *J. Chem. Technol. Metall.* **2019**, *54*, 454–462.
24. Alrabie, K.; Saidan, M.N. A preliminary solar-hydrogen system for Jordan: Impacts assessment and scenarios analysis. *Int. J. Hydrog. Energy* **2018**, *43*, 9211–9223. [CrossRef]
25. Khasawneh, H.; Saidan, M.; Al-Addous, M. Utilization of hydrogen as clean energy resource in chlor-alkali process. *Energy Explor. Exploit.* **2019**, *37*, 1053–1072. [CrossRef]
26. Saidan, M. Cross-sectional survey of non-hazardous waste composition and quantities in industrial sector and potential recycling in Jordan. *Environ. Nanotech. Monitor. Manag.* **2019**, *12*, 100227. [CrossRef]
27. Al-Addous, M.; Saidan, M.; Bdour, M.; Dalalah, Z.; Albatayneh, A.; Class, C.B. Key aspects and feasibility assessment of a proposed wind farm in Jordan. *Int. J. Low-Carbon Technol.* **2019**. [CrossRef]
28. Saidan, M.N.; Abu Drais, A.; Linton, C.; Hamdan, S. Solid waste Characterization and recycling in Syrian refugees hosting communities in Jordan. In *Waste Management in MENA Regions*, 1st ed.; Negm, A.M., Noama, E., Eds.; Earth and Environmental Sciences Series, Applied Environmental Science and Engineering for a Sustainable Future; Springer International Publishing AG, Part of Springer Nature: Cham, Switzerland, 2019.
29. Saidan, M.N.; Ansour, L.M.; Saidan, H. Management of Plastic Bags Waste: An Assessment of Scenarios in Jordan. *J. Chem. Technol. Metall.* **2017**, *52*, 148–154.
30. Hararah, M.A.; Saidan, M.N.; Alhamamre, Z.; Abu-Jrai, A.M.; Alsawair, J.; Damra, R.A. The PCDD/PCDF emission inventory in Jordan: Aqaba city. *J. Chem. Technol. Metall.* **2016**, *51*, 112–120.
31. Saidan, M.; Tarawneh, A. Estimation of potential E-waste generation in Jordan. *Ekoloji* **2015**, *24*, 60–64. [CrossRef]
32. Tarawneh, A.; Saidan, M. Households awareness, behaviors, and willingness to participate in E-waste management in Jordan. *Int. J. Ecosys* **2013**, *3*, 124–131.
33. Saidan, M.; Khasawneh, H.J.; Tayyem, M.; Hawari, M. Getting energy from poultry waste in Jordan: Cleaner production approach. *J. Chem. Technol. Metall.* **2017**, *52*, 595–601.

34. Hindiyeh, M.; Altalafha, T.; Al-Naerat, M.; Saidan, H.; Al-Salaymeh, A.; Sbeinati, L.A.; Saidan, M.N. Process modification of pharmaceutical tablet manufacturing operations: An Ecoefficiency approach. *Processes* **2018**, *6*, 15. [CrossRef]
35. METAP-World Bank. Water Quality Management Jordan—World Bank. Available online: http://siteresources.worldbank.org/EXTMETAP/Resources/WQM.JordanP.pdf (accessed on 2 October 2019).
36. Salahat, M.A.; Al-Qinna, M.I.; Badran, R.A. Potential of Treated Wastewater Usage for Adaptation to Climate Change: Jordan as a Success Story. In *Water Resources in Arid Areas: The Way Forward. Springer Water*; Abdalla, O., Kacimov, A., Chen, M., Al-Maktoumi, A., Al-Hosni, T., Clark, I., Eds.; Springer: Cham, Switzerland, 2017.
37. Carr, G.; Potter, R.B.; Nortcliff, S. Water reuse for irrigation in Jordan: Perceptions of water quality among farmers. *Agric. Water Manag.* **2011**, *98*, 847–854. [CrossRef]
38. Scott, C.A.; El-Naser, H.; Hagan, R.E.; Hijazi, A. Facing Water Scarcity in Jordan Reuse, Demand Reduction, Energy, and Transboundary Approaches to Assure Future Water Supplies. *Water Int.* **2003**, *28*, 209–216. [CrossRef]
39. The World Bank. *Water Reuse in The Arab World, from Principle to Practice, A Summary of Proceedings, Expert Consultation Wastewater Management in the Arab World 22–24 May 2011*; The World Bank: Dubai, UAE, 2011.
40. Hussain, M.I.; Muscolo, A.; Farooq, M.; Ahmad, W. Sustainable use and management of non-conventional water resources for rehabilitation of marginal lands in arid and semiarid environments. *Agric. Water Manag.* **2019**, *221*, 462–476. [CrossRef]
41. Mateo-Sagasta, J.; Medlicott, K.; Qadir, M.; Raschid-Sally, L.; Drechsel, P.; Liebe, J. *UN-Water Project on the Safe Use of Wastewater in Agriculture*; Jens Liebe, J., Ardakanian, K., Eds.; UN: Now York, NY, USA, 2013.
42. Hussein, I.; Abu-Sharar, T.M.; Al-Jayyousi, O. The use of treated wastewater for irrigation in Jordan: Opportunities and constraints. *Chart. Inst. Water Environ. J.* **2003**, *17*, 232–238.
43. Qadir, M.; Bahri, A.; Sato, T.; Al-Karadsheh, E. Wastewater production, treatment, and irrigation in Middle East and North Africa. *Irrig. Drain. Syst.* **2010**, *24*, 37–51. [CrossRef]
44. Maldonado, P.M.A. Water Resources Assessment under Semi-arid Conditions—Modelling Applications in a Complex Surface-groundwater System in Jordan. Ph.D. Thesis, KIT-Bibliothek, Karlsruhe, Germany, 2017.
45. Mohsen, M.; Jaber, J.O. Potential of industrial wastewater reuse. *Desalination* **2002**, *152*, 281–289. [CrossRef]
46. Schyns, J.F.; Hamaideh, A.; Hoekstra, A.Y.; Mekonnen, M.M.; Schyns, M. Mitigating the risk of extreme water scarcity and dependency: The case of Jordan. *Water* **2015**, *7*, 5705–5730. [CrossRef]
47. SIWI: World Water Week. *Water and Waste: Reduce and Reuse*; SIWI: Stockholm, Sweden, 2017.
48. Pintilie, L.; Torres, C.M.; Teodosiu, C.; Castells, F. Urban wastewater reclamation for industrial reuse: An LCA case study. *J. Clean. Prod.* **2016**, *139*, 1–14. [CrossRef]
49. Lyu, S.; Chen, W.; Zhang, W.; Fan, Y.; Jiao, W. Wastewater reclamation and reuse in China: Opportunities and challenges. *J. Environ. Sci.* **2016**, *39*, 86–96. [CrossRef]
50. Mo, W.; Zhang, Q. Energy-nutrients-water-nexus: Integrated resource recovery in municipal wastewater treatment plants. *J. Environ. Manag.* **2013**, *127*, 255–267. [CrossRef]
51. Husnain, T.; Liu, Y.; Riffat, R.; Mi, B. Integration of forward osmosis and membrane distillation for sustainable wastewater reuse. *Sep. Purif. Technol.* **2015**, *156*, 424–431. [CrossRef]
52. Geissen, V.; Mol, H.; Klumpp, E.; Umlauf, G.; Nadal, M.; van der Ploeg, M.; van de Zee, S.E.A.T.M.; Ritsema, C.J. Emerging pollutants in the environment: A challenge for water resource management. *Int. Soil Water Conserv. Res.* **2015**, *3*, 57–65. [CrossRef]
53. USAID. *National Strategic Wastewater Master Plan for Jordan*; USAID-ISSP: Washington, DC, USA, 2013.
54. Hussein, H. Tomatoes, tribes, bananas, and businessmen: An analysis of the shadow state and of the politics of water in Jordan. *Environ. Sci. Policy* **2018**, *84*, 170–176. [CrossRef]
55. Water Authority of Jordan. *Groundwater Basins Directorate Open Files*; Water Authority of Jordan: Amman, Jordan, 2016.
56. Tong, L.; Liu, X.; Liu, X.; Zengwei, Y.; Zhang, Q. Life cycle assessment of water reuse systems in an industrial park. *J. Environ. Manag.* **2013**, *129*, 471–478. [CrossRef]
57. Chang, D.; Ma, Z. Wastewater reclamation and reuse in Beijing: Influence factors and policy implications. *Desalination* **2012**, *297*, 72–78. [CrossRef]
58. Yang, H.; Abbaspour, H.C. Analysis of wastewater reuse potential in Beijing. *Desalination* **2007**, *212*, 238–250. [CrossRef]

59. Smith, K.; Liu, S.; Hu, H.Y.; Dong, X.; Wen, X. Water and energy recovery: The future of wastewater in China. *Sci. Total Environ.* **2018**, *637*, 1466–1470. [CrossRef]
60. Li, Y.; Xiong, W.; Zhang, W.; Wang, C.; Wang, P. Life cycle assessment of water supply alternatives in water-receiving areas of the South-to-NorthWater Diversion Project in China. *Water Res.* **2016**, *89*, 9–19. [CrossRef]
61. Anne DE PAZZIS. As Samra Wastewater Treatment Plant A Major Asset for Sustainability in Jordan. Presented at the Stockholm World Water Week, Stockholm, Sweden, 31 August 2017.
62. SAMRA WASTEWATER TREATMENT PLANT JORDAN. Available online: http://www.fogevaporator.com/pdf/samra-wastewater-treatment-plant.pdf (accessed on 18 November 2019).
63. McCarty, P.L.; Bae, J.; Kim, J. Domestic wastewater treatment as a net energy producer–can this be achieved? *Environ. Sci. Technol.* **2011**, *45*, 7100–7106. [CrossRef]
64. Saidan, M.; Khasawneh, H.J.; Aboelnga, H.; Meric, S.; Kalavrouziotis, I.; Hayek, B.O.; Al-Momany, S.; Al Malla, M.; Porro, J.C. Baseline carbon emission assessment in water utilities in Jordan using ECAM tool. Baseline carbon emission assessment in water utilities in Jordan using ECAM tool. *J. Water Supply Res. Technol.* **2019**, *68*, 460–473. [CrossRef]
65. Mehariya, S.; Patel, A.K.; Obulisamy, P.K.; Punniyakotti, E.; Wong, J.W.C. Co-digestion of food waste and sewage sludge for methane production: Current status and perspective. *Bioresour. Technol.* **2018**, *265*, 519–531. [CrossRef] [PubMed]
66. Agyeman, F.O.; Tao, W. Anaerobic co-digestion of food waste and dairy manure: Effects of food waste particle size and organic loading rate. *J. Environ. Manag.* **2014**, *133*, 268–274. [CrossRef]
67. Appels, L.; Lauwers, J.; Degrève, J.; Helsen, L.; Lievens, B.; Willems, K.; Van Impe, J.; Dewil, R. Anaerobic digestion in global bio-energy production: Potential and research challenges. *Renew. Sustain. Energy Rev.* **2011**, *15*, 4295–4301. [CrossRef]
68. Awasthi, M.K.; Pandey, A.K.; Bundela, P.S.; Wong, J.W.C.; Li, R.; Zhang, Z. Co-composting of gelatin industry sludge combined with organic fraction of municipal solid waste and poultry waste employing zeolite mixed with enriched nitrifying bacterial consortium. *Bioresour. Technol.* **2016**, *213*, 181–189. [CrossRef] [PubMed]
69. Chiu, S.L.H.; Lo, I.M.C. Reviewing the anaerobic digestion and co-digestion process of food waste from the perspectives on biogas production performance and environmental impacts. *Environ. Sci. Pollut. Res.* **2016**, *23*, 24435–24450. [CrossRef]
70. Al-Addous, M.; Saidan, M.N.; Bdour, M.; Alnaief, M. Evaluation of Biogas Production from the Co-Digestion of Municipal Food Waste and Wastewater Sludge at Refugee Camps Using an Automated Methane Potential Test System. *Energies* **2019**, *12*, 32. [CrossRef]
71. Dai, X.; Li, X.; Zhang, D.; Chen, Y.; Dai, L. Simultaneous enhancement of methane production and methane content in biogas from waste activated sludge and perennial ryegrass anaerobic co-digestion: The effects of pH and C/N ratio. *Bioresour. Technol.* **2016**, *216*, 323–330. [CrossRef]
72. Wan, J.; Gu, J.; Zhao, Q.; Liu, Y. COD capture: A feasible option towards energy self sufficient domestic wastewater treatment. *Sci. Rep.* **2016**, *6*, 25054. [CrossRef]
73. Gu, Y.; Li, Y.; Li, X.; Luo, P.; Wang, H.; Robinson, Z.P.; Wang, X.; Wu, J.; Li, F. The feasibility and challenges of energy self-sufficient wastewater treatment plants. *Appl. Energy* **2017**, *204*, 1463–1475. [CrossRef]
74. Hagos, K.; Zong, J.; Li, D.; Liu, C.; Lu, X. Anaerobic co-digestion process for biogas production: Progress, challenges and perspectives. *Renew. Sustain. Energy Rev.* **2017**, *76*, 1485–1496. [CrossRef]
75. Ronteltap, M.; Maurer, M.; Gujer, W. The behaviour of pharmaceuticals and heavy metals during struvite precipitation in urine. *Water Res.* **2007**, *41*, 1859–1868. [CrossRef] [PubMed]
76. Uysal, A.; Yilmazel, Y.D.; Demirer, G. The determination of fertilizer quality of the formed struvite from effluent of a sewage sludge anaerobic digester. *J. Hazard. Mater.* **2010**, *181*, 248–254. [CrossRef]
77. Antakyali, D.; Kuch, B.; Preyl, V.; Steinmetz, H. Effect of Micropollutants in Wastewater on Recovered Struvite, Proc. *Water Environ. Fed. Nutr. Recovery Manag.* **2011**, *1*, 575–582. [CrossRef]
78. Jabr, G.; Saidan, M.; Al-Hmoud, N. Phosphorus Recovery by Struvite Formation from Al Samra Municipal Wastewater Treatment Plant in Jordan. *Desalin. Water Treat.* **2019**. [CrossRef]
79. EPA. *Guidelines for Water Reuse, U.S.*; Environmental Protection Agency: Washington, DC, USA, 2004.
80. Saidan, M.N.; Al-Yazjeen, H.; Abdalla, A.; Khasawneh, H.J.; Al-Naimat, H.; Al Alami, N.; Adawy, M.; Jaber, M.S.; Sowan, N. Assessment of on-site treatment process of institutional building's wastewater. *Processes* **2018**, *6*, 26. [CrossRef]

81. Bixio, D.; Wintgens, T. *Water Reuse System Management—Manual AQUAREC. Directorate-General for Research*; European Commission: Brussels, Belgium, 2006.
82. Huber, M.M.; Gobel, A.; Joss, A.; Hermann, N.; Loffler, D.; McArdell, C.S.; Ried, A.; Siegrist, H.; Ternes, T.A.; von Gunten, U. Oxidation of pharmaceuticals during ozonation of municipal wastewater effluents: A pilot study. *Environ. Sci. Technol.* **2005**, *39*, 4290–4299. [CrossRef] [PubMed]
83. Feigin, A.; Ravina, I.; Shalhevet, J. *Irrigation with Treated Sewage Effluent*, 1st ed.; Springer: Berlin/Heidelberg, Germany, 1991.
84. Brissaud, F. Low technology systems for wastewater treatment: Perspectives. *Water Sci. Technol.* **2007**, *55*, 1–9. [CrossRef] [PubMed]
85. Ghermandi, A.; Bixio, D.; Traverso, P.; Cersosimo, I.; Thoeye, C. The removal of pathogens in surface-flow constructed wetlands and its implications for water reuse. *Water Sci. Technol.* **2007**, *56*, 207–216. [CrossRef] [PubMed]
86. Oron, G.; Gillerman, L.; Buriakovsky, N.; Bick, A.; Gargir, M.; Dolan, Y.; Manor, Y.; Katz, L.; Hagin, J. Membrane technology for advanced wastewater reclamation for sustainable agriculture production. *Desalination* **2008**, *218*, 170–180. [CrossRef]
87. Hyun, K.S.; Lee, S.J. Biofilm/membrane filtration for reclamation and reuse of rural wastewaters. *Water Sci. Technol.* **2009**, *59*, 2145–2152. [CrossRef]
88. Xu, P.; Drewes, J.E.; Bellona, C.; Amy, G.; Kim, T.U.; Adam, M.; Heberer, T. Rejection of emerging organic micropollutants in nanofiltration-reverse osmosis membrane applications. *Water Environ. Res.* **2005**, *77*, 40–48. [CrossRef] [PubMed]
89. Norton-Brandao, D.; Scherrenberg, S.M.; van Lier, J.B. Reclamation of used urban waters for irrigation purposes—A review of treatment technologies. *J. Environ. Manag.* **2013**, *122*, 85–98. [CrossRef] [PubMed]
90. Cano, A.; Cañizares, P.; Barrera, C.; Sáez, C.; Rodrigo, M.A. Use of low current densities in electrolyses with conductive-diamond electrochemical e oxidation to disinfect treated wastewaters for reuse. *Electrochem. Commun.* **2011**, *13*, 1268–1270. [CrossRef]
91. Wei, X.Z.; Gu, P.; Zhang, G.H. Research progress in the new model of pollutants in water and their removing methods. *Ind. Water Treat.* **2014**, *34*, 8–12.
92. Bernabeu, A.; Vercher, R.F.; Santos-Juanes, L.; Simon, P.J.; Lardin, C.; Martinez, M.A.; Vicente, J.A.; Gonzalez, R.; llosa, C.; Arques, A.; et al. Solar photocatalysis as a tertiary treatment to remove emerging pollutants from wastewater treatment plant effluents. *Catal. Today* **2011**, *161*, 235–240. [CrossRef]
93. Pedrero, F.; Albuquerque, A.; Amado, L.; Marecos do Monte, H.; Alarcón, J. Analysis of the reclamation treatment capability of a constructed wetland for reuse. *Water Pract. Technol.* **2011**, *6*, wpt2011050. [CrossRef]
94. Cirelli, G.L.; Consoli, S.; Di Grande, V. Long-term storage of reclaimed water: The case studies in Sicily (Italy). *Desalination* **2008**, *218*, 62–73. [CrossRef]
95. Yi, L.; Jiao, W.; Chen, X.; Chen, W. An overview of reclaimed water reuse in China. *J. Environ. Sci.* **2011**, *23*, 1585–1593. [CrossRef]
96. Waterwise Queensland. Manual for Recycled Water Agreements in Queensland. The State of Queensland. Environmental Protection Agency. 2005. Available online: http://www.epa.qld.gov.au/waterrecyclingagreement (accessed on 12 November 2019).
97. COAG. *Communique August*; Council of Australian Governments: Canberra, Australia, 2003.
98. Di Carlo, A.; Sherman, A. The Reclaimed Water Agreement Manual, Russell Kennedy Solicitors, Melbourne. 2004. Available online: http://www.clw.csiro.au/priorities/urban/awcrrp/stage1files/AWCRRP_8%20Final_27Apr2004.pdf (accessed on 12 November 2019).
99. Moore, W.; Gould, J. Shoalhaven Reclaimed Water Management Scheme—How a small region built a large water recycling scheme. In Proceedings of the Water Recycling Australia, 2nd National Conference, Brisbane, Australia, 1–2 September 2003.

# 12

# Degradation of Trace Organic Contaminants by a Membrane Distillation—Enzymatic Bioreactor

**Muhammad B. Asif [1], Faisal I. Hai [1,*], Jinguo Kang [1,2], Jason P. van de Merwe [3], Frederic D. L. Leusch [3], Kazuo Yamamoto [4], William E. Price [2] and Long D. Nghiem [1]**

1 Strategic Water Infrastructure Lab, School of Civil, Mining and Environmental Engineering, University of Wollongong, Wollongong NSW 2522, Australia; mba409@uowmail.edu.au (M.B.A.); jkang@uow.edu.au (J.K.); longn@uow.edu.au (L.D.N.)

2 Strategic Water Infrastructure Lab, School of Chemistry, University of Wollongong, Wollongong NSW 2522, Australia; wprice@uow.edu.au

3 Australian Rivers Institute and Griffith School of Environment, Griffith University, Gold Coast QLD 4222, Australia; j.vandemerwe@griffith.edu.au (J.P.v.d.M.); f.leusch@griffith.edu.au (F.D.L.L.)

4 Environmental Science Centre, Department of Urban Engineering, University of Tokyo, Tokyo 113-0033, Japan; yamamoto@esc.u-tokyo.ac.jp

* Correspondence: faisal@uow.edu.au.

**Abstract:** A high retention enzymatic bioreactor was developed by coupling membrane distillation with an enzymatic bioreactor (MD-EMBR) to investigate the degradation of 13 phenolic and 17 non-phenolic trace organic contaminants (TrOCs). TrOCs were effectively retained (90–99%) by the MD membrane. Furthermore, significant laccase-catalyzed degradation (80–99%) was achieved for 10 phenolic and 3 non-phenolic TrOCs that contain strong electron donating functional groups. For the remaining TrOCs, enzymatic degradation ranged from 40 to 65%. This is still higher than those reported for enzymatic bioreactors equipped with ultrafiltration membranes, which retained laccase but not the TrOCs. Addition of three redox-mediators, namely syringaldehyde (SA), violuric acid (VA) and 1-hydroxybenzotriazole (HBT), in the MD-EMBR significantly broadened the spectrum of efficiently degraded TrOCs. Among the tested redox-mediators, VA (0.5 mM) was the most efficient and versatile mediator for enhanced TrOC degradation. The final effluent (i.e., membrane permeate) toxicity was below the detection limit, although there was a mediator-specific increase in toxicity of the bioreactor media.

**Keywords:** enzymatic membrane bioreactor (EMBR); laccase; membrane distillation; redox-mediators; trace organic contaminants (TrOCs)

---

## 1. Introduction

Laccase (EC 1.10.3.2), a copper-containing oxidoreductase enzyme, has been studied extensively for the degradation of recalcitrant compounds such as phenols and aromatic hydrocarbons [1–5]. In recent years, laccase-catalyzed degradation of trace organic contaminants (TrOCs) such as pharmaceuticals, pesticides, personal care products, industrial chemicals and steroid hormones has gained significant attention [6,7]. These TrOCs occur ubiquitously in municipal wastewater and have the potential to adversely affect aquatic ecosystems and human health [8–10].

TrOC degradation by laccase depends on a number of factors including pH, temperature, chemical structure of TrOCs and laccase properties [11–13]. In general, effective laccase-catalyzed degradation of TrOCs containing electron donating functional groups (EDGs) such as amine ($-NH_2$), alkoxy (–OR) or hydroxyl (–OH) was observed. On the other hand, degradation of TrOCs containing electron withdrawing functional groups (EWGs) such as halogen (–X), amide ($-CONR_2$) or nitro

($-NO_2$) has been reported to be poor or unstable [11,14]. Degradation of TrOCs can be improved by adding different natural and synthetic redox-mediators that are low molecular weight compounds capable of exchanging electrons between laccase and TrOCs [15–17].

Initial studies have assessed the performance of laccase-catalyzed TrOC degradation in batch enzymatic bioreactors due to the concern of enzyme washout in a continuous flow system. In an attempt to prevent enzyme washout, an enzymatic membrane bioreactor (EMBR) was developed by coupling an ultrafiltration (UF) membrane to an enzymatic bioreactor [18,19]. Interestingly, during the operation of the EMBR, adsorption of some hydrophobic TrOCs (e.g., amitriptyline, oxybenzone and octocrylene) onto the enzyme gel layer over the membrane surface resulted in enhanced degradation of the adsorbed compounds [18]. In another study, removal of four non-phenolic TrOCs, namely atrazine, sulfamethoxazole, diclofenac and carbamazepine was improved by 15–25% following the addition of granular activated carbon (GAC) in EMBR. This was probably because simultaneous adsorption of laccase and TrOCs on GAC promoted the interaction of TrOCs with the active sites of laccase [20]. Results from previous studies indicate the complementarity of simultaneous laccase and TrOC retention within EMBR in contrast to only laccase retention by UF membranes utilized in the previously developed EMBRs. Hence, in this study, it is postulated that the integration of an enzymatic bioreactor with a high retention membrane could facilitate the degradation of resistant TrOCs by retaining both laccase and TrOCs.

Different configurations of conventional activated sludge-based high retention membrane bioreactors (HR-MBR), employing membrane distillation (MD), forward osmosis (FO) or nanofiltration (NF) membranes, have been investigated for advanced wastewater treatment [21–24]. Complete TrOC retention in HR-MBR improved the membrane permeate quality, but the poor removal of certain groups of TrOCs such as those containing EWGs led to their accumulation in the bioreactor. This indicates the necessity of formulating means to enhance biodegradation of TrOCs. In this context, it is noteworthy that recent reports confirm enhanced laccase-catalyzed degradation of selected TrOCs that are not amenable to degradation by conventional activated sludge [25,26]. However, the performance of a high retention—enzymatic membrane bioreactor for the removal of a wide range of TrOCs remains to be elucidated.

Among the high retention membrane systems, in MD, a vapor-liquid interface is developed around a hydrophobic micro-porous membrane that allows the water to pass through the membrane via diffusion due to vapor pressure gradient. Compared to conventional distillation processes such as fractional distillation, the MD process requires low temperature and could be operated by using low grade heat or solar energy [27,28]. Since the mass transfer in the MD process occurs in gaseous phase, it can theoretically achieve 100% retention of all non-volatile compounds [29]. The standalone MD process has been investigated for seawater desalination [30], industrial wastewater treatment [31], municipal wastewater treatment [32] and TrOC removal [29,33]. Thus, the MD process was selected for coupling to an enzymatic bioreactor in this study.

The aim of this study was to assess the performance of a laccase based membrane distillation—enzymatic membrane bioreactor (MD-EMBR) for the removal of TrOCs having diverse physicochemical properties (e.g., EDGs/EWGs, hydrophobicity and phenolic/non-phenolic moieties). A special focus was given to the improvement in TrOC degradation due to the addition of three redox-mediators, namely syringaldehyde (SA), violuric acid (VA) and 1-hydroxybenzotriazole (HBT) at different concentrations. In addition, performance of laccase-mediator systems was systematically compared based on TrOC degradation, enzyme stability and effluent toxicity.

## 2. Materials and Methods

### 2.1. Trace Organic Contaminants (TrOCs), Laccase and Mediators

A set of 30 TrOCs comprising 10 pharmaceuticals, four personal care products, six pesticides, four industrial chemicals, five steroid hormones and one phytoestrogen was selected based on their widespread occurrence in surface water bodies (see Supplementary Data Table S1). Key physicochemical properties of the TrOCs including molecular weight, water solubility, hydrophobicity (log D) and volatility ($pK_H$) are presented in Table 1. All TrOCs were purchased from Sigma Aldrich (Sydney, NSW, Australia), and were of analytical grade. A stock solution (25 mg/L) containing the mixture of 30 TrOCs was prepared in pure methanol, and kept in dark at −18 °C.

Laccase from genetically modified *Aspergillus oryzae* was supplied by Novozymes Australia Pty Ltd (Sydney, NSW, Australia). According to the supplier, the enzyme has a molecular weight, purity, activity and density of 56 KDa, 10% (*w*/*w*), 150,000 $\mu M_{(DMP)}$/min (measured using 2,6-dimethoxy phenol, DMP, as substrate) and 1.12 g/mL, respectively.

Two N–OH type redox-mediators, namely 1-hydroxybenzotriazole (HBT) and violuric acid (VA), and one phenolic redox-mediator, namely syringaldehyde (SA), were used in this study (see Supplementary Data Table S2). The selected mediators all follow hydrogen atom transfer (HAT) pathway for TrOC degradation [34], but the oxidation of phenolic and N–OH type redox-mediators by laccase produces highly reactive phenoxyl and aminoxyl radicals, respectively. The mediators were also purchased from Sigma Aldrich (Sydney, NSW, Australia). A separate stock solution (50 mM) of each redox-mediator was prepared, and stored at 4 °C in dark.

**Table 1.** Physicochemical properties of the selected TrOCs.

| TrOCs | Chemical Formula | Molecular Weight | Log D at pH = 7 | Water Solubility at 25 °C | Vapor Pressure | $pK_H$ at pH 7 |
|---|---|---|---|---|---|---|
| | | g/mole | | mg/L | (mmHg) | |
| **Non-Phenolic Compounds** | | | | | | |
| Primidone | $C_{12}H_{14}N_2O$ | 218.25 | 0.83 | 1500 | $6.08 \times 10^{-11}$ | 13.93 |
| Ketoprofen | $C_{16}H_{14}O_3$ | 254.28 | 0.19 | 554,000 | $3.32 \times 10^{-8}$ | 13.70 |
| Naproxen | $C_{14}H_{14}O_3$ | 230.26 | 0.73 | 435,000 | $3.01 \times 10^{-7}$ | 12.68 |
| Gemfibrozil | $C_{15}H_{22}O_3$ | 250.33 | 2.07 | 263,000 | $6.13 \times 10^{-7}$ | 12.11 |
| Metronidazole | $C_6H_9N_3O_3$ | 171.15 | −0.14 | 29,000 | $2.67 \times 10^{-7}$ | 11.68 |
| Diclofenac | $C_{14}H_{11}Cl_2NO_2$ | 296.15 | 1.77 | 20,000 | $1.59 \times 10^{-7}$ | 11.51 |
| Fenoprop | $C_9H_7Cl_3O_3$ | 269.51 | −0.13 | 230,000 | $2.13 \times 10^{-6}$ | 11.48 |
| Ibuprofen | $C_{13}H_{18}O_2$ | 206.28 | 0.94 | 928,000 | $1.39 \times 10^{-4}$ | 10.39 |
| Ametryn | $C_9H_{17}N_5S$ | 27.33 | 2.97 | 140 | $1.72 \times 10^{-6}$ | 9.35 |
| Clofibric acid | $C_{10}H_{11}ClO_3$ | 214.65 | −1.06 | 100,000 | $1.03 \times 10^{-4}$ | 9.54 |
| Carbamazepine | $C_{15}H_{12}N_2O$ | 236.27 | 1.89 | 220 | $5.78 \times 10^{-7}$ | 9.09 |
| Octocrylene | $C_{24}H_{27}N$ | 361.48 | 6.89 | 0.36 | $2.56 \times 10^{-9}$ | 8.47 |
| Amitriptyline | $C_{20}H_{23}N$ | 277.40 | 2.28 | 83 | $1.50 \times 10^{-6}$ | 8.18 |
| Atrazine | $C_8H_{14}ClN_5$ | 215.68 | 2.64 | 69 | $1.27 \times 10^{-5}$ | 7.28 |
| Propoxur | $C_{11}H_{15}NO_3$ | 209.24 | 1.54 | 800 | $1.53 \times 10^{-3}$ | 6.28 |
| Benzophenone | $C_{13}H_{10}O$ | 182.22 | 3.21 | 150 | $8.23 \times 10^{-4}$ | 5.88 |
| DEET | $C_{12}H_{17}NO$ | 191.3 | 2.42 | 1000 | $5.6 \times 10^{-3}$ | 5.85 |
| **Phenolic Compounds** | | | | | | |
| Enterolactone | $C_{18}H_{18}O_4$ | 288.38 | 2.53 | 200 | $3.29 \times 10^{-13}$ | 15.20 |
| Estriol | $C_{18}H_{24}O_3$ | 298.33 | 1.89 | 32 | $1.34 \times 10^{-9}$ | 10.78 |
| 17α-Ethinylestradiol | $C_{20}H_{24}O_2$ | 269.40 | 4.11 | 3.9 | $3.74 \times 10^{-9}$ | 9.47 |
| Oxybenzone | $C_{14}H_{12}O_3$ | 228.24 | 3.89 | 2700 | $5.26 \times 10^{-6}$ | 9.23 |
| Estrone | $C_{18}H_{22}O_2$ | 270.37 | 3.62 | 5.9 | $1.54 \times 10^{-8}$ | 9.03 |
| 17β-Estradiol | $C_{18}H_{24}O_2$ | 272.38 | 4.15 | 3 | $9.82 \times 10^{-9}$ | 8.93 |
| 17β-Estradiol-17-acetate | $C_{20}H_{26}O_3$ | 314.42 | 5.11 | 1.9 | $9.88 \times 10^{-9}$ | 8.67 |
| Bisphenol A | $C_{15}H_{16}O_2$ | 228.29 | 3.64 | 73 | $5.34 \times 10^{-7}$ | 8.66 |
| Salicylic acid | $C_7H_6O_3$ | 138.12 | −1.13 | 2240 | $8.2 \times 10^{-5}$ | 8.18 |
| Pentachlorophenol | $C_6HCl_5O$ | 266.34 | 2.85 | 4800 | $3.49 \times 10^{-4}$ | 7.59 |
| Triclosan | $C_{12}H_7Cl_3O_2$ | 289.54 | 5.28 | 19 | $3.26 \times 10^{-5}$ | 6.18 |
| 4-tert-Butylphenol | $C_{10}H_{14}O$ | 150.22 | 3.40 | 1000 | 0.0361 | 5.15 |
| 4-tert-Octylphenol | $C_{14}H_{22}O$ | 206.32 | 5.18 | 62 | $1.98 \times 10^{-3}$ | 5.06 |

## 2.2. *Experimental Setup*

The laboratory scale MD-EMBR setup comprised a glass enzymatic bioreactor (1.5 L) and an external direct contact membrane distillation system (Figure 1). The glass enzymatic bioreactor covered with aluminum foil was placed in a water bath, and the temperature of the water bath was maintained at 30 ± 0.2 °C using an immersion heating unit (Julabo, Seelbach, Germany). The enzymatic bioreactor was equipped with an air pump (ACO-002, Zhejiang Sensen Industry Co. Ltd., Zhoushan, China) to maintain a dissolved oxygen concentration of above 3 mg/L.

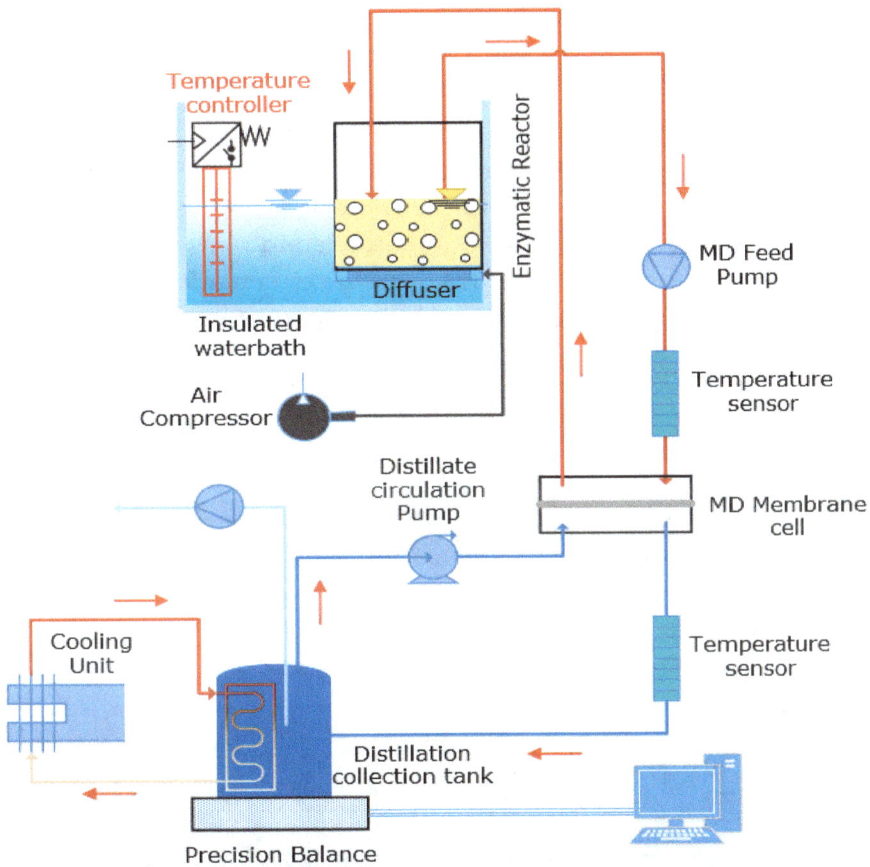


**Figure 1.** Schematic representation of the membrane distillation—enzymatic membrane bioreactor (MD-EMBR).

The external direct contact membrane distillation system contained an acrylic glass membrane cell, two circulation pumps (Micropump Inc., Vancouver, WA, USA) and a glass permeate tank (Figure 1). Feed and permeate flow channels were engraved on each block of the membrane cell. Length, width and height of each flow channel were 145, 95 and 3 mm, respectively.

A hydrophobic microporous polytetrafloroethylene (PTFE) membrane (GE, Minnetonka, MN, USA) was used during each experiment. The PTFE membrane has a nominal pore size of 0.22 μm, thickness of 175 μm, porosity of 70% and an active layer thickness of 5 μm [35].

### 2.3. Experimental Protocol

A series of experiments was carried out to evaluate the performance of MD-EMBR for TrOC degradation. At the start of the experiment, a mixture of the selected TrOCs (each at 20 μg/L) in Milli-Q water was added to the bioreactor. Laccase was added to the bioreactor for achieving an initial enzymatic activity of 95–100 $\mu M_{(DMP)}$/min. The media from the glass enzymatic bioreactor and water from the permeate tank were recirculated in their respective flow channels separated by the membrane. A chiller (SC100-A10, Thermo Scientific, Waltham, MA, USA) was used to regulate the temperature of the permeate tank at 10 ± 0.1 °C. The permeate tank was also placed on a precision balance (Mettler Toledo Inc., Columbus, OH, USA) to monitor permeate flux. The recirculation flow rate of both feed and the distillate was controlled at 1 L/min (corresponding to the cross flow velocity of 9 cm/s) using two rotameters.

Duplicate samples from the enzymatic bioreactor (100 mL each) and permeate tank (500 mL each) were taken after operating the MD-EMBR for 12 h. After evaluating the laccase-catalyzed degradation of TrOCs in MD-EMBR, the possible improvement in TrOC degradation was assessed with the addition of three redox-mediators (HBT, VA and SA) at two different concentrations (0.25 and 0.5 mM) via separate runs. Again duplicate samples from the enzymatic bioreactor and permeate tank were collected for the quantification of TrOCs.

Samples collected from the enzymatic bioreactor were diluted to 500 mL with Milli-Q water and were filtered through 0.45 μm glass fiber filter paper (Filtech, Wollongong, NSW, Australia). The pH of samples was adjusted to 2–2.5 using 4 M $H_2SO_4$ before solid phase extraction (SPE) and GC/MS analysis. For toxicity analysis, undiluted samples from the enzymatic bioreactor and permeate tank were collected in 2 mL amber vials at the end of each experiment, and stored at 4 °C until analysis.

## 2.4. Analytical Methods

### 2.4.1. TrOC Analysis

The concentration of TrOCs was measured using an analytical method involving SPE derivatization and quantitative determination by a Shimadzu GC/MS (QP5000) system as described by Hai et al. [36]. Limit of detection (LOD) for this method was compound specific and ranged from 1 to 20 ng/L (see Supplementary Data Table S1). Removal efficiencies by the enzymatic bioreactor ($R_1$) and the MD-EMBR ($R_2$) were calculated using Equations (1) and (2), respectively:

$$R_1 = 100 \times (1 - C_f/C_o) \quad (1)$$

$$R_2 = 100 \times (1 - C_p/C_o) \quad (2)$$

where, $C_o$ and $C_f$ are the concentration (ng/L) of specific TrOC in the enzymatic bioreactor at the beginning ($t$ = 0 h) and end ($t$ = 12 h) of experiment, respectively, while $C_p$ is the concentration of specific TrOC in permeate at $t$ = 12 h. The enzymatic transformation/degradation of TrOCs in the MD-EMBR was calculated using Equation (3):

$$C_o \times V_o = (C_f \times V_f) + (C_p \times V_p) + \text{biodegradation} \quad (3)$$

where, $V_o$, $V_f$ and $V_p$ represents the volume of feed (at $t$ = 0 h), supernatant ($t$ = 12 h) and permeate ($t$ = 12 h), respectively.

### 2.4.2. Enzymatic Activity, ORP and Toxicity Assay

Laccase activity and effluent toxicity were examined as described elsewhere [18]. Laccase activity was measured by recording the change in absorbance at 468 nm due to the oxidation of 2,6-dimethoxyl phenol (DMP) in the presence of 100 mM sodium citrate (pH 4.5). Laccase activity expressed as $\mu M_{(DMP)}$/min was then calculated from the molar extinction coefficient of 49.6/mM cm. Oxidation reduction potential (ORP) was measured at the start and end of each experiment using an ORP meter (WP-80D dual pH-mV meter, Thermo Fisher Scientific, Scoresby, VIC, Australia). Samples for toxicity analysis were collected from the enzymatic bioreactor and permeate tank at end of each experiment. Toxicity, expressed as a relative toxicity unit (rTU), was analyzed by measuring the inhibition of luminescence in the naturally bioluminescent bacteria, *Photobacterium leiognathi*, as previously described [37,38].

## 3. Results and Discussion

### 3.1. Overall Removal of TrOCs

In theory, MD membranes can retain all but the volatile organic compounds. In this study, the concentration of non-volatile ($pK_H > 9$) TrOCs in the permeate of the MD-EMBR was below the limit of detection of GC/MS. This is consistent with the observation in a previous study, where an MD membrane was coupled with an activated sludge bioreactor [22]. On the other hand, the MD system achieved 90–99% removal of relatively volatile TrOCs having $pK_H < 9$ (Figure 2). This compares favorably to their previously reported moderate to high removal (54–99%) by a standalone MD system [29]. In particular, removal of octocrylene ($pK_H = 8.47$), benzophenone ($pK_H = 5.88$), 4-tert-butylphenol ($pK_H = 5.15$), 4-tert-octylphenol ($pK_H = 5.06$) by the MD-EMBR was above 99%, compared to their 55–70% removal by the MD only [29]. These results suggest that the coupling of enzymatic degradation process to the MD system was favorable for achieving high TrOC removal.

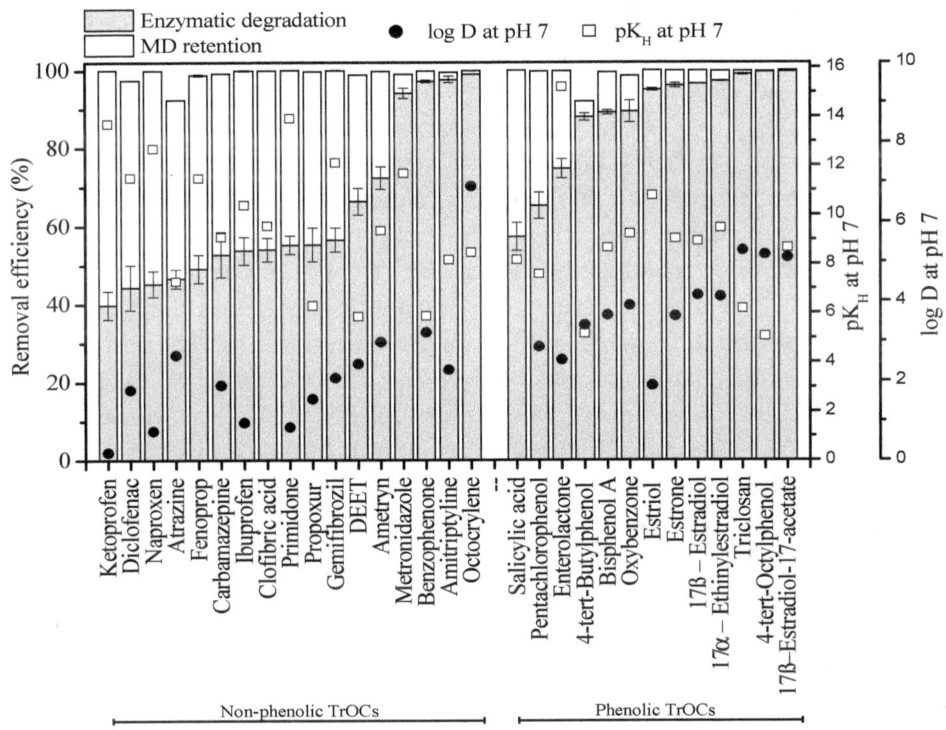


**Figure 2.** Overall removal and enzymatic degradation of 30 TrOCs in the MD-EMBR. Error bars indicate the standard deviation of duplicate samples. MD-EMBR operating conditions: the initial TrOC concentration and laccase activity was 20 μg/L and 95–100 $\mu M_{(DMP)}$/min, respectively; temperature of the enzymatic bioreactor and the permeate tank were kept at 30 and 10 °C, respectively; and cross-flow rate of media from the enzymatic bioreactor and distillate was 1 L/min (corresponding to a cross-flow velocity of 9 cm/s).

### 3.2. TrOC Degradation in Enzymatic Bioreactor

Degradation of a substrate by laccase involves the transfer of an electron from the substrate to laccase with concomitant reduction of atmospheric oxygen to water. The extent of degradation depends on, among others, the molecular properties (e.g., EWGs, EDGs or phenolic moiety) of the target substrate [11,39]. In this study, high degradation (87–99%) of 10 out 13 phenolic TrOCs was achieved by the MD-EMBR (Figure 2). These included five steroid hormones (estriol, estrone, 17β–estradiol, 17α–ethinylestradiol and 17β-estradiol-17-acetate (95–99%)), two industrial chemicals (4-tert-butylphenol, and 4-tert-octylphenol (87–99%)) and two personal care products (oxybenzone and triclosan (89–98%)). On the other hand, enzymatic degradation of some phenolic compounds, namely pentachlorophenol, enterolactone and salicylic acid, ranged from 55 to 75%. Their incomplete degradation, despite the presence of a strong EDG (i.e., hydroxyl group), can be attributed to the concomitant presence of an EWG (e.g., halogen) in their molecular structure (see Supplementary Data Table S1) [39].

Depending on the medium ORP, laccase can also degrade non-phenolic compounds. However, the reaction kinetics can be slow [17,39]. In this study, the enzymatic degradation of 17 non-phenolic TrOCs varied from 40–99% (Figure 2). Laccase catalyzed degradation of 13 compounds fell in the range of 40–65%, while the degradation of the remaining four TrOCs was in the range of 94–98%. The well degraded non-phenolic TrOCs include metronidazole, benzophenone, amitriptyline and octocrylene. High laccase-catalyzed degradation (80–99%) in continuous flow UF-EMBR has been previously reported [18,37] for benzophenone, amitriptyline and octocrylene, but not for metronidazole. Metronidazole contains both EWGs (i.e., $-NO_2$) and EDGs (i.e., methyl and hydroxyethyl) in its molecule (see Supplementary Data Table S1). High enzymatic degradation of metronidazole following its complete retention by the MD membrane in MD-EMBR can be attributed to the prolonged contact time that may have promoted the interaction of laccase with the EDGs of metronidazole.

An overall degradation of only 40–65% was achieved by the MD-EMBR for a number of non-phenolic TrOCs (Figure 2), however, these removal efficiencies in fact compare favorably with those reported in the literature [12,15,37]. For instance, laccase catalyzed degradation of carbamazepine, clofibric acid, fenoprop and atrazine has been reported to be less than 10% in both batch and continuous flow ultrafiltration based enzymatic bioreactors [15,18,40]. By contrast, 40–45% degradation of these TrOCs by the MD-EMBR was observed in this study. Since most of the selected non-phenolic TrOCs contain both EWGs and EDGs in their structure (see Supplementary Data Table S1), complete retention of these TrOCs in the enzymatic bioreactor may have facilitated the interaction of EDGs with nearby redox centers, thereby providing higher possibility of electron transfer to enzyme [17]. Previously, Nguyen et al. [20] reported that dosing of GAC into an UF-EMBR led to simultaneous adsorption of laccase and TrOCs on GAC, yielding significant improvement in the degradation of four non-phenolic TrOCs, namely atrazine, sulfamethoxazole, diclofenac and carbamazepine. Although our approach was different, it is conceivable that prolonged retention of TrOCs in the enzymatic bioreactor can improve their degradation.

It is noteworthy that phenolic TrOCs (e.g., triclosan, oxybenzone, bisphenol A and steroid hormones) can act as redox-mediators, and the fragments of phenoxyl radicals formed following their degradation by laccase can oxidize non-phenolic compounds [39]. Indeed Margot et al. [12] observed that degradation of diclofenac by laccase was significantly higher in the mixture of TrOCs containing diclofenac, bisphenol A and mefenamic acid than its degradation as a single compound. It is possible that complete retention of phenoxyl radicals formed due to the degradation of phenolic TrOCs aided better degradation of non-phenolic TrOCs by MD-EMBR as compared to previously developed UF-EMBR [18,37]. Further investigation would be required to substantiate this hypothesis but that is beyond the scope of this study.

This study confirms for the first time the improvement in TrOC degradation in an enzymatic bioreactor by coupling with it a high retention membrane (such as membrane distillation) as compared to a conventional ultrafiltration membrane. We used a direct contact membrane distillation module, but there may be case-specific scope of choice between different formats of membrane distillation. Future studies are recommended to assess the commercial viability of different configurations of MD such as vacuum MD and air gap MD, but that is beyond the scope of this study.

### 3.3. Impact of Mediator Addition on TrOC Degradation

As noted in Section 3.2, of the 30 TrOCs tested, MD-EMBR achieved high degradation (85–99%) for 14 compounds (10 phenolic and 4 non-phenolic compounds) but the degradation efficiency varied widely (40–70%) for the rest of the compounds. To improve the degradation of the latter group, three redox-mediators, namely SA, VA and HBT, were added at 0.25 and 0.5 mM concentrations each in separate runs. Depending on the redox-mediator type and concentration, degradation of phenolic compounds and non-phenolic compounds by the MD-EMBR was improved by 20–30% and 10–50%, respectively (Figure 3) as explained in the following sections.

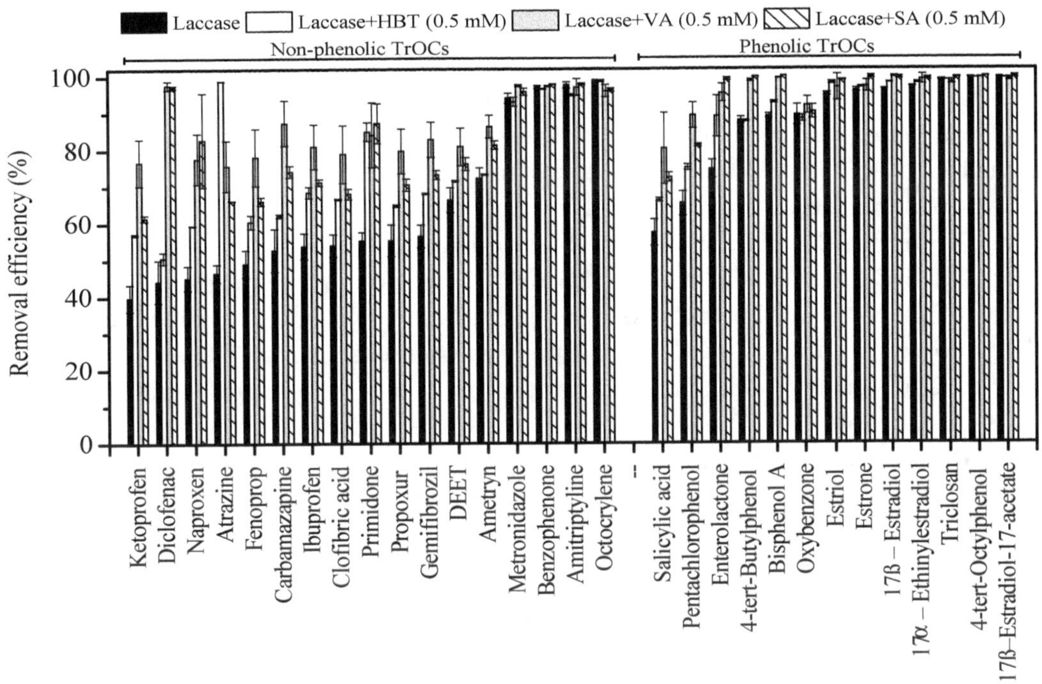


**Figure 3.** Enzymatic degradation of 30 TrOCs in the presence of three mediators, namely HBT, VA and SA (separately at 0.5 mM) in the MD-EMBR. Error bars indicate the standard deviation of duplicate samples. Operating conditions of the MD-EMBR are given in the caption of Figure 2.

### 3.3.1. Comparison of Redox-Mediators

To date, the impact of redox-mediator type on the improvement of TrOC degradation has been assessed mainly in small scale and batch tests [34,41,42]. For instance, Ashe et al. [34] investigated the performance of seven different redox-mediators including SA, HBT and VA for the degradation of four resistant TrOCs, namely atrazine, naproxen, oxybenzone and pentachlorophenol in 10 mL batch reactors. They achieved significant improvement (40–90%) at a concentration of 1 mM. Nguyen et al. [18] achieved enhanced (10–90%) removal of TrOCs in UF-EMBR using SA and HBT. However, this is the first study investigating the efficacy of SA, VA and HBT for enhanced degradation of a broad spectrum of TrOCs by an MD-EMBR.

All the tested redox-mediators enhanced the degradation of TrOCs. However, the best overall performance was shown by VA (Figure 3). In line with the findings of Nguyen et al. [37], degradation of the phenolic TrOCs that were already highly degraded by laccase (Figure 2) remained almost the same after the addition of redox-mediators. For the remaining phenolic TrOCs, VA (at 0.5 mM), compared to HBT and SA achieved better removal for two compounds, namely salicylic acid (80%) and pentachlorophenol (90%). Both VA and SA achieved above 95% degradation of enterolactone, which compares favorably with 45–70% degradation achieved in absence of mediators (Figure 3).

Of the 17 non-phenolic compounds, degradation of four compounds viz metronidazole, benzophenone, amitriptyline and octocrylene, was at least 90%, regardless of the mediator type (Figure 3). For the remaining compounds, VA (at 0.5 mM) achieved better degradation for 10 compounds compared to SA and HBT. SA (at 0.5 mM) performed the best for the degradation of two compounds, namely naproxen and primidone. It is well-known that the herbicide atrazine is resistant to laccase catalyzed degradation [18]. Compared to other redox-mediators, HBT was particularly efficient (>99%) for the degradation of atrazine. Although a superior ability of VA compared to other mediators for the degradation of non-phenolic TrOCs has been reported previously in a batch enzymatic bioreactor spiked with four TrOCs [34], the effectiveness of VA for the degradation of a broad spectrum of non-phenolic TrOCs is demonstrated for the first time in this study.

### 3.3.2. Impact of Mediator Concentration

Redox-mediator dose can affect TrOC degradation by changing the abundance, stability and reversibility of the generated radicals [43]. Therefore, the impact of two mediator concentrations (0.25 and 0.5 mM) on ORP, TrOC degradation, and enzyme stability was investigated.

Concentration-dependent improvement in the degradation of 18 TrOCs (5 phenolic and 13 non-phenolic compounds, Figure 4) was observed in MD-EMBR. The highest improvement in the degradation of TrOCs was achieved at 0.5 mM. Notably, increasing the concentration of SA, HBT and VA from 0.25 to 0.5 mM improved TOC degradation by up to 7%, 15% and 25%, respectively (Figure 4). This corresponds well with the respective increase of 2%, 5% and 15% of the reaction media ORP (Figure 5). On the other hand, degradation of 8 phenolic and 4 non-phenolic TrOCs in MD-EMBR was comparable at all the tested mediator concentrations (Supplementary data Figure S3). For instance, HBT achieved over 99% degradation of atrazine in MD-EMBR irrespective of the mediator concentration. This is consistent with HBT performance reported in case of UF-EMBR [18].

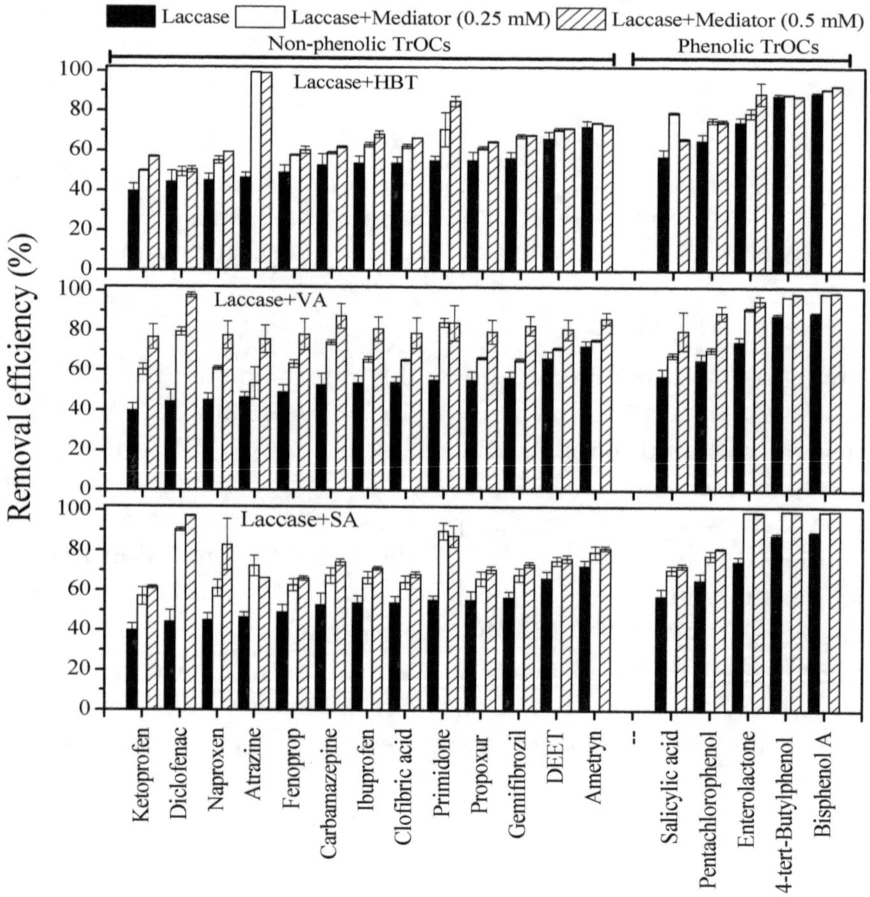


**Figure 4.** Impact of mediator concentration (0.25 and 0.5 mM) on the degradation of TrOCs in the MD-EMBR. Error bars indicate the standard deviation of duplicate samples. Operating conditions of the MD-EMBR are given in the caption of Figure 2. Only those TrOCs showing mediator concentration-dependent improvement in their degradation are shown here. For remaining TrOCs, results are given in Supplementary Data Figure S3.

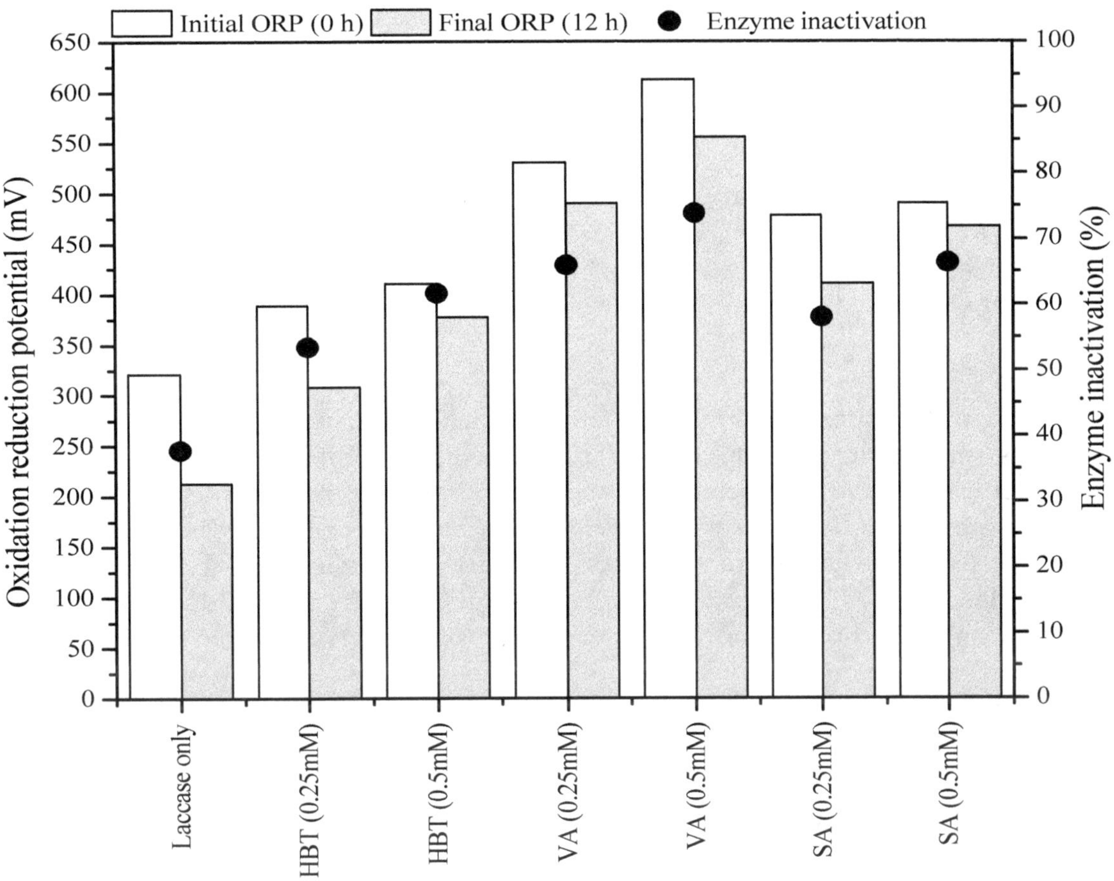


**Figure 5.** Effect of mediator type and concentration on oxidation reduction potential (ORP) and laccase inactivation in the MD-EMBR. Operating conditions of the MD-EMBR are given in the caption of Figure 2.

In general, the degradation of TrOCs that are easily amenable to laccase (Supplementary data Figure S3) does not improve significantly (less than 5% in this study), while the degradation of resistant TrOCs (Figure 4) may improve with the increase in mediator concentration, and may reach a plateau beyond a certain mediator concentration. However, the mediator concentration beyond which no improvement occurs may depend on the type of mediators as well as the target TrOCs [41,44].

### 3.3.3. Effect of Mediators on Enzyme Stability

In this study, a gradual inactivation of laccase was observed despite the absence of any known chemical inhibitors in the synthetic wastewater (Figure 5). In the absence of redox-mediators, a 37% laccase inactivation was observed over a period of 12 h. This was possibly due to the blockage of the active enzyme sites by the charged metabolites and/or hydraulic stress during membrane filtration [25,41]. Since the MD membrane can conceptually retain all nonvolatile organics including the transformation products/radicals, laccase inactivation with or without the presence of redox-mediators can be expected. The extent of laccase inactivation increased further when the mediators were added (61%, 66% and 73% for HBT, SA and VA, respectively, each at a concentration of 0.5 mM). The highly reactive radicals generated from mediators can enhance the degradation of TrOCs but at the same time may inactivate laccase [45]. Purich [17] suggested that the metabolites from the oxidation of substrate and/or mediators could react with enzyme to form non-productive complexes, thereby inactivating the enzyme.

The extent of laccase inactivation also depends on the concentration of redox-mediators. For instance, Khlifi-Slama et al. [45] observed a gradual increase in the inactivation of laccase from *Trametes trogii* following a stepwise increase in the concentration of HBT from 0.1 to 10 mM. In another study, increasing SA concentration from 0.1 to 1 mM resulted in aggravated inactivation of laccase from *Trametes versicolor* [42]. These results suggest that the degree of laccase inactivation is strongly influenced by redox-mediator concentration. Indeed, loss in laccase activity was increased by 7%, 9% and 11% in MD-EMBR due to the increase in the concentration of HBT, SA and VA, respectively, from 0.25 to 0.5 mM (Figure 5). Although laccase activity was greatly affected in the presence of redox-mediators, it was compensated by the improvement in TrOC degradation (Figure 3). For example, the highest drop in laccase activity was observed in the presence of VA (Figure 5), but it outperformed SA and HBT in terms of enhanced TrOC degradation (Figure 3).

### 3.4. Effluent Toxicity

The charged metabolites and highly reactive radicals produced following the oxidation of redox-mediators may improve TrOC degradation [18,46], but these can also cause an increase in effluent toxicity [18,47]. In this study, it was not possible to relate individual metabolites to specific parent compounds because we investigated a mixture of 30 TrOCs. Hence, the overall bacterial toxicity of the reaction mixture and permeate was evaluated at the end of each run. Of the three mediators tested, SA significantly increased the toxicity of the solution in the enzymatic bioreactor, whereas HBT and VA showed no effect on toxicity levels (Table 2). Compared to the background toxicity of the mixture of laccase and TrOCs in the enzymatic bioreactor of MD-EMBR (<1 to 1.8 rTU; $n = 2$), toxicity in the enzymatic bioreactor due to addition of HBT, VA and SA ranged from <1 to 1.7 rTU ($n = 2$), 3.3 to 3.9 rTU ($n = 2$) and 109 to 116 rTU ($n = 2$), respectively. Notably, the final effluent (i.e., membrane permeate) was not toxic to bacteria (<1 rTU) for any of the enzyme/mediator combinations, indicating that MD not only retained TrOCs and laccase but also the transformation byproducts and radicals responsible for inducing bacterial toxicity. This is an added advantage of coupling a high retention membrane to the enzymatic bioreactor.

**Table 2.** Toxicity of the reactor mixture and permeate following treatment of TrOCs with different mediators in MD-EMBR, expressed as relative toxic unit (rTU). Mediators were added separately at a concentration of 0.5 mM. The limit of detection of the toxicity assay was 10% inhibition of luminescence (i.e., 1 rTU). Toxicity in all permeate samples was below the limit of detection ($n = 2$).

| Reaction Mixture | Toxicity of the Reactor Mixture (rTU) | Toxicity of the Permeate (rTU) |
|---|---|---|
| TrOCs + Laccase | <1–1.8 | <1 |
| TrOCs + Laccase + HBT (0.5 mM) | <1–1.7 | <1 |
| TrOCs + Laccase + VA (0.5 mM) | 3.3–3.9 | <1 |
| TrOCs + Laccase + SA (0.5 mM) | 109–116 | <1 |

### 3.5. Permeate Flux

The driving force of permeate flux in MD is the difference between feed and distillate temperature. Ideally, feed and distillate temperature is maintained at over 50 and 20–25 °C, respectively to obtain a permeate flux of approximately 10 L/m$^2$ h [27,48]. In this study, however, to avoid thermal inhibition of laccase [46], temperature of the enzymatic reactor and permeate tank was kept at 30 and 10 °C, respectively. A stable permeate flux of around 4 L/m$^2$ h was observed during all experiments (Supplementary Data Figure S4), suggesting that membrane fouling did not occur during the operation period. This level of flux is consistent with the feed temperature employed. Notably, the average permeate flux (Figure 6) for laccase only, laccase-HBT, laccase-VA and laccase-SA variations was 3.69 $\pm$ 0.44 L/m$^2$ h ($n = 150$), 3.89 $\pm$ 0.63 L/m$^2$ h ($n = 283$), 3.92 $\pm$ 0.62 L/m$^2$ h ($n = 291$) and 3.86 $\pm$ 0.66 L/m$^2$ h ($n = 288$) LMH, respectively, confirming negligible impact of different type of mediator addition on membrane flux. In this study, the mass transfer coefficient ($K_m$) of the DCMD, which was calculated based on the method described by Nghiem et al. [49], ranged from 1.22 to 1.28 ($\times 10^{-3}$) L/m$^2$ h Pa. This value is in good agreement with that in previous studies [48,50].

Thus, this study shows both stable membrane hydraulic performance and improved enzymatic degradation of TrOCs following their complete retention by the MD membrane.

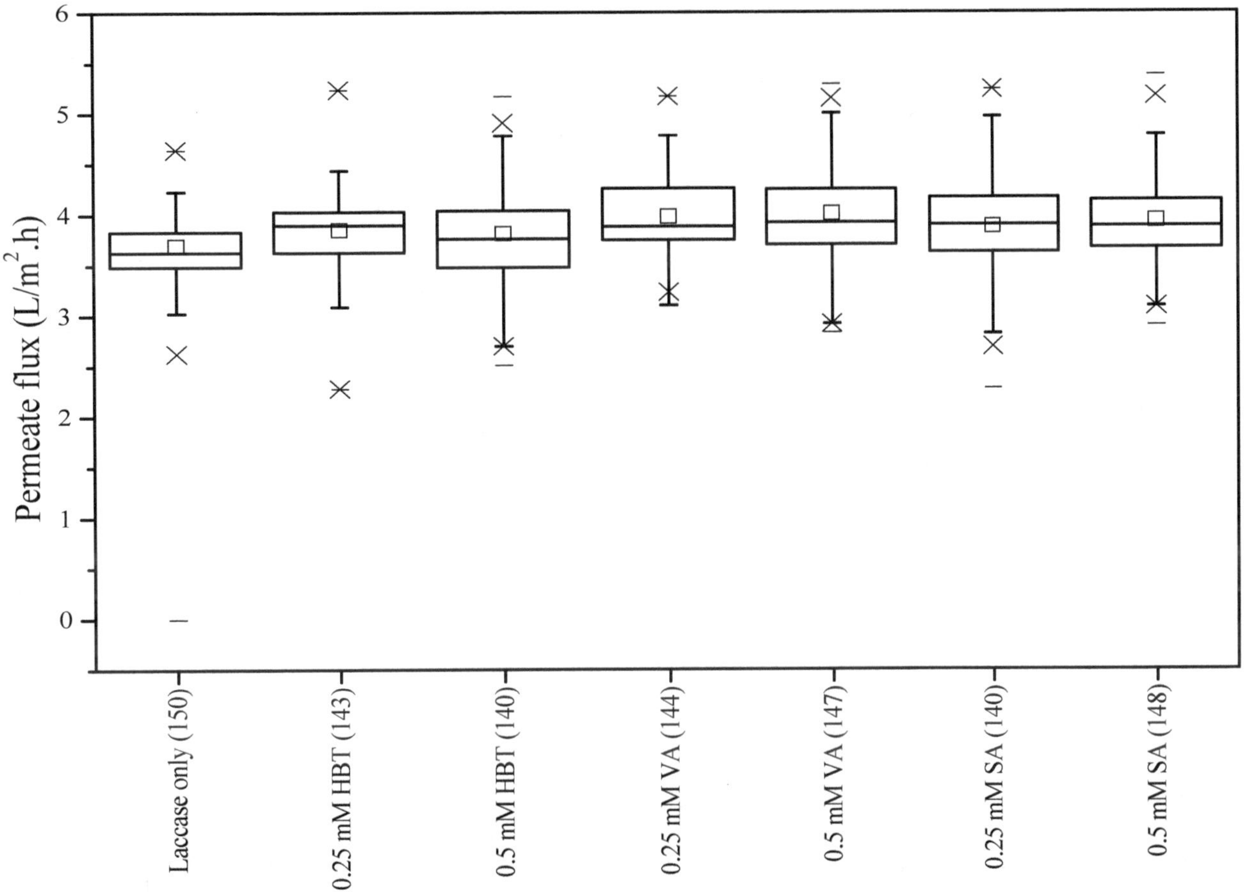


**Figure 6.** Permeate flux during the operation of MD-EMBR with and without the presence of redox-mediators. Box-and-whisker plot shows the interquartile range; median (horizontal line in the box); min and max (whiskers); average (small square in the box); and 1 and 99% percentiles (cross above and below the whiskers). Operating conditions for MD-EMBR: Temperature of the enzymatic bioreactor and the permeate tank were kept at 30 and 10 °C, respectively; cross-flow rate of water from enzymatic bioreactor and distillate was 1 L/min (corresponding to cross-flow velocity of 9 cm/s); the initial TrOC concentration and laccase activity was 20 µg/L and 95–100 $\mu M_{(DMP)}$/min, respectively; and each mediator was added at 0.25 or 0.5 mM concentration in separate runs.

## 4. Conclusions

Performance of an enzymatic bioreactor integrated with the MD system (MD-EMBR) was examined for the removal of 13 phenolic and 17 non-phenolic compounds. Based on permeate quality, MD-EMBR achieved 90–99% TrOC retention. Degradation of TrOCs varied (40–99%) depending on their molecular properties (electron withdrawing functional groups electron donating functional groups and phenolic moiety). High degradation (above 90%) of TrOCs containing EDGs in their chemical structure was observed in the MD-EMBR, while those containing EWGs in their molecular structure were moderately degraded (40–75%). Degradation of TrOCs was further improved by adding three redox-mediators, namely syringaldehyde (SA), violuric acid (VA) and 1-hydroxybenzotriazole (HBT). VA at 0.5 mM concentration was found to be the most effective mediator for improving the degradation of phenolic and non-phenolic TrOCs. Moreover, it was observed that the degradation of non-phenolic compounds in laccase-mediator system was strongly influenced by the tested concentration of the redox-mediators. Despite an increase in the toxicity of the reaction mixture caused by SA, the final effluent of the MD-EMBR was nontoxic.

**Supplementary Materials:** Table S1: Physicochemical properties of the selected trace organic contaminants (TrOCs), Table S2: Physicochemical properties of the selected redox-mediators, Figure S3: Impact of mediator concentration (0.25 and 0.5 mM) on the degradation of after an incubation time of 12 h in the MD-EMBR. Error bars indicate the standard deviation of duplicate samples. Degradation of these TrOCs did not improve by increasing mediator concentration, Figure S4: Permeate flux obtained during the operation of enzymatic membrane distillation (MD-EMBR) with and without the addition of redox mediators.

**Acknowledgments:** This research has been conducted with the support of the Australian Government Research Training Program Scholarship. Novozymes Pty. Ltd., Australia is thanked for the provision of enzyme solution. This study was partially funded by the GeoQuEST Research Centre, University of Wollongong, Australia.

**Author Contributions:** F.I.H. conceived and led the project. F.I.H. and M.B.A. planned the experiments in consultation with the coauthors. M.B.A. conducted the experiments. J.K., F.D.D.L. and J.P.M. analyzed TrOC and toxicity samples. F.I.H. and M.B.A. analyzed the data and prepared the manuscript with the contribution of K.Y., W.E.P., L.D.N., F.D.D.L. and J.P.M. to specific sections.

## References

1. Hai, F.I.; Yamamoto, K.; Nakajima, F.; Fukushi, K. Application of a gac-coated hollow fiber module to couple enzymatic degradation of dye on membrane to whole cell biodegradation within a membrane bioreactor. *J. Membr. Sci.* **2012**, *389*, 67–75. [CrossRef]
2. Hai, F.I.; Yamamoto, K.; Fukushi, K. Hybrid treatment systems for dye wastewater. *Crit. Rev. Environ. Sci. Technol.* **2007**, *37*, 315–377. [CrossRef]
3. Liu, Y.; Yan, M.; Geng, Y.; Huang, J. Laccase immobilization on poly(p-phenylenediamine)/$Fe_3O_4$ nanocomposite for reactive blue 19 dye removal. *Appl. Sci.* **2016**, *6*, 232. [CrossRef]
4. Hai, F.I.; Yamamoto, K.; Fukushi, K. Development of a submerged membrane fungi reactor for textile wastewater treatment. *Desalination* **2006**, *192*, 315–322. [CrossRef]
5. Hai, F.I.; Yamamoto, K.; Nakajima, F.; Fukushi, K.; Nghiem, L.D.; Price, W.E.; Jin, B. Degradation of azo dye acid orange 7 in a membrane bioreactor by pellets and attached growth of *Coriolus versicolour*. *Bioresour. Technol.* **2013**, *141*, 29–34. [CrossRef] [PubMed]
6. Asif, M.B.; Hai, F.I.; Singh, L.; Price, W.E.; Nghiem, L.D. Degradation of pharmaceuticals and personal care products by white-rot fungi—A critical review. *Curr. Pollut. Rep.* **2017**, *3*, 88–103. [CrossRef]
7. Asif, M.B.; Hai, F.I.; Hou, J.; Price, W.E.; Nghiem, L.D. Impact of wastewater derived dissolved interfering compounds on growth, enzymatic activity and trace organic contaminant removal of white rot fungi—A critical review. *J. Environ. Manag.* **2017**, *201*, 89–109. [CrossRef] [PubMed]
8. Luo, Y.; Guo, W.; Ngo, H.H.; Nghiem, L.D.; Hai, F.I.; Zhang, J.; Liang, S.; Wang, X.C. A review on the occurrence of micropollutants in the aquatic environment and their fate and removal during wastewater treatment. *Sci. Total Environ.* **2014**, *473*, 619–641. [CrossRef] [PubMed]
9. Alexander, J.T.; Hai, F.I.; Al-aboud, T.M. Chemical coagulation-based processes for trace organic contaminant removal: Current state and future potential. *J. Environ. Manag.* **2012**, *111*, 195–207. [CrossRef] [PubMed]
10. Pal, A.; Gin, K.Y.-H.; Lin, A.Y.-C.; Reinhard, M. Impacts of emerging organic contaminants on freshwater resources: Review of recent occurrences, sources, fate and effects. *Sci. Total Environ.* **2010**, *408*, 6062–6069. [CrossRef] [PubMed]
11. Yang, S.; Hai, F.I.; Nghiem, L.D.; Price, W.E.; Roddick, F.; Moreira, M.T.; Magram, S.F. Understanding the factors controlling the removal of trace organic contaminants by white-rot fungi and their lignin modifying enzymes: A critical review. *Bioresour. Technol.* **2013**, *141*, 97–108. [CrossRef] [PubMed]
12. Margot, J.; Maillard, J.; Rossi, L.; Barry, D.A.; Holliger, C. Influence of treatment conditions on the oxidation of micropollutants by trametes versicolor laccase. *New Biotechnol.* **2013**, *30*, 803–813. [CrossRef] [PubMed]
13. Hai, F.I.; Nghiem, L.D.; Khan, S.J.; Price, W.E.; Yamamoto, K. Wastewater reuse: Removal of emerging trace organic contaminants. In *Membrane Biological Reactors: Theory, Modeling, Design, Management and Applications to Wastewater Reuse*; Hai, F.I., Yamamoto, K., Lee, C., Eds.; IWA Publishing: London, UK, 2014; pp. 165–205, ISBN 9781780400655.

14. Hai, F.I.; Tadkaew, N.; McDonald, J.A.; Khan, S.J.; Nghiem, L.D. Is halogen content the most important factor in the removal of halogenated trace organics by mbr treatment? *Bioresour. Technol.* **2011**, *102*, 6299–6303. [CrossRef] [PubMed]
15. Tran, N.H.; Urase, T.; Kusakabe, O. Biodegradation characteristics of pharmaceutical substances by whole fungal culture trametes versicolor and its laccase. *J. Water Environ. Technol.* **2010**, *8*, 125–140. [CrossRef]
16. Cañas, A.I.; Camarero, S. Laccases and their natural mediators: Biotechnological tools for sustainable eco-friendly processes. *Biotechnol. Adv.* **2010**, *28*, 694–705. [CrossRef] [PubMed]
17. Purich, D.L. *Enzyme Kinetics: Catalysis & Control: A Reference of Theory and Best-Practice Methods*; Elsevier: Amsterdam, The Netherlands, 2010; p. 759, ISBN 9780123809247.
18. Nguyen, L.N.; Hai, F.I.; Price, W.E.; Kang, J.; Leusch, F.D.; Roddick, F.; van de Merwe, J.P.; Magram, S.F.; Nghiem, L.D. Degradation of a broad spectrum of trace organic contaminants by an enzymatic membrane reactor: Complementary role of membrane retention and enzymatic degradation. *Int. Biodeterior. Biodegrad.* **2015**, *99*, 115–122. [CrossRef]
19. Lloret, L.; Eibes, G.; Feijoo, G.; Moreira, M.T.; Lema, J.M. Degradation of estrogens by laccase from myceliophthora thermophila in fed-batch and enzymatic membrane reactors. *J. Hazard. Mater.* **2012**, *213–214*, 175–183. [CrossRef] [PubMed]
20. Nguyen, L.N.; Hai, F.I.; Price, W.E.; Leusch, F.D.; Roddick, F.; Ngo, H.H.; Guo, W.; Magram, S.F.; Nghiem, L.D. The effects of mediator and granular activated carbon addition on degradation of trace organic contaminants by an enzymatic membrane reactor. *Bioresour. Technol.* **2014**, *167*, 169–177. [CrossRef] [PubMed]
21. Luo, W.; Hai, F.I.; Price, W.E.; Guo, W.; Ngo, H.H.; Yamamoto, K.; Nghiem, L.D. High retention membrane bioreactors: Challenges and opportunities. *Bioresour. Technol.* **2014**, *167*, 539–546. [CrossRef] [PubMed]
22. Wijekoon, K.C.; Hai, F.I.; Kang, J.; Price, W.E.; Guo, W.; Ngo, H.H.; Cath, T.Y.; Nghiem, L.D. A novel membrane distillation–thermophilic bioreactor system: Biological stability and trace organic compound removal. *Bioresour. Technol.* **2014**, *159*, 334–341. [CrossRef] [PubMed]
23. Luo, W.; Hai, F.I.; Kang, J.; Price, W.E.; Nghiem, L.D.; Elimelech, M. The role of forward osmosis and microfiltration in an integrated osmotic-microfiltration membrane bioreactor system. *Chemosphere* **2015**, *136*, 125–132. [CrossRef]
24. Choi, J.-H.; Dockko, S.; Fukushi, K.; Yamamoto, K. A novel application of a submerged nanofiltration membrane bioreactor (NF MBR) for wastewater treatment. *Desalination* **2002**, *146*, 413–420. [CrossRef]
25. Hai, F.I.; Nghiem, L.D.; Modin, O. Biocatalytic membrane reactors for the removal of recalcitrant and emerging pollutants from wastewater. In *Handbook of Membrane Reactors: Reactor Types and Industrial Applications*; Basile, A., Ed.; Woodhead Publishing: Cambridge, UK, 2013; Volume 2, pp. 763–807, ISBN 9780857094155.
26. Modin, O.; Hai, F.I.; Nghiem, L.D.; Basile, A.; Fukushi, K. Gas-diffusion, extractive, biocatalytic and electrochemical membrane biological reactors. In *Membrane Biological Reactors: Theory, Modeling, Design, Management and Applications to Wastewater Reuse*; Hai, F.I., Yamamoto, K., Lee, C., Eds.; IWA Publishing: London, UK, 2014; pp. 299–334, ISBN 9781780400655.
27. Alkhudhiri, A.; Darwish, N.; Hilal, N. Membrane distillation: A comprehensive review. *Desalination* **2012**, *287*, 2–18. [CrossRef]
28. Qtaishat, M.R.; Banat, F. Desalination by solar powered membrane distillation systems. *Desalination* **2013**, *308*, 186–197. [CrossRef]
29. Wijekoon, K.C.; Hai, F.I.; Kang, J.; Price, W.E.; Cath, T.Y.; Nghiem, L.D. Rejection and fate of trace organic compounds (TrOCs) during membrane distillation. *J. Membr. Sci.* **2014**, *453*, 636–642. [CrossRef]
30. Duong, H.C.; Duke, M.; Gray, S.; Cooper, P.; Nghiem, L.D. Membrane scaling and prevention techniques during seawater desalination by air gap membrane distillation. *Desalination* **2016**, *397*, 92–100. [CrossRef]
31. Li, J.; Wu, J.; Sun, H.; Cheng, F.; Liu, Y. Advanced treatment of biologically treated coking wastewater by membrane distillation coupled with pre-coagulation. *Desalination* **2016**, *380*, 43–51. [CrossRef]
32. Phattaranawik, J.; Fane, A.G.; Pasquier, A.C.; Bing, W. A novel membrane bioreactor based on membrane distillation. *Desalination* **2008**, *223*, 386–395. [CrossRef]
33. Asif, M.B.; Nguyen, L.N.; Hai, F.I.; Price, W.E.; Nghiem, L.D. Integration of an enzymatic bioreactor with membrane distillation for enhanced biodegradation of trace organic contaminants. *Int. Biodeterior. Biodegrad.* **2017**, in press. [CrossRef]

34. Ashe, B.; Nguyen, L.N.; Hai, F.I.; Lee, D.-J.; van de Merwe, J.P.; Leusch, F.D.; Price, W.E.; Nghiem, L.D. Impacts of redox-mediator type on trace organic contaminants degradation by laccase: Degradation efficiency, laccase stability and effluent toxicity. *Int. Biodeterior. Biodegrad.* **2016**, *113*, 169–176. [CrossRef]
35. Nghiem, L.D.; Cath, T. A scaling mitigation approach during direct contact membrane distillation. *Sep. Purif. Technol.* **2011**, *80*, 315–322. [CrossRef]
36. Hai, F.I.; Tessmer, K.; Nguyen, L.N.; Kang, J.; Price, W.E.; Nghiem, L.D. Removal of micropollutants by membrane bioreactor under temperature variation. *J. Membr. Sci.* **2011**, *383*, 144–151. [CrossRef]
37. Nguyen, L.N.; van de Merwe, J.P.; Hai, F.I.; Leusch, F.D.; Kang, J.; Price, W.E.; Roddick, F.; Magram, S.F.; Nghiem, L.D. Laccase-syringaldehyde-mediated degradation of trace organic contaminants in an enzymatic membrane reactor: Removal efficiency and effluent toxicity. *Bioresour. Technol.* **2016**, *200*, 477–484. [CrossRef] [PubMed]
38. Van de Merwe, J.P.; Leusch, F.D. A sensitive and high throughput bacterial luminescence assay for assessing aquatic toxicity—The blt-screen. *Environ. Sci. Process. Impacts* **2015**, *17*, 947–955. [CrossRef] [PubMed]
39. D'Acunzo, F.; Galli, C.; Gentili, P.; Sergi, F. Mechanistic and steric issues in the oxidation of phenolic and non-phenolic compounds by laccase or laccase-mediator systems. The case of bifunctional substrates. *New J. Chem.* **2006**, *30*, 583–591. [CrossRef]
40. Hata, T.; Shintate, H.; Kawai, S.; Okamura, H.; Nishida, T. Elimination of carbamazepine by repeated treatment with laccase in the presence of 1-hydroxybenzotriazole. *J. Hazard. Mater.* **2010**, *181*, 1175–1178. [CrossRef] [PubMed]
41. Lloret, L.; Eibes, G.; Moreira, M.; Feijoo, G.; Lema, J. On the use of a high-redox potential laccase as an alternative for the transformation of non-steroidal anti-inflammatory drugs (NSAIDs). *J. Mol. Catal. B Enzym.* **2013**, *97*, 233–242. [CrossRef]
42. Nguyen, L.N.; Hai, F.I.; Kang, J.; Leusch, F.D.; Roddick, F.; Magram, S.F.; Price, W.E.; Nghiem, L.D. Enhancement of trace organic contaminant degradation by crude enzyme extract from trametes versicolor culture: Effect of mediator type and concentration. *J. Taiwan Inst. Chem. Eng.* **2014**, *45*, 1855–1862. [CrossRef]
43. D'Acunzo, F.; Baiocco, P.; Galli, C. A study of the oxidation of ethers with the enzyme laccase under mediation by two N–OH–type compounds. *New J. Chem.* **2003**, *27*, 329–332. [CrossRef]
44. Mizuno, H.; Hirai, H.; Kawai, S.; Nishida, T. Removal of estrogenic activity of iso-butylparaben and *n*-butylparaben by laccase in the presence of $_1$-hydroxybenzotriazole. *Biodegradation* **2009**, *20*, 533–539. [CrossRef] [PubMed]
45. Khlifi-Slama, R.; Mechichi, T.; Sayadi, S.; Dhouib, A. Effect of natural mediators on the stability of trametes trogii laccase during the decolourization of textile wastewaters. *J. Microbiol.* **2012**, *50*, 226–234. [CrossRef] [PubMed]
46. Kim, Y.-J.; Nicell, J.A. Impact of reaction conditions on the laccase-catalyzed conversion of bisphenol A. *Bioresour. Technol.* **2006**, *97*, 1431–1442. [CrossRef] [PubMed]
47. Marco-Urrea, E.; Pérez-Trujillo, M.; Vicent, T.; Caminal, G. Ability of white-rot fungi to remove selected pharmaceuticals and identification of degradation products of ibuprofen by trametes versicolor. *Chemosphere* **2009**, *74*, 765–772. [CrossRef]
48. Duong, H.C.; Hai, F.I.; Al-Jubainawi, A.; Ma, Z.; He, T.; Nghiem, L.D. Liquid desiccant lithium chloride regeneration by membrane distillation for air conditioning. *Sep. Purif. Technol.* **2017**, *177*, 121–128. [CrossRef]
49. Nghiem, L.D.; Hildinger, F.; Hai, F.I.; Cath, T. Treatment of saline aqueous solutions using direct contact membrane distillation. *Desalination Water Treat.* **2011**, *32*, 234–241. [CrossRef]
50. Duong, H.C.; Cooper, P.; Nelemans, B.; Cath, T.Y.; Nghiem, L.D. Optimising thermal efficiency of direct contact membrane distillation by brine recycling for small-scale seawater desalination. *Desalination* **2015**, *374*, 1–9. [CrossRef]

# 13

# Suppressing Salt Transport through Composite Pervaporation Membranes for Brine Desalination

**Lin Li [1], Jingwei Hou [1,*], Yun Ye [1], Jaleh Mansouri [1,2], Yatao Zhang [3] and Vicki Chen [1]**

1 The United Nations Educational, Scientific and Cultural Organization (UNESCO) Centre for Membrane Science and Technology, School of Chemical Engineering, University of New South Wales, Sydney 2052, Australia; lin.li@student.unsw.edu.au (L.L.); yun.ye@unsw.edu.au (Y.Y.); j.mansouri@unsw.edu.au (J.M.); v.chen@unsw.edu.au (V.C.)
2 Cooperative Research Centre for Polymers, Notting Hill 3168, Australia
3 School of Chemical Engineering and Energy, Zhengzhou University, Zhengzhou 450001, China; zhangyatao@zzu.edu.cn
* Correspondence: Jingwei.hou@unsw.edu.au

**Featured Application: The pervaporation membranes fabricated in this study can be potentially used for brine treatment.**

**Abstract:** Pervaporation membranes have gained renewed interest in challenging feedwaters desalination, such as reverse osmosis (RO) concentrated brine wastewater. In this study, composite polyvinyl alcohol (PVA)/polyvinylidene fluoride (PVDF) pervaporation membranes were prepared for brine treatment. The composite membrane was firstly studied by adjusting the cross-linking density of PVA by glutaraldehyde: the membrane with higher cross-linking density exhibited much higher salt rejection efficiency for long-term operation. A trace of salt on the permeate side was found to diffuse through the membrane in the form of hydrated ions, following solution-diffusion mechanism. To further suppress the salt transport and achieve long-term stable operation, graphene oxide (GO) was incorporated into the PVA layer: the addition of GO had minor effects on water permeation but significantly suppressed the salt passage, compared to the pure PVA/PVDF membranes. In terms of brine wastewater containing organic/inorganic foulant, improved anti-fouling performance was also observed with GO-containing membranes. Furthermore, the highest flux of 28 $L/m^2h$ was obtained for the membrane with 0.1 wt. % of GO using 100 g/L NaCl as the feed at 65 °C by optimising the pervaporation rig, with permeate conductivity below 1.2 μS/cm over 24 h (equivalent to a salt rejection of >99.99%).

**Keywords:** brine wastewater treatment; pervaporation; composite PVA/PVDF membrane; graphene oxide; anti-fouling properties

---

## 1. Introduction

Desalination has been widely utilized to relieve the shortage of fresh water in many parts of the world. However, a large amount of brine wastewater is also produced as a by-product of desalination. For example, seawater reverse osmosis (RO) retentate may contain concentrated salt, humic acid and other dissolved solids. The disposal of brine wastewater into the ocean or inland could lead to environmental and ecological problems [1]. Due to high osmotic pressures, RO cannot be utilized to treat such wastewater. On the other hand, the treatment efficiency of thermally driven processes like membrane distillation (MD) is less dependent on the feed solution concentration, and can be regarded as a promising candidate in brine wastewater treatment [2,3]. However, the major problems for the industrial application of MD are membrane fouling/scaling and pore wetting.

More recently, to address the pore wetting, a dense hydrophilic pervaporation membrane has been explored for brine treatment. Pervaporation is also a thermally driven process, and the main water transport mechanism is solution-diffusion [4]. It has been extensively studied for organic dehydration [5,6]. The purpose of both desalination and organic solvents dehydration is to separate water from the bulk feed solution; however, feed properties vary significantly. For organic solvent dehydration, water usually only accounts for less than 10% of the bulk solution [7], while water is the major component in feed for desalination (more than 90%). Thus, the major problem of applying many conventional pervaporation membranes for desalination is that the membranes are not stable in the solution with both high water fractions and elevated temperatures.

Polyvinyl alcohol (PVA) pervaporation membranes have high water vapor permeability, satisfactory membrane-forming capability and excellent anti-fouling property. Recently, PVA membranes have been applied for pervaporation desalination, but a certain amount of salt can still diffuse through the polymeric layers in the form of hydrated salt ions [8,9]. This unfavorable effect can be partially mitigated by polymer cross-linking. However, a higher salt rejection would be always compromised by a reduced permeability [10,11]. Aside from the organic cross-linker, the addition of inorganic nanofillers can also enhance the stability of the composite pervaporation membranes. For example, the PVA/maleic acid/silica freestanding membranes exhibited a reduced swelling degree with higher silica concentration [12,13]. Furthermore, graphene oxide (GO), a unique 2D inorganic material, has good interfacial compatibility with PVA due to the formation of hydrogen bonds [14], showing good potential to enhance PVA stability. GO has been extensively investigated for gas and liquid separation membranes [15,16]. For porous filtration membranes, the incorporation of GO can improve the permeation flux, salt rejection and anti-fouling properties due to improved hydrophilicity [17–23]. For a porous hydrophobic membrane for MD desalination applications, the immobilization of GO by PVDF as the binder material on the surface of polytetrafluoroethylene (PTFE) membranes showed enhanced flux due to selective sorption, nanocapillary effect, reduced temperature polarization and polar functional groups in GO [24]. It can also improve the thermal, mechanical and electrical properties of the composite membranes [25,26]. Considering its 2D structure, the incorporation of GO can potentially suppress the passage of hydrated salt ions within the PVA polymer. In addition, stacked GO laminates can also form highly efficient molecular sieving channels within polymeric matrix [27,28]. However, these aspects have yet to be explored for pervaporation membranes.

Furthermore, most studies on pervaporation desalination so far only use single monovalent salt solutions as feed, and the systems are only evaluated for a relatively short period. The long-term performance of composite pervaporation membranes is crucial for evaluating their feasibility for practical application, especially for complex brine treatment. Thus, in this study, a series of composite PVA/PVDF membranes were fabricated by coating a thin layer of PVA onto commercial hydrophobic PVDF membranes. Different cross-linking density was investigated to understand their effect on the membrane performance. In addition, GO nanosheets were blended into the PVA layer to further suppress salt transport through the membrane. To better understand the properties of the GO/PVA composite membrane, freestanding GO/PVA membranes were fabricated to investigate the interactions between GO and PVA, as well as the water/salt diffusion process within the composite layer. Furthermore, the composite membrane's anti-fouling performance was explored using highly concentrated brine solutions containing humic and calcium salts. Lastly, the pervaporation rig optimization was carried out to promote the operational flux for the composite membranes.

## 2. Materials and Methods

### *2.1. Materials*

Commercial hydrophobic microfiltration PVDF flat sheet membrane (Millipore, GVHP 0.22 μm, and thickness 125 μm, Billerica, MA, USA) was used in this study. Polyvinyl alcohol (PVA, Mw 89k–98k, 99+ % hydrolysed) was purchased from Sigma-Aldrich (Sigma-Aldrich Corp., St. Louis, MO, USA).

Sodium chloride (NaCl) and potassium hydroxide (KOH) pellets were supplied by Ajax Finechem (Cheltenham, VIC, Australia). Dextran (Mw 9k–11k) and glutaraldehyde (25% in water) were obtained from Alfa Aesar (Thermo Fisher Scientific, Heysham, UK). Hydrochloric acid (HCl, 32% aqueous solution) was obtained from RCl Labscan (RCl Labscan Limited, Bangkok, Thailand). Graphene oxide nanosheets were fabricated from graphene with a modified Hummer's method. The detailed process can be found in our previous publication [28].

### 2.2. *Composite Membrane Preparation*

The commercial supporting membrane was firstly pre-treated with 1 M KOH solution at 65 °C water bath for different length (0.5, 1, 2, 6 and 8 h) [29,30]. The KOH treatment led to reduction in water contact angle due to the hydrophilization effect. In this work, 2 h pre-treatment time was selected as longer treatment time did not lead to the increase of surface hydrophilicity (Figure S1). Subsequently, the membrane was rinsed with Milli-Q water and dried at room temperature. For the casting solution, aqueous PVA solution (10 wt. %) was prepared by dissolving PVA powder in Milli-Q water at 80 °C water bath with mechanical stirring for at least 6 h. Then the PVA solution was cooled down to room temperature. A certain amount of glutaraldehyde, and quencher methanol were added to the aqueous PVA solution and the mixture was magnetically stirred for another 30 min, which was followed by the addition of catalyst HCl and stirring for another 5 min. The molar ratio of glutaraldehyde/PVA repeat unit (denoted as MR value) was varied, while the ratio of methanol/HCl/PVA was maintained invariant, i.e., 1 g PVA was mixed with 2 mL of 10% methanol and 0.4 mL of 1 M HCl. To prepare the composite membrane with MR value of 0.2, for 1 g PVA prepared for the casting solution, the glutaraldehyde used (25% in water) was 1.72 mL. The PVA concentration in the final casting solution was maintained at 5 wt. % by adjusting the amount of Milli-Q water in the casting solution. The casting machine (Sheen 1133N automatic film applicator, Sheen Instruments, Surry, UK) was used to cast the PVA membrane on PVDF supports with a casting speed of 50 mm/s under 50% humidity. Then the coated membrane was dried at room temperature overnight, followed by oven drying at 80 °C for 30 min.

For the GO-containing PVA composite membrane fabrication, a certain amount of GO was evenly dispersed into the PVA casting solution containing glutaraldehyde and methanol. To initiate the cross-linking, HCl was then added to the above casting solution and stirred for 5 min. The membrane casting and drying procedures are identical to the previous pure PVA composite membranes. The weight ratio of GO to PVA (0.1, 0.2, 0.3 wt. %) was used to denote different samples, e.g., PVA0.1GO/PVDF denoted composite membrane with 0.1 wt. % of GO in PVA matrix. For the PVA-GO composite membranes, the molar ratio of glutaraldehyde to PVA repeat unit was kept invariant of 0.2.

In order to understand the interfacial interactions between GO and PVA, the freestanding PVA-GO membranes were prepared by pouring the above-mentioned casting solution in Petri dishes, followed by the same drying and post-heat treatment procedure as the cast coated membrane.

### 2.3. *Pervaporation Desalination*

Pervaporation desalination experiments were conducted by using a laboratory scale pervaporation unit, shown in Figure 1 A membrane with an effective surface area of 40 $cm^2$ was placed in the middle of the module, where four thermocouples were installed to monitor the inlet and outlet temperature of feed and permeate, in order to have an accurate measurement of the temperature difference on both feed and permeate sides across the membrane. During the pervaporation test, the feed solution was preheated in a water bath and pumped to the membrane module with a cross-flow velocity (CFV) of 0.625 m/s, while maintaining feed inlet temperature to the module at 65 ± 1 °C. The permeate water vapor was withdrawn by a pump on the permeate side with vacuum pressure around 24 kPa (unless otherwise stated), and collected in the beaker on the balance after condensing (cold water maintained at ~10 °C). Both the feed pump and permeate pump used in this experiment were Masterflex L/S variable speed digital peristaltic pump (Cole-Parmer, Vernon Hills, USA). Spacers (commercial RO module spacer, with filament thickness of 0.66 mm, mesh length of 3.1 mm and spacer thickness of 1.18 mm, Synderfiltration, Vacaville, USA) were used in both of the membrane feed and permeate sides to support the membrane. During the operation, the feed tank was topped up using the solution with the same concentration as the original feed.

The salt rejection (R) was calculated by the following equation:

$$R = \frac{C_f - C_p}{C_f} \times 100\% = \frac{\kappa_f - \kappa_p}{\kappa_f} \times 100\% \quad (1)$$

where $C_f$ and $C_p$ referred to the feed and permeate salt concentration, $\kappa_f$ and $\kappa_p$ referred to the feed and permeate conductivity.

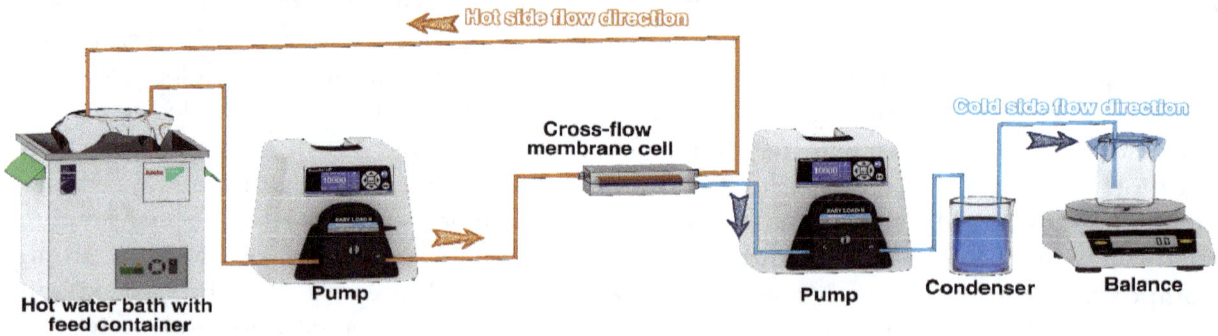


**Figure 1.** Schematic diagram of cross-flow pervaporation setup with flat-sheet membrane module.

### *2.4. Membrane Characterization*

#### 2.4.1. Scanning Electron Microscopy (SEM)

The surface and cross-section of the membranes were characterized by field emission scanning electron microscopy (FE-SEM, FEI Nova NanoSEM, Hillsboro, OR, USA). The cross-sectional SEM images of the composite membrane were obtained by snapping the membrane in liquid nitrogen. Samples for FE-SEM were prepared by sputter coating a thin layer of chromium under vacuum to generate conductivity. The qualitative surface chemistry of the membrane after desalination was investigated by energy dispersive X-ray spectroscopy (EDX) (FEI Nova NanoSEM and Hitachi S3400, Hitachi Ltd, Tokyo, Japan) to detect the presence of salt.

#### 2.4.2. Membrane Hydrophobicity/Hydrophilicity Characterization

Static contact angles were measured using a contact angle goniometer (KSV CAM 200, Biolin Scientific, Gothenburg, Sweden) by the sessile drop method. Reported values were the average of at least 5 measurements.

#### 2.4.3. Equilibrium Water Content (EWC)

EWC was measured to assess the cross-linked PVA's swelling property, as the ratio of the weight of water in the hydrogel to the weight of the hydrogel at equilibrium hydration. A freestanding hydrogel film was weighted and then immersed in 50 mL Milli-Q water for at least 24 h at room temperature, in order to achieve the equilibrium hydration state. The surface of the film was then blotted with an absorbent tissue paper (Kimwipe, Holcomb Bridge Road Roswell, GA, Canada) to remove excess water present on the surface and weighed again. The EWC value was calculated as follows:

$$EWC = \frac{m_w - m_d}{m_w} \quad (2)$$

where $m_w$ and $m_d$ represented the weight of cross-linked PVA at equilibrium hydration and dry state.

2.4.4. Salt Desorption Test

Salt transport in hydrogel film was measured using kinetic desorption experiment at room temperature. Dense freestanding membrane was prepared following the same procedure as for the EWC test. A hydrogel film with known thickness was immersed in 50 mL of 100 g/L NaCl for 24 h, which was long enough for the film to absorb an equilibrium amount of NaCl from the solution. Then the surface of the film was dried by Kimwipe and placed in a beaker with 50 mL Milli-Q water and stirred by magnetic stirring at around 500 rpm speed. Solution conductivity increased as the salt diffused out of the film into the Milli-Q water and the solution conductivity was recorded. Calculation of the NaCl diffusivity in the hydrogels was done using a Fickian analyses of NaCl desorption from a planar film.

2.4.5. Pore size Characterization

Liquid entry pressure (LEP) and mean pore size of the composite membrane were measured using capillary flow porometer from Porous Materials Inc. (PMI, Ithaca, NY, USA), based on wet/dry flow method [31]. A membrane sample with a diameter of 13 mm was placed in the chamber between two O-rings. In addition, then the membrane surface was covered by the wetting agent (Galwick®, PMI, Ithaca, NY, USA) with defined surface tension (15.9 dynes/cm). The gas flow rate through the membrane was measured as a function of the differential pressure. All porometry tests were performed at room temperature.

2.4.6. Pressurized Dead-End Filtration Test

For the membrane pore size smaller than the porometer instrument capability, a pressurized filtration test was used to determine the membrane pore size with a dead-end filtration membrane cell. Membrane resistance and permeability were also investigated using the same technique. Before the test, membrane sample was soaked in Milli-Q water overnight. Permeate flux over time was monitored with different feed solutions, i.e., Milli-Q water, 30 g/L NaCl, and 100 mg/L dextran. In addition, the dextran content of both feed and permeate was measured using the total organic carbon analyser (TOC, Shimadzu V-CSH, Kyoto, Japan). The entire filtration test was performed at room temperature.

2.4.7. X-ray Diffraction (XRD)

To determine whether GO sheets were dispersed as separated sheets in PVA matrix, XRD measurements (Empyrean X-ray diffraction system, PANalytical, 7602 EA Almelo, the Netherlands) was carried out by using Cu $K_\alpha$ radiation.

2.4.8. Fourier Transform Infrared Spectroscopy (FT-IR)

FT-IR (Spotlight 400, PerkinElmer, Waltham, MA, USA) was used to characterize the surface properties of composite membranes.

2.4.9. Differential Scanning Calorimetry (DSC)

Thermal analysis of the freestanding membrane was conducted using a DSC Q20 (TA Instruments, Inc., New Castle, DE, USA). The samples were heated from −30 °C to 300 °C at a rate of 10 °C/min in a nitrogen atmosphere. Approximately 6–8 mg of sample is used for each DSC measurement.

## 3. Results

### *3.1. Investigation on Cross-linking of PVA by Glutaraldehyde*

#### 3.1.1. Pervaporation Performance Using Single Salt Brine

PVA/PVDF membranes fabricated with different molar ratio of glutaraldehyde to PVA repeat unit (MR values of 0.025, 0.1, 0.2) were investigated regarding their pervaporation desalination performance using 100 g/L NaCl as the feed (Figure 2). In this work, each pervaporation test was carried out with at least two membrane samples, and the difference in permeation flux and salt rejection was less than 10% throughout the whole testing process.

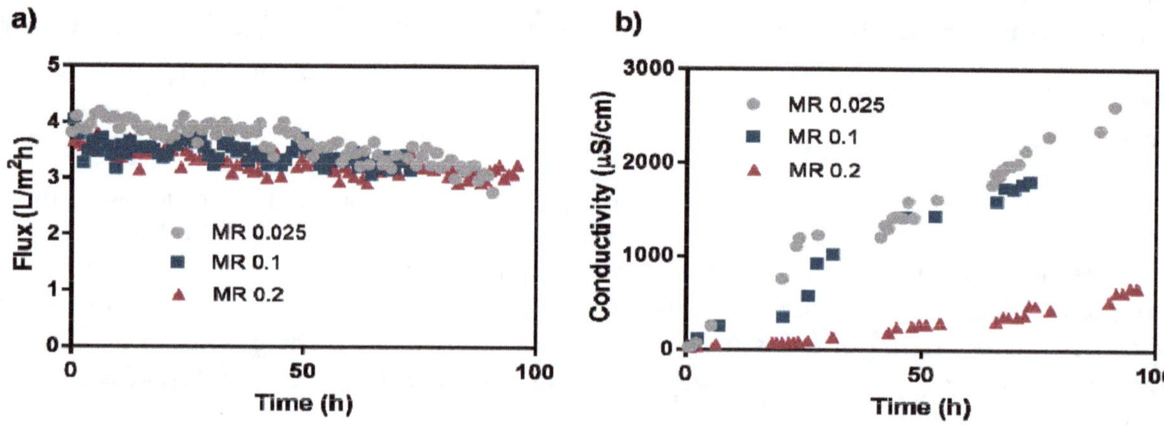


**Figure 2.** (**a**) Flux and (**b**) permeate conductivity profile of composite PVA/PVDF membrane with different molar ratio of glutaraldehyde to PVA repeat unit (MR 0.025, 0.1, 0.2) using a feed solution of 100 g/L NaCl at 65 °C.

As shown in Figure 2a, all membrane samples had comparable water flux under the same operating condition although a slightly lower flux was observed for membranes with higher MR value. Besides, all the membranes experienced continuous flux declines during the whole testing process due to the gradual increase of the feed concentration. In terms of the permeate conductivity, all membrane samples showed an increased conductivity with a longer operation time, and the membrane with the highest MR value (0.2) showed the most stable salt rejection after 96 h operation. This trend suggested that the cross-linking of PVA can benefit the operational stability of the membrane, which can be attributed to more acetal/ether linkages with higher MR value, leading to the improved structural stability with reduced free volume and swelling tendency [32,33]. Comparatively, the virgin supporting membrane, hydrophobic PVDF membrane with nominal pore size of 0.22 µm, showed a permeate conductivity of 2 mS/cm after 10 min of operation under the same operating condition. This indicated that the salt rejection efficiency of the pervaporation process was not attributed to the supporting membrane used in the composite membrane.

The water flux results in this work were comparable to the values reported in previous literatures of pervaporation membranes, when taking the operating temperature and feed solution properties into consideration. For example, composite membranes containing 100–1000 nm PVA surface coating on polysulfone membrane yielded fluxes of 4.6–7.4 $L/m^2h$ with 30 g/L NaCl solution at 70 °C [34]. In another line of research, a slightly higher water flux (5.57 $L/m^2h$) was obtained with a PVA/polyacrylonitrile (PAN)/polyethylenimine (PEI) triple-layered membrane when using 50 g/L NaCl solution as feed under room temperature [35]. In terms of the salt rejection, the PVA membrane in this work exhibited higher salt rejections even with more concentrated feed solution compared with the literatures [34,35].We also explored the membrane stability by recycling the membrane and testing the membrane using different feed concentrations and salt types, as shown in Figures S2 and S3.

However, considering the permeate flux was relatively invariant, the gradual increase of permeate conductivity for all the PVA/PVDF membranes indicated a more rapid transport of salt ions as the pervaporation test proceeded. The following characterizations and discussion were aimed to better understand the transport mechanism of salt through the composite PVA/PVDF membrane, to provide guidance for the further modification of such membranes.

3.1.2. Transport Mechanism of Salt

The salt transport through the membrane was further confirmed by the EDX results (Figure 3): the presence of salt crystals was observed on the membrane feed side as well as within the porous substrate after 96 h pervaporation test. The transport of salt ions through the PVA layer could be originated via the free volume between PVA polymeric chains. In this work, the gradual increase of the conductivity suggested a change of PVA chain structure overtime, possibly due to the swelling effect by hot water feed during the pervaporation process.

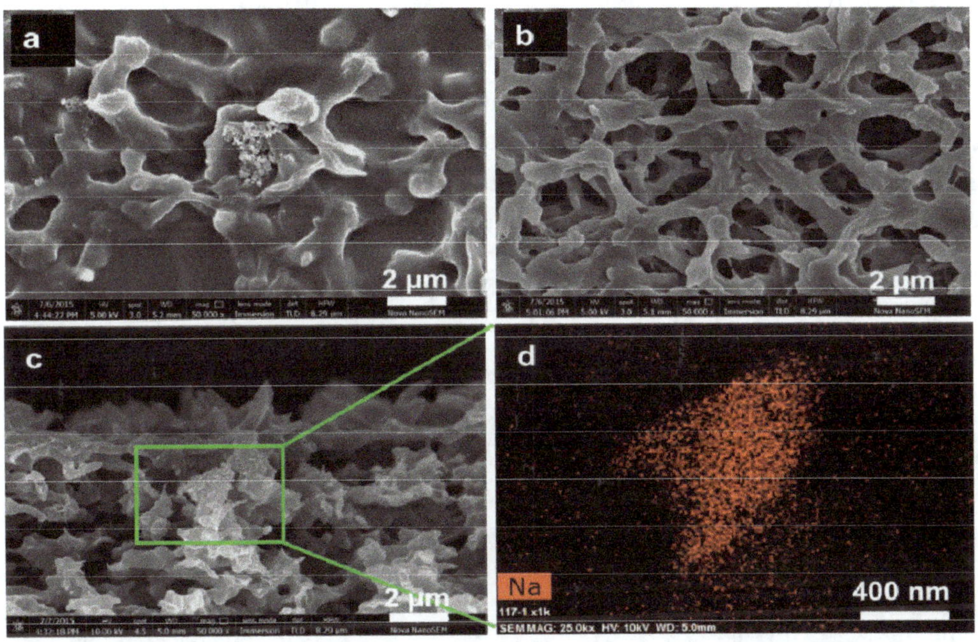


**Figure 3.** SEM and EDX autopsy of the composite membrane after 96 h pervaporation test (membrane with MR of 0.2). (**a**) Feed side, (**b**) permeate side and (**c**) cross-section of the composite membrane; and (**d**) EDX mapping the membrane cross-section beneath the PVA coating layer.

To further explore the PVA membrane pore structure, a porometer test was carried out with the composite membrane (both fresh membranes and membranes after 96 h permeation test) using Galwick as a wetting agent at room temperature. However, no direct liquid penetration through the membrane was observed for all the membranes with different MR values under maximum feed of 10 bar. To further explore the pore size (free volume) range of the composite membrane, we conducted the dead-end filtration test with 15 bar feed pressure using fresh composite membranes at room temperature. As suggested in Table 1, the resistance for MR 0.2 membrane was in the range of RO membranes, with a pure water flux of 1.4 ± 0.4 L/m$^2$h. The rejection of dextran (1–2 nm) can remain ~83.4% over 10 h operation. In comparison, the rejection of NaCl (diameter of hydrated $Na^+$ 0.716 nm, hydrated $Cl^-$ 0.664 nm) was much lower at around 33%, which was lower than the reported PVA composite nanofiltration (NF) membrane [36,37]. Thus, it was believed that the pore size (free volume) of the composite membrane (MR of 0.2) was in the range of NF membrane (below 1 nm). Deng et al. also reported that the diameter of casted/cross-linked PVA membrane's free volume was around 0.426 nm tested by bulk positron annihilation lifetime spectroscopy (PALS) [38].

**Table 1.** Pressurized filtration test results of composite PVA/PVDF membrane using feed solutions of Milli-Q water, 30 g/L NaCl or 100 mg/L dextran, under TMP of 15 bar at room temperature. The flux was recorded after the stabilization of 30 min. The membrane resistance ($R_m$) was calculated based on $R_m = \frac{\Delta P}{\mu J}$, where ΔP, μ and J indicated the trans-membrane pressure, viscosity of Milli-Q water and corresponded pure water flux, respectively.

| Pressurized Filtration Test Results | Unit | Commercial PVDF | MR0.025 | MR0.1 | MR0.2 |
|---|---|---|---|---|---|
| Pure water flux | $L/m^2h$ | >15,000 | $240 \pm 56$ | $5.5 \pm 0.27$ | $1.4 \pm 0.4$ |
| Permeability | $L/m^2\,h{\cdot}bar$ | >1000 | $16 \pm 3.5$ | $0.37 \pm 0.018$ | $0.093 \pm 0.027$ |
| Average membrane resistance | $m^{-1}$ | $<3.9 \times 10^{11}$ | $\sim 2.5 \times 10^{13}$ | $\sim 1.1 \times 10^{15}$ | $4.3 \times 10^{15}$ |
| Flux using 30 g/L NaCl as the feed | $L/m^2h$ | / | $129 \pm 16.5$ | $3.74 \pm 0.23$ | $0.54 \pm 0.1$ |
| Salt rejection using 30 g/L NaCl as the feed | % | / | ~0 | ~9 | ~33 |
| Flux using 100 mg/L Dextran as the feed | $L/m^2h$ | / | $142 \pm 21$ | $4.69 \pm 0.42$ | $0.99 \pm 0.18$ |
| Rejection using 100 mg/L Dextran as the feed | % | / | ~0 | 53.9% | 83.4 |

It also indicated that the change of MR value had a significant effect on the PVA layer pore (free volume) structure. With the increase of MR value, the membrane resistance gradually increased, accompanied with the increase of rejections for salt ions and dextran. In this work, the lowest resistance was observed for the membrane with MR 0.025: it fell into the range of UF or even MF membranes.

Figure 4 showed the SEM and atomic force microscope (AFM) analysis of membrane with different MRs. With the increase of MR value, the PVA layer thickness became more homogeneous, accompanied with the loss of surface roughness. The surface roughness of PVDF membrane after KOH pre-treatment was ~200 nm, which was comparable to the composite membrane with MR of 0.025. This indicated that the coating of PVA with low MR value (0.025) did not significantly change the membrane's surface roughness. Furthermore, the change of MR value did not only alter the membrane pore (free volume) structure; it can also affect the chemical component within the selective layer. Theoretically, the cross-linking of PVA with glutaraldehyde can consume the hydroxyl groups for the polymer. It can also increase the polymer chain structure rigidity. Both aspects can lead to the reduced EWC value for the membrane as shown in Table 2, which well aligns other studies [32,33].

**Table 2.** Characterization results of composite PVA/PVDF membrane (Diffusivity of NaCl in pure water is $14.7 \times 10^{-6}$ $cm^2/s$). MR: Molar Ratio. EWC: Equilibrium Water Content.

| Composite Membrane | Surface Roughness (nm) | EWC (%) in Water | EWC (%) in 100 g/L NaCl | NaCl Diffusivity ($10^{-6}$ $cm^2/s$) |
|---|---|---|---|---|
| MR0.025 | $281.5 \pm 7.78$ | $184.5 \pm 35.3$ | $129.7 \pm 16.8$ | $2.02 \pm 0.97$ |
| MR0.1 | $140 \pm 31.1$ | $60.4 \pm 5.2$ | $56.6 \pm 2.5$ | $1.42 \pm 0.36$ |
| MR0.2 | $106.4 \pm 13.58$ | $49.6 \pm 7.0$ | $38.1 \pm 6.3$ | $0.64 \pm 0.13$ |

This could well explain the observation of slightly lower permeate flux but improved salt rejection for the membrane with higher MR values (Figure 2): the transport rate of bulkier penetrant (hydrated salt ions) was more sensitive to the changes in free volume than those of smaller penetrant (water vapor) [39,40]. Therefore, the decreased diffusivity of salt through the membrane with higher MR (Table 2) can be attributed to its lower free volume.

During the pervaporation process, salt ions could diffuse through the PVA layer in the form of hydrated ions and condensed together with water due to the temperature drop in the permeate side. The increase of cross-linking density can retard but not completely block the passage of salt ions. At the same time, the gradual up-take of water can lead to the swelling of PVA due to the preferable adsorption of water molecules by the hydroxyl groups on PVA chains. In the swollen PVA especially the one with large free volume, more water was adsorbed by the hydroxyl groups of PVA chains, where a small fraction of hydrated salt ions could fit into such enlarged free volume, allowing their gradual passage. As a result, the increased free volume allows fast dissolution and diffusion of bulkier hydrated salt ions through the membrane [34,41,42], which explained the gradual loss of the salt rejection during the extended pervaporation test (Figure 5). Thus, in the following session, GO nanosheets, impermeable hydrophilic nanosheets, were incorporated into the PVA matrix to limit the free volume increase with time in PVA layer and further suppress salt transport through the membrane. MR value of 0.2 was applied to fabricate the membranes with GO because of its highest salt rejection efficiency compared to membranes with lower MR values.

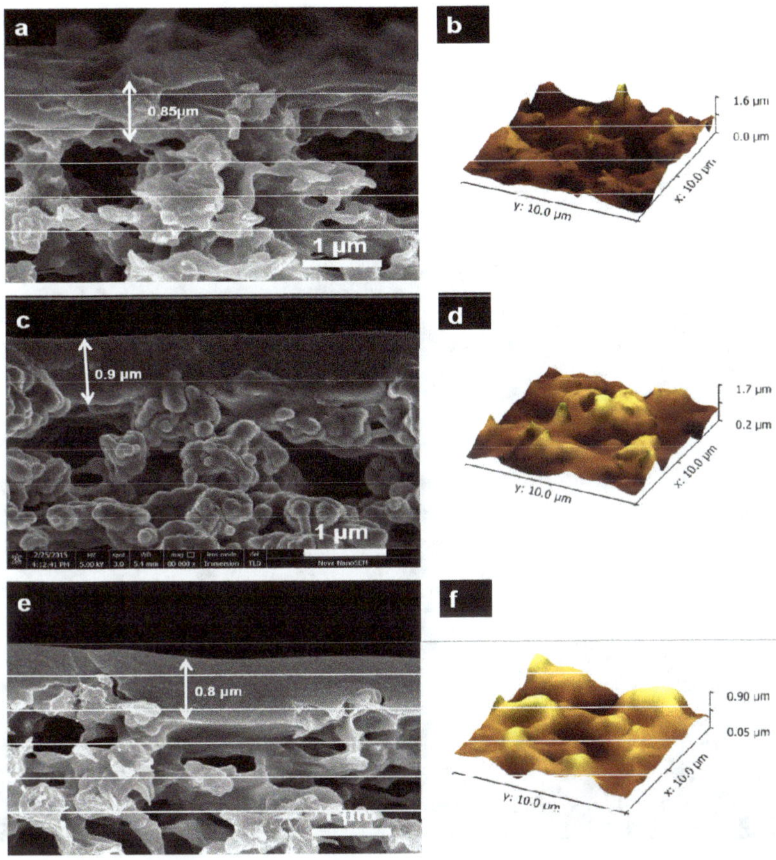


**Figure 4.** Cross-sectional SEM and AFM surface height images of PVA/PVDF composite membrane with MR of (**a**) and (**b**) 0.025; (**c**) and (**d**) 0.1; (**e**) and (**f**) 0.2.

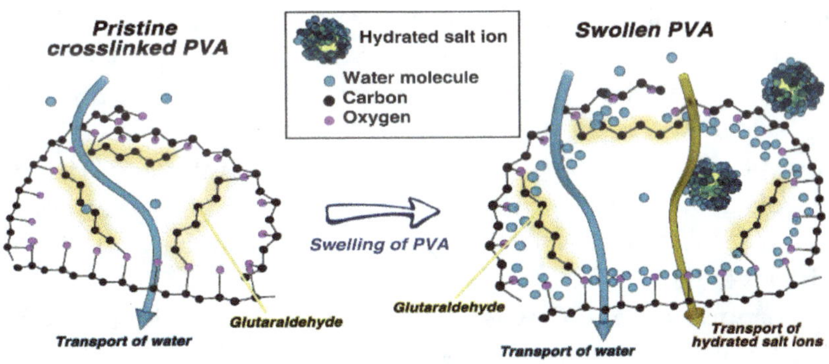


**Figure 5.** Schematic diagram of salt passage through the swollen PVA membrane (not to scale).

## *3.2. Incorporation of GO into PVA Matrix*

### 3.2.1. Characterizations

The addition of a small amount of GO into PVA formed small protrusions on the membrane surface, and this became more significant with the increase of GO loading within the coating layer

(Figure 6). Similar surface morphology was observed after incorporating GO into polyether block amide (Pebax) matrix using cast coating method [43]. In terms of the cross-sectional images (Figure S4), they showed good agreement with the surface SEM images. The pure PVA coating was even with a thickness of around 0.8 μm. The incorporation of GO reduced the smoothness for the PVA-GO coating layer, especially when the GO loading was relatively high. During the formation of PVA-GO surface coating, the gradual evaporation of the solvent (water) allowed the cross-linking of the PVA chains, forming the final coating layer of 0.5–1 μm. In terms of the GO, the evaporation-induced capillary flow during the drying process, together with the viscous PVA solution, disrupted the entropy-driven phase transition of GO, leading to the formation of surface protrusions [44].

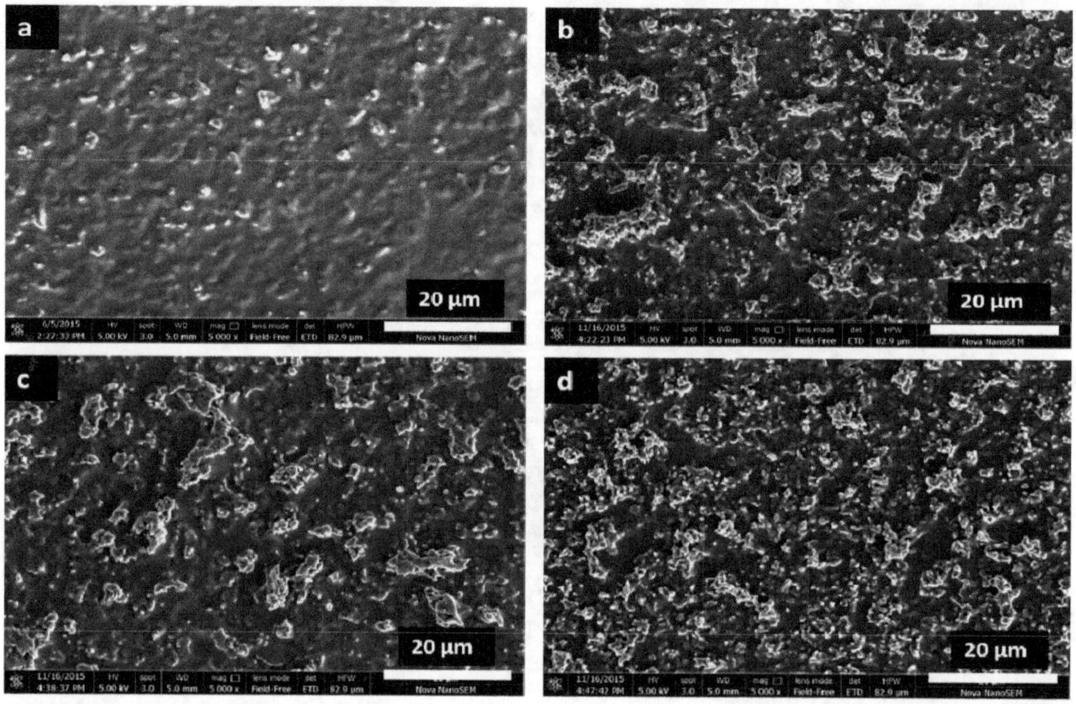


**Figure 6.** Surface SEM images of composite membranes fabricated with cast coating method: (**a**) pure PVA coating; and PVA-GO hybrid coating with (**b**) 0.1 wt. %, (**c**) 0.2 wt. %, (**d**) 0.3 wt. % GO loading.

The XRD pattern can reveal the GO nanosheets arrangement within the hybrid coating layer. As shown in Figure 7, for the pure GO coated membrane, it had a clear peak at 10.6°, suggesting the GO nanosheets were piled together with an average interlayer distance of 0.83 nm. After incorporating the GO into PVA matrix, no clear GO peak can be observed. This observation was in good agreement with the previous researches [14,45]. This can be attributed to the good interfacial compatibility between GO and PVA: the presence of carboxyl groups and hydroxyl groups on GO can form hydrogen bonding with hydroxyl groups on PVA. As a result, the GO nanosheets can be well dispersed within the PVA matrix at the molecular level [26].

The successful incorporation of GO was further evidenced by the FT-IR patterns, as shown in Figure 8 The characteristic peaks of GO powder at around 3280 $cm^{-1}$ and 1725 $cm^{-1}$ corresponded to O–H and C=O stretch, which indicated the existence of hydroxyl groups and carbonyl groups on GO. Besides, the peak at around 2950 $cm^{-1}$ confirmed a large amount of C–H stretch on GO. Compared to pure PVA coating, the emerging peaks for the PVA-GO composite coating layer at the region of 3100–3010 $cm^{-1}$ can be ascribed to alkenyl C–H stretch due to the presence of GO. In addition, the right shift of the O-H stretching peaks were detected with the addition of GO in PVA matrix (from 3375 $cm^{-1}$ for pure PVA to 3345 $cm^{-1}$ for PVA with GO of 0.3 wt. %). This observation confirmed the formation of hydrogen bonding between hydroxyl/carboxylate groups on GO and hydroxyl groups on PVA chains, which was consistent with those of other studies [14,45,46].

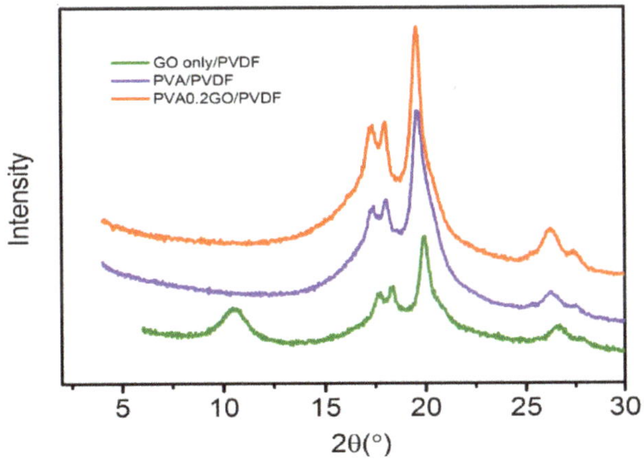


**Figure 7.** XRD curves of composite membranes.

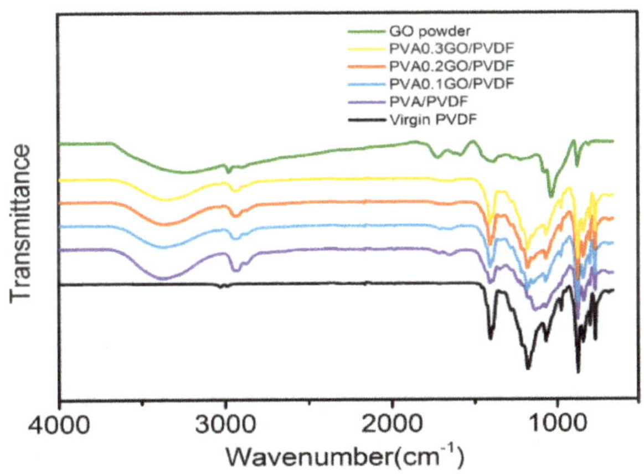


**Figure 8.** FT-IR curves of PVA-GO/PVDF membranes with GO loading of 0.1, 0.2 and 0.3 wt. %.

Regarding the contact angle shown in Table S1, the addition of GO into PVA matrix slightly increased the contact angle of the membrane surface. For example, the pure PVA surface coating has a contact angle of $30.2 \pm 3.5°$, compared with the $37.1 \pm 2.3°$ for the composite membrane containing 0.2 wt. % GO. This can be attributed to the presence of a large amount non-polar benzene rings on GO surface, as well as the formation of hierarchical surface morphology after the addition of GO [47].

To better understand the effect of GO incorporation on the coating layer properties, freestanding PVA-GO films were also prepared and characterized to explore their surface morphology, thermal and swelling properties. The thickness of the freestanding films was around 100 μm. From the digital photo (Figure S5), with higher GO loading, the freestanding film color changed from almost transparent to yellowish brown. Similar to the composite membrane fabricated via cast coating, the addition of GO introduced wrinkles and protrusions onto the membrane surface, as shown in Figure S6. As the freestanding films were fabricated via the solvent evaporation method, without the shear force induced alignment during the casting process, the GO nanosheets randomly oriented within the original membrane casting solution, and eventually led to the wrinkled paper-like structure on the surface due to the capillary flow during the solvent evaporation process. Based on the SEM images, GO was homogeneously dispersed within the PVA matrix. Same as the composite membranes, no GO peak was observed in freestanding films after the incorporation of GO in PVA matrix from the

XRD patterns (Figure S7), indicating a good interfacial compatibility between the GO nanosheets and the PVA matrix.

As shown in Table 3, the DSC results exhibited an increased Tg for the freestanding films after the incorporation of GO nanosheets (from around 48 °C to around 60 °C). This further confirmed the formation of hydrogen bonding between GO nanosheets and PVA matrix, which reduced the polymer chain mobility. This could be also lead to the rigidification of polymer chains near the GO sheets [15]. Similar trends were also observed by incorporating a small amount of GO into PVA matrix [45,48].

**Table 3.** Thermal and swelling property of the freestanding films (PVA0.1GO denoted membrane with 0.1 wt. % of GO in PVA matrix). PVA: polyvinyl alcohol. GO: graphene oxide.

| Freestanding Films | Tg (°C) | EWC in Milli-Q Water (%) | EWC in 100 g/L NaCl (%) | NaCl Diffusivity ($10^{-6}$ $cm^2/s$) |
|---|---|---|---|---|
| **PVA only** | 48 | 49.6 ± 7 | 38.1 ± 6.3 | 0.64 ± 0.13 |
| **PVA0.1GO** | 59 | 25.35 ± 1.41 | 28.06 ± 1.07 | 0.712 ± 0.001 |
| **PVA0.2GO** | 60 | 38.15 ± 2.05 | 34.98 ± 1.66 | 0.549 ± 0.066 |
| **PVA0.3GO** | 60 | 35.35 ± 4.37 | 30.68 ± 0.54 | 0.526 ± 0.005 |

The freestanding films' swelling properties were also studied as it is closely related to the membrane's cross-linking density and free volume [12]. The equilibrium water content (EWC) values of freestanding films (Table 3) in both Mill-Q water and 100 g/L NaCl aqueous solution were investigated. Among different membranes, the pure PVA freestanding film exhibited the highest EWC values for both Milli-Q water and salt water. The initial addition of GO (0.1 wt. %) significantly reduced the EWC value. This can be attributed to the impermeable feature of GO and formation of hydrogen bonding between GO and PVA, which reduced the solubility of water in the PVA polymer chains. However, with the increase of GO loading (0.2 and 0.3 wt. %), the EWC values increased again due to the water adsorption on the nanosheet surface, but they were still lower than the pure PVA films. Another potential reason could be the aggregation of GO sheets in PVA matrix created new pathways for water adsorption and diffusion. Smaller free volume fraction was also obtained by adding GO into PVA matrix from other studies [45].

By comparing the EWC value in Milli-Q water and salt water, it was found that for pure PVA films the EWC value in salt water was significantly lower than that in Milli-Q water. While for the films with GO incorporated, the difference is less significant. This indicated that the swelling ability of PVA with GO was less affected by the salt content, compared to pure PVA. Furthermore, with the addition of GO into the PVA matrix, the salt diffusivity tended to decrease. This was caused by much longer and more torturous transport pathways for salt to pass through the film, since GO is an impermeable nanosheet to salt ions. It is worth mentioning that in the salt diffusivity test, with the addition of GO, the amount of salt absorbed by the membrane decreases significantly. After immersing the hydrated films into Milli-Q water, the equilibrium solution conductivity dropped from around 120 µS/cm for pure PVA films down to 20 µS/cm for PVA with 0.1 wt. % GO. It took similar time (~200 s) for both films to achieve equilibrium during the immersion process. This clearly indicates that less salt is adsorbed by the membrane with GO incorporated: i.e., the GO incorporation considerably decreased the membrane's salt solubility in the PVA films.

From the above characterizations on both composite membranes and freestanding films with GO incorporated into PVA matrix, a good dispersion of GO in PVA was observed by SEM and XRD results; and the hydrogen bonding formed between PVA and GO is detected from FT-IR and DSC analysis. The EWC and NaCl diffusivity results showed that the incorporation of impermeable GO decreased the size of the free volume in PVA matrix and also the water and salt solubility in the matrix.

### 3.2.2. Pervaporation Performance Using Single Salt Brine

The composite membranes with different GO loadings in PVA matrix (0.1, 0.2, and 0.3 wt. %) were tested in the pervaporation setup, which was then compared with the pure PVA/PVDF membrane (MR value of 0.2). For each membrane, Milli-Q water permeation tests were firstly conducted for at least 4 h. Compared with pure PVA/PVDF membrane, the incorporation of 0.1 wt. % GO showed the negligible effect on the Milli-Q water flux (4.38 L/$m^2$h). However, further increase of the GO loading can lead to a marginal loss of the flux: 94.4 and 85.5% of the original PVA/PVDF flux can be obtained for composite membranes containing 0.2 and 0.3 wt. % GO (results not shown). In highly cross-linked PVA polymer chains, the free volume was organized in a tortuous manner, which led to the random and non-directional water passage pathway [49]. After the addition of GO, the presence of a large amount of hydrophobic aromatic rings on its surface can facilitate the rapid "sliding" of water molecules along its surface [50]. This aspect would facilitate the water transport through the membrane. On the other hand, the incorporation of impermeable GO sheets also increased the transport pathway length of water vapor. Meanwhile, lower free volume, measured as the EWC value, was obtained as a result of hydrogen bonding between GO and PVA chains. As a result, the initial addition of GO (0.1 wt. %) had a negligible effect on the Milli-Q water flux and a further increase of the GO loading (0.2 and 0.3 wt. %) slightly reduced the permeation flux.

In terms of the brine desalination tests with 100 g/L NaCl solution, similar flux profiles (Figure 9) were observed as to the Milli-Q results: the initial addition of 0.1 wt. % GO had a negligible effect on the brine solution flux, and a further increase of the GO loading can lead to more obvious loss of the permeation flux. In terms of permeate conductivity, it constantly increased to over 650 μS/cm (salt rejection of over 99.6%) after 96 h operation for the pure PVA/PVDF membrane, while the composite membrane containing 0.1 wt. % GO had a much more stable performance: the permeate conductivity gradually increased to less than 150 μS/cm (salt rejection of over 99.9%) for 120 h operation while maintaining lower than 50 μS/cm in the first 50 h. As discussed above, the presence of GO nanosheets within the PVA matrix can facilitate the selective water molecule transport along with its "slippery" surface, which suppressed the transport of salt ions through the composite membrane. The reduced free volume of PVA matrix had a more significant blocking effect on the larger hydrated salt ions compared with smaller water molecules. As a result, the addition of GO can improve the salt rejection for the composite membrane.

When 0.2 wt. % GO was added into the PVA matrix, the permeate conductivity was more stable (less than 30 μS/cm in 82 h operation). However, in this case, the permeate flux was much lower than the PVA0.1GO/PVDF membrane, and a significant flux decline was observed after 10 h operation. Due to the reduced free volume after GO addition, the blockage effect by the hydrated salt ions to the smaller water molecules can be more significant. This can also explain the more significant flux loss for membrane containing 0.3 wt. % GO. In addition, for the membrane containing 0.3 wt. % GO, the permeate conductivity increased more rapidly compared with a membrane containing 0.1 and 0.2 wt. % GO. The possible reason was that some defects or cracks may exist due to strong filler interactions and agglomerations. Similar results were also observed with a composite membrane containing silica nanoparticles and ordered mesoporous carbon (OMC), less water uptake and permeable flux were detected with higher loading of nanoparticles in the thin-film layer, and the overdose of nanofillers can also lead to the loss of membrane selectivity [51–53].

Upon the completion of the long-term brine desalination test, the membranes were re-tested with Milli-Q water to investigate their permeation flux recovery. The results suggested even though the initial flux was much lower, it can gradually recover to the original Milli-Q water flux prior to the brine desalination test in less than 2 h, shown in Figure S8. This observation confirmed flux decline during the brine desalination test was mainly due to the blockage of water transport pathway by the bulkier hydrated salt ions. In all, with the increase of GO loading in PVA matrix, the water flux tended to decrease for both Milli-Q water and brine, while the salt rejection firstly increased then decreased. Similar membrane behavior has been reported for the PVA-GO-based composite pervaporation membrane for organic solution dehydration [54].

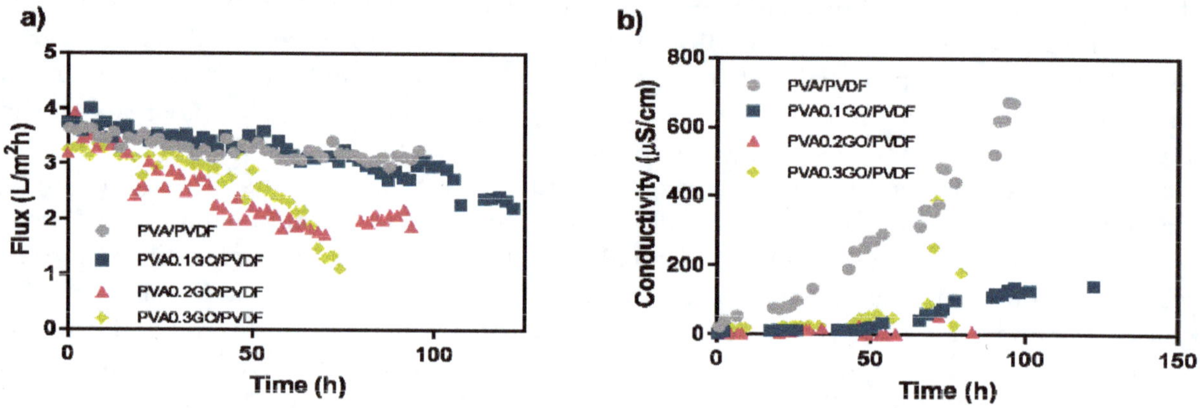


**Figure 9.** (**a**) Flux and (**b**) permeate conductivity profiles of the composite membrane using a feed solution of 100 g/L NaCl at 65 °C.

### *3.3. Anti-Fouling Performance*

GO nanosheets are known to improve the anti-fouling property for the mixed matrix filtration membranes [18,20–23]. This is ascribed to the improved membrane's hydrophilicity due to the hydrophilic groups on GO. However, whether GO can promote the anti-fouling performance of pervaporation membranes has not been explored yet. The fouling study on pure PVA/PVDF composite membranes (Figure 10) showed that the presence of 10 mg/L humic acid in 100 g/L NaCl aqueous solution had a negligible effect on the flux, but increased the permeate conductivity significantly (from 370 µS/cm to 2.8 mS/cm at the end of 72 h operation). This may indicate that the attachment of humic acid on the surface of PVA layer and the binding between humic acid and salt ions accelerated the salt transport through the membrane [55]. With GO incorporated in the PVA coating layer, increased permeate conductivity was also observed (from 140 to 355 µS/cm at the end of 120 h operation), which well aligned with the membrane without GO. However, the increase was less significant. This indicated that the incorporation of GO in PVA matrix stabilized the composite membrane's performance when processing brine containing organic foulant (humic acid).

We further investigated the effect of $CaCl_2$ on the membrane performance by adding 1.26 g/L $CaCl_2$ in the mixture of 100 g/L NaCl and 10 mg/L humic acid: the presence of the inorganic salt ions had a negligible effect on the permeation flux. It is interesting that the presence of calcium slightly reduced the permeation conductivity for both pure PVA/PVDF and PVA-GO/PVDF composite membranes, which can be attributed to the formation of the bulkier calcium-humic complex [55]. The results showed that a much lower permeate conductivity was obtained for the composite membrane containing GO than the pure PVA/PVDF membrane, which clearly demonstrated the improved anti-fouling property for the composite membrane after the addition of GO. This observation also suggested better operational stability for the pervaporation desalination process compared with the MD process: the presence of $CaCl_2$ can lead to severe fouling and flux decline for virgin hydrophobic PVDF (flux dropping to 0 with permeate conductivity rising to 1 mS/cm at the end of 20 h' operation) and modified superhydrophobic membrane (flux dropping to 0 with permeate conductivity rising to 3 mS/cm at the end of 40 h' operation) during the vacuum MD process, as suggested by our previous work [56].

The permeate solution composition at the different operating time was analyzed by ICP when using feed solution containing 100 g/L NaCl, 10 mg/L humic acid and 1.26 g/L $CaCl_2$ with the composite membrane containing 0.1 wt. % GO (Table 4). With the progress of the pervaporation operation, the permeate conductivity increased due to the gradual diffusion of the sodium ions and calcium ions through the membrane, and the diffusion rate of the smaller Na ions was higher than the bulkier Ca ions. In all, the incorporation of small amount of GO in PVA matrix efficiently slowed down the transport of salt ions through the membrane during the pervaporation process, although this was not completely prevented.

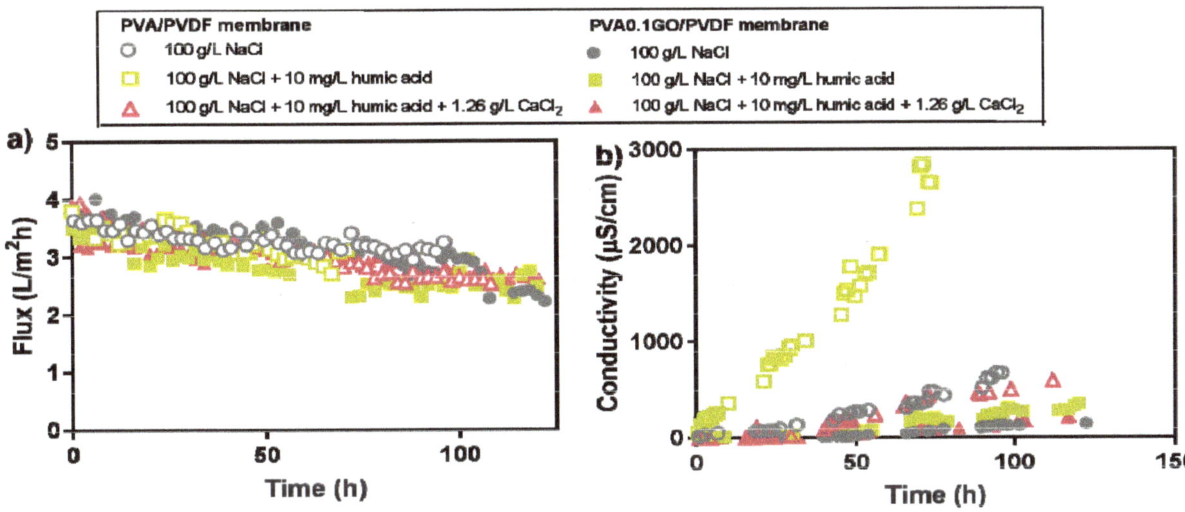


**Figure 10.** (**a**) Flux and (**b**) permeate conductivity profiles of the composite membrane PVA/PVDF, and PVA0.1GO/PVDF prepared by cast coating method using different feed solution.

**Table 4.** Permeate property of membrane PVA0.1GO/PVDF at different operation time by using feed solution of 100 g/L NaCl, 10 mg/L humic acid and 1.26 g/L $CaCl_2$, where initial feed conductivity is around 155 mS/cm, with $Na^+$ 39.4 g/L and $Ca^{2+}$ 0.164 g/L.

| Operation Time (h) | Conductivity (μS/cm) | $Na^+$ (mg/L) | $Ca^{2+}$ (mg/L) |
|---|---|---|---|
| 19.2 | 3.7 | 0.55 | 0.00 |
| 66.5 | 78.8 | 13.6 | 0.03 |
| 116.4 | 205 | 37.1 | 0.18 |
| 164.25 | 493 | 127 | 1.07 |

### 3.4. *Effect of Permeate Pressure on Pervaporation Performance*

As the pervaporation process is a thermal-driven process, the vapor pressure difference across the membrane plays a major role in the membrane water vapor permeability. While the main goal of this study was to suppress the salt transport through the composite membrane by modifications on PVA layer, there is still a need to suggest how the permeate pressure influenced on the membranes pervaporation performance, especially the membrane productivity. Conducting a thorough study could be an independent endeavor, however, in this study a limited number of tests were conducted to explore the effect of permeate pressure on membrane performance. A secondary condensation unit was applied next to the membrane cell in the permeate side, in order to maintain the permeate pressure at 3 kPa (instead of the initial pressure of 24 kPa). The membrane PVA/PVDF and PVA0.1GO/PVDF (both with 0.2 MR value and effective membrane area of 14.7 $cm^2$) were further tested, with both Milli-Q water and 100 g/L NaCl as the feeds. For the Milli-Q water feed, PVA/PVDF membrane and PVA0.1GO/PVDF membrane had very comparable flux (average flux 39.7 and 39.2 $L/m^2h$ for 4 h operation after stabilization, results not shown). The pervaporation performance using highly concentrated NaCl as the feed was shown in Figure 11 an initial volume of 2 L feed solution with a concentration of 100 g/L was used without top-up during a 24 h operation. Both membranes showed an initial flux of 27–28 $L/m^2h$, which was followed by obvious flux decline (around 25%) as a result of continuously increased feed concentration. After the 24 h pervaporation test using salt water, a flux of around 39 L/m$^2$h was restored using Milli-Q water as the feed for both membranes (results not shown). In terms of permeate conductivity, the membrane PVA0.1GO/PVDF obtained a constant value of 1–1.2 μS/cm, while the membrane PVA/PVDF showed a value of 15 μS/cm at the end of 24 h testing. This also confirmed that the addition of 0.1 wt. % GO effectively suppressed salt diffusion through the membrane. More importantly, the flux value using the same feed concentration was 9 times to that of using permeate pressure of 24 kPa. This suggested the further direction for optimization for such pervaporation wastewater treatment processes.

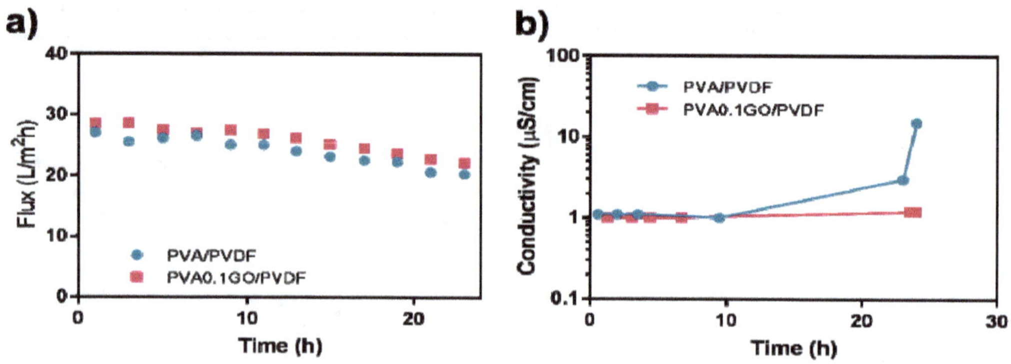


**Figure 11.** (**a**) Flux and (**b**) permeate conductivity profiles of composite PVA/PVDF membrane (MR value of 0.2) and PVA0.1GO/PVDF membrane using 2 L of 100 g/L NaCl as the feed without top-up at operating temperature of 65 °C.

## 4. Conclusions

Initially, in this work, composite membrane desalination performance was optimized by varying the cross-linking density of PVA. The results suggested the transport of salt ions through the membrane followed the solution-diffusion mechanism in the form of hydrated ions. Then, by incorporating a small amount of GO (0.1 wt. %) into PVA matrix, a high salt rejection can be maintained during the extended desalination operation, while the flux was relatively unaffected. This can be attributed to the impermeable properties of GO to salt ions and fast water molecule transport along its surface. In addition, the presence of polar groups on GO can also form hydrogen bonds with PVA, which rigidified the PVA chains and suppressed the transport of bulkier salt ions. The GO/PVA composite membrane also exhibited satisfactory anti-fouling properties during the treatment of calcium and humic-containing brine wastewater. Lastly, the study on permeate pressure indicated its crucial role in affecting the membrane productivity for such PVA-based pervaporation membranes.

**Supplementary Materials:**
Figure S1: Contact angle after membrane pre-treatment, Figure S2: Flux and conductivity profile of composite PVDF membrane with KOH/MR 0.2/5 wt. % by one layer casting, where the membrane was cleaned with in-situ cleaning method: (top) "96 h 100 g/L NaCl + 24 h Milli-Q water cycle; (bottom) "23 h 100 g/L NaCl + 1 h Milli-Q water cycle". The feed inlet temperature is 65 ± 1 °C, Figure S3: Flux and conductivity profile of composite PVA/PVDF membrane with KOH/MR 0.2/5 wt. % by one layer casting using a feed solution of 30 g/L, 100 g/L NaCl and PAC solution with the same conductivity as 30 g/L NaCl (~55 mS/cm). The feed inlet temperature was 65 ± 1 °C, Figure S4: Cross-sectional SEM images of the composite membrane fabricated with the cast coating: (a,b) PVA/PVDF, (c,d) PVA0.1GO/PVDF, Figure S5: Digital photo of freestanding PVA samples with different GO loadings: (a) 0, (b) 0.1 wt. %, (c) 0.2 wt. %, and (d) 0.3 wt. %, Figure S6: SEM images of freestanding PVA membrane with GO loading of (a) 0, (b) 0.1 wt. %, (c) 0.2 wt. %, (d) magnified 0.1 wt. %, Figure S7: XRD curves of GO only/PVDF, freestanding PVA, and freestanding PVA with GO loading of 0.1 wt. %, Figure S8: Flux profile of composite membrane PVA0.1GO/PVDF using Milli-Q water, 100 g/L NaCl and Milli-Q water as the feed, sequentially, Table S1: Contact angles of the composite membrane samples with different GO loadings.

**Acknowledgments:** This research was supported under Australian Research Council's Discovery Projects funding scheme (DP130104048).

**Author Contributions:** Lin Li, Jingwei Hou and Yatao Zhang conceived the idea of suppressing salt transport by GO incorporation. The experimental work was carried out by Lin Li. Yun Ye and Jingwei Hou built the pervaporation setup. Lin Li, Jingwei Hou, Yun Ye, Jaleh Mansouri, Yatao Zhang and Vicki Chen contributed to the manuscript writing and editing.

## References

1. Lattemann, S.; Höpner, T. Environmental impact and impact assessment of seawater desalination. *Desalination* **2008**, *220*, 1–15. [CrossRef]
2. Mariah, L.; Buckley, C.A.; Brouckaert, C.J.; Curcio, E.; Drioli, E.; Jaganyi, D.; Ramjugernath, D. Membrane distillation of concentrated brines—Role of water activities in the evaluation of driving force. *J. Membr. Sci.* **2006**, *280*, 937–947. [CrossRef]
3. Mericq, J.-P.; Laborie, S.; Cabassud, C. Vacuum membrane distillation of seawater reverse osmosis brines. *Water Res.* **2010**, *44*, 5260–5273. [CrossRef] [PubMed]
4. Khayet, M.; Matsuura, T. Pervaporation and vacuum membrane distillation processes: Modeling and experiments. *AIChE J.* **2004**, *50*, 1697–1712. [CrossRef]
5. Feng, X.; Huang, R.Y. Pervaporation with chitosan membranes. I. Separation of water from ethylene glycol by a chitosan/polysulfone composite membrane. *J. Membr. Sci.* **1996**, *116*, 67–76. [CrossRef]
6. Chapman, P.D.; Oliveira, T.; Livingston, A.G.; Li, K. Membranes for the dehydration of solvents by pervaporation. *J. Membr. Sci.* **2008**, *318*, 5–37. [CrossRef]
7. Shah, D.; Kissick, K.; Ghorpade, A.; Hannah, R.; Bhattacharyya, D. Pervaporation of alcohol–water and dimethylformamide–water mixtures using hydrophilic zeolite NaA membranes: Mechanisms and experimental results. *J. Membr. Sci.* **2000**, *179*, 185–205. [CrossRef]
8. Quiñones-Bolaños, E.; Zhou, H.; Soundararajan, R.; Otten, L. Water and solute transport in pervaporation hydrophilic membranes to reclaim contaminated water for micro-irrigation. *J. Membr. Sci.* **2005**, *252*, 19–28. [CrossRef]
9. Wang, Q.; Li, N.; Bolto, B.; Hoang, M.; Xie, Z. Desalination by pervaporation: A review. *Desalination* **2016**, *387*, 46–60. [CrossRef]
10. Bolto, B.; Tran, T.; Hoang, M.; Xie, Z. Crosslinked poly (vinyl alcohol) membranes. *Prog. Polym. Sci.* **2009**, *34*, 969–981. [CrossRef]
11. Praptowidodo, V.S. Influence of swelling on water transport through pva-based membrane. *J. Mol. Struct.* **2005**, *739*, 207–212. [CrossRef]
12. Xie, Z.; Hoang, M.; Duong, T.; Ng, D.; Dao, B.; Gray, S. Sol–gel derived poly (vinyl alcohol)/maleic acid/silica hybrid membrane for desalination by pervaporation. *J. Membr. Sci.* **2011**, *383*, 96–103. [CrossRef]
13. Xie, Z.; Ng, D.; Hoang, M.; Duong, T.; Gray, S. Separation of aqueous salt solution by pervaporation through hybrid organic–inorganic membrane: Effect of operating conditions. *Desalination* **2011**, *273*, 220–225. [CrossRef]
14. Huang, H.-D.; Ren, P.-G.; Chen, J.; Zhang, W.-Q.; Ji, X.; Li, Z.-M. High barrier graphene oxide nanosheet/poly (vinyl alcohol) nanocomposite films. *J. Membr. Sci.* **2012**, *409*, 156–163. [CrossRef]
15. Tan, B.; Thomas, N.L. A review of the water barrier properties of polymer/clay and polymer/graphene nanocomposites. *J. Membr. Sci.* **2016**, *514*, 595–612. [CrossRef]
16. Cui, Y.; Kundalwal, S.I.; Kumar, S. Gas barrier performance of graphene/polymer nanocomposites. *Carbon* **2016**, *98*, 313–333. [CrossRef]
17. Ganesh, B.; Isloor, A.M.; Ismail, A.F. Enhanced hydrophilicity and salt rejection study of graphene oxide-polysulfone mixed matrix membrane. *Desalination* **2013**, *313*, 199–207. [CrossRef]
18. Zinadini, S.; Zinatizadeh, A.A.; Rahimi, M.; Vatanpour, V.; Zangeneh, H. Preparation of a novel antifouling mixed matrix pes membrane by embedding graphene oxide nanoplates. *J. Membr. Sci.* **2014**, *453*, 292–301. [CrossRef]
19. Yin, J.; Zhu, G.; Deng, B. Graphene oxide (GO) enhanced polyamide (PA) thin-film nanocomposite (TFN) membrane for water purification. *Desalination* **2016**, *379*, 93–101. [CrossRef]
20. Bano, S.; Mahmood, A.; Kim, S.-J.; Lee, K.-H. Graphene oxide modified polyamide nanofiltration membrane with improved flux and antifouling properties. *J. Mater. Chem. A* **2015**, *3*, 2065–2071. [CrossRef]

21. Zhao, H.; Wu, L.; Zhou, Z.; Zhang, L.; Chen, H. Improving the antifouling property of polysulfone ultrafiltration membrane by incorporation of isocyanate-treated graphene oxide. *Phys. Chem. Chem. Phys.* **2013**, *15*, 9084–9092. [CrossRef] [PubMed]
22. Wu, H.; Tang, B.; Wu, P. Development of novel $SiO_2$–GO nanohybrid/polysulfone membrane with enhanced performance. *J. Membr. Sci.* **2014**, *451*, 94–102. [CrossRef]
23. Zhao, C.; Xu, X.; Chen, J.; Yang, F. Effect of graphene oxide concentration on the morphologies and antifouling properties of pvdf ultrafiltration membranes. *J. Environ. Chem. Eng.* **2013**, *1*, 349–354. [CrossRef]
24. Bhadra, M.; Roy, S.; Mitra, S. Desalination across a graphene oxide membrane via direct contact membrane distillation. *Desalination* **2016**, *378*, 37–43. [CrossRef]
25. Yin, J.; Deng, B. Polymer-matrix nanocomposite membranes for water treatment. *J. Membr. Sci.* **2015**, *479*, 256–275. [CrossRef]
26. Yang, H.-C.; Hou, J.; Chen, V.; Xu, Z.-K. Surface and interface engineering for organic–inorganic composite membranes. *J. Mater. Chem. A* **2016**, *4*, 9716–9729. [CrossRef]
27. Shen, J.; Liu, G.; Huang, K.; Jin, W.; Lee, K.-R.; Xu, N. Membranes with fast and selective gas-transport channels of laminar graphene oxide for efficient $CO_2$ capture. *Angew. Chem. Int. Ed.* **2015**, *54*, 578–582.
28. Zhang, Y.; Shen, Q.; Hou, J.; Sutrisna, P.D.; Chen, V. Shear-aligned graphene oxide laminate/pebax ultrathin composite hollow fiber membranes using a facile dip-coating approach. *J. Mater. Chem. A* **2017**, *5*, 7732–7737. [CrossRef]
29. Liang, B.; Zhan, W.; Qi, G.; Nan, Q.; Liu, Y.; Lin, S.; Cao, B.; Pan, K. High performance graphene oxide/polyacrylonitrile composite pervaporation membranes for desalination applications. *J. Mater. Chem. A* **2015**. [CrossRef]
30. Li, X.; Chen, Y.; Hu, X.; Zhang, Y.; Hu, L. Desalination of dye solution utilizing PVA/PVDF hollow fiber composite membrane modified with $TiO_2$ nanoparticles. *J. Membr. Sci.* **2014**, *471*, 118–129. [CrossRef]
31. Li, L.; Hashaikeh, R.; Arafat, H.A. Development of eco-efficient micro-porous membranes via electrospinning and annealing of poly (lactic acid). *J. Membr. Sci.* **2013**, *436*, 57–67. [CrossRef]
32. Kim, K.-J.; Lee, S.-B.; Han, N.W. Effects of the degree of crosslinking on properties of poly (vinyl alcohol) membranes. *Polym. J.* **1993**, *25*, 1295–1302. [CrossRef]
33. Zhang, L.; Yu, P.; Luo, Y. Dehydration of caprolactam–water mixtures through cross-linked pva composite pervaporation membranes. *J. Membr. Sci.* **2007**, *306*, 93–102. [CrossRef]
34. Chaudhri, S.G.; Rajai, B.H.; Singh, P.S. Preparation of ultra-thin poly (vinyl alcohol) membranes supported on polysulfone hollow fiber and their application for production of pure water from seawater. *Desalination* **2015**, *367*, 272–284. [CrossRef]
35. Liang, B.; Pan, K.; Li, L.; Giannelis, E.P.; Cao, B. High performance hydrophilic pervaporation composite membranes for water desalination. *Desalination* **2014**, *347*, 199–206. [CrossRef]
36. Peng, F.; Huang, X.; Jawor, A.; Hoek, E.M. Transport, structural, and interfacial properties of poly (vinyl alcohol)–polysulfone composite nanofiltration membranes. *J. Membr. Sci.* **2010**, *353*, 169–176. [CrossRef]
37. Baroña, G.N.B.; Choi, M.; Jung, B. High permeate flux of PVA/PSf thin film composite nanofiltration membrane with aluminosilicate single-walled nanotubes. *J. Colloid Interface Sci.* **2012**, *386*, 189–197. [CrossRef] [PubMed]
38. Deng, Y.H.; Chen, J.T.; Chang, C.H.; Liao, K.S.; Tung, K.L.; Price, W.E.; Yamauchi, Y.; Wu, K.C.W. A drying-free, water-based process for fabricating mixed-matrix membranes with outstanding pervaporation performance. *Angew. Chem. Int. Ed.* **2016**, *55*, 12793–12796. [CrossRef] [PubMed]
39. Ju, H.; Sagle, A.C.; Freeman, B.D.; Mardel, J.I.; Hill, A.J. Characterization of sodium chloride and water transport in crosslinked poly (ethylene oxide) hydrogels. *J. Membr. Sci.* **2010**, *358*, 131–141. [CrossRef]
40. Xie, Z. *Hybrid Organic-Inorganic Pervaporation Membranes for Desalination*; Victoria University: Toronto, ON, Canada, 2012.
41. Kusumocahyo, S.P.; Sano, K.; Sudoh, M.; Kensaka, M. Water permselectivity in the pervaporation of acetic acid–water mixture using crosslinked poly (vinyl alcohol) membranes. *Sep. Purif. Technol.* **2000**, *18*, 141–150. [CrossRef]
42. Xu, R.; Lin, P.; Zhang, Q.; Zhong, J.; Tsuru, T. Development of ethenylene-bridged organosilica membranes for desalination applications. *Ind. Eng. Chem. Res.* **2016**, *55*, 2183–2190. [CrossRef]

43. Zhao, D.; Ren, J.; Qiu, Y.; Li, H.; Hua, K.; Li, X.; Deng, M. Effect of graphene oxide on the behavior of poly (amide-6-b-ethylene oxide)/graphene oxide mixed-matrix membranes in the permeation process. *J. Appl. Polym. Sci.* **2015**, *132*. [CrossRef]
44. Qin, L.; Zhao, Y.; Liu, J.; Hou, J.; Zhang, Y.; Wang, J.; Zhu, J.; Zhang, B.; Lvov, Y.; Van der Bruggen, B. Oriented clay nanotube membrane assembled on microporous polymeric substrates. *ACS Appl. Mater. Interfaces* **2016**, *8*, 34914–34923. [CrossRef] [PubMed]
45. Sharma, S.; Prakash, J.; Pujari, P. Effects of the molecular level dispersion of graphene oxide on the free volume characteristics of poly (vinyl alcohol) and its impact on the thermal and mechanical properties of their nanocomposites. *Phys. Chem. Chem. Phys.* **2015**, *17*, 29201–29209. [CrossRef] [PubMed]
46. Loryuenyong, V.; Saewong, C.; Aranchaiya, C.; Buasri, A. The improvement in mechanical and barrier properties of poly(vinyl alcohol)/graphene oxide packaging films. *Packag. Technol. Sci.* **2015**, *28*, 939–947. [CrossRef]
47. Hou, J.; Dong, G.; Ye, Y.; Chen, V. Enzymatic degradation of bisphenol-a with immobilized laccase on tio 2 sol–gel coated pvdf membrane. *J. Membr. Sci.* **2014**, *469*, 19–30. [CrossRef]
48. Liang, J.; Huang, Y.; Zhang, L.; Wang, Y.; Ma, Y.; Guo, T.; Chen, Y. Molecular-level dispersion of graphene into poly (vinyl alcohol) and effective reinforcement of their nanocomposites. *Adv. Funct. Mater.* **2009**, *19*, 2297–2302. [CrossRef]
49. Ma, H.; Burger, C.; Hsiao, B.S.; Chu, B. Highly permeable polymer membranes containing directed channels for water purification. *ACS Macro Lett.* **2012**, *1*, 723–726. [CrossRef]
50. Joshi, R.; Carbone, P.; Wang, F.-C.; Kravets, V.G.; Su, Y.; Grigorieva, I.V.; Wu, H.; Geim, A.K.; Nair, R.R. Precise and ultrafast molecular sieving through graphene oxide membranes. *Science* **2014**, *343*, 752–754. [CrossRef] [PubMed]
51. Yin, J.; Kim, E.-S.; Yang, J.; Deng, B. Fabrication of a novel thin-film nanocomposite (TFN) membrane containing MCM-41 silica nanoparticles (NPs) for water purification. *J. Membr. Sci.* **2012**, *423*, 238–246. [CrossRef]
52. Kim, E.-S.; Deng, B. Fabrication of polyamide thin-film nano-composite (PA-TFN) membrane with hydrophilized ordered mesoporous carbon (H-OMC) for water purifications. *J. Membr. Sci.* **2011**, *375*, 46–54. [CrossRef]
53. Wang, X.; Chen, X.; Yoon, K.; Fang, D.; Hsiao, B.S.; Chu, B. High flux filtration medium based on nanofibrous substrate with hydrophilic nanocomposite coating. *Environ. Sci. Technol.* **2005**, *39*, 7684–7691. [CrossRef] [PubMed]
54. Wang, N.; Ji, S.; Li, J.; Zhang, R.; Zhang, G. Poly (vinyl alcohol)–graphene oxide nanohybrid "pore-filling" membrane for pervaporation of toluene/n-heptane mixtures. *J. Membr. Sci.* **2014**, *455*, 113–120. [CrossRef]
55. Meng, S.; Ye, Y.; Mansouri, J.; Chen, V. Fouling and crystallisation behaviour of superhydrophobic nano-composite pvdf membranes in direct contact membrane distillation. *J. Membr. Sci.* **2014**, *463*, 102–112. [CrossRef]
56. Meng, S.; Ye, Y.; Mansouri, J.; Chen, V. Crystallization behavior of salts during membrane distillation with hydrophobic and superhydrophobic capillary membranes. *J. Membr. Sci.* **2015**, *473*, 165–176. [CrossRef]

# 14

# Advanced Treatment of Real Grey Water by SBR Followed by Ultrafiltration—Performance and Fouling Behavior

**Gabriela Kamińska and Anna Marszałek** *

Department of Water and Wastewater Engineering, Silesian University of Technology, Konarskiego 18, 44-100 Gliwice, Poland; gabriela.liszczyk@gmail.com
* Correspondence: anna.marszalek@polsl.pl.

**Abstract:** Grey water has been identified as a potential source of water in a number of applications e.g., toilet flushing, laundering in first rinsing, floor cleaning, and irrigation. The major obstacle to the reuse of grey water relates to pathogens, nutrients, and organic matter found in grey water. Therefore, much effort has been put to treat grey water, in order to yield high-quality water deprived of bacteria and with an appropriate value in a wide range of quality parameters (Total Organic Carbon (TOC), nitrate, phosphate, ammonium, pH, and absorbance), similar to the values for tap water. The aim of this study was to treat the real grey water, and turn it into high-quality, safe water. For this purpose, the real grey water was treated by means of a sequential biological reactor (SBR) followed by ultrafiltration. Initially, grey water was treated in a laboratory SBR reactor with a capacity of 3 L, operated in a 24 h cycle. Then, SBR effluent was purified in a cross-flow ultrafiltration setup. Treatment efficiency in SBR and ultrafiltration was assessed using extended physicochemical and microbiological analyses (pH, conductivity, color, absorbance, Chemical Oxygen Demand (COD), Biological Oxygen Demand ($BOD_5$), nitrate, phosphate, ammonium, total nitrogen, phenol index, nonionic and anionic surfactants, TOC, *Escherichia coli*, and enterococci). Additionally, ultrafiltration was evaluated in terms of fouling behavior for three polymer membranes with different MWCO (molecular weight cut-off). The values of quality parameters (pH, conductivity, COD, $BOD_5$, TOC, $N\text{-}NH_4^+$, $N\text{-}NO_3^-$, $N_{tot}$, and $P\text{-}PO_4^{3-}$) measured in SBR effluent did not exceed permissible values for wastewater discharged to soil and water. Ultrafiltration provided the high-quality water with very low values of COD (5.8–18.1 mg/L), TOC (0.47–2.19 mg/L), $absorbance_{UV254}$ (0.015–0.048 1/cm), color (10–29 mgPt/L) and concentration of nitrate (0.18–0.56 mg/L), phosphate (0.9–2.1 mg/L), ammonium (0.03–0.11 mg/L), and total nitrogen (3.3–4.7 mg/L) as well as lack of *E. coli* and enterococci. Membrane structural and surface properties did not affect the treatment efficiency, but did influence the fouling behavior.

**Keywords:** grey water; SBR; ultrafiltration; fouling; zeta potential

---

## 1. Introduction

Water is emerging as one of the single most important resources of Planet Earth for the prosperity of the economy and human life. However, freshwater resources have been increasingly polluted and depleted globally. The constant increase in water usage and climate change—such as altered weather-patterns (including droughts or floods), deforestation, and increased pollution—are the main driving forces for the rising global scarcity of water. It is experienced especially by European countries, due to an improvement of living standards and economic development in recent years. Along with economic development, the total water use in Europe has been sharply increasing in the last decades [1].

Grey water recycling is the solution to the growing challenges associated with water shortages [2]. Quantitatively, grey water represents around 65% of the total volume of domestic sewage, making it the largest stream for water reuse [3]. Grey water contains nitrate, phosphate, organic matter, surfactants, pharmaceuticals, oils, and pathogens [4]. Therefore, the treatment of grey water before its reuse is necessary. Purified grey water could be reused for cleaning, car washing, concrete production, and irrigation [5,6]. The guidelines of the World Health Organization (WHO) point to four criteria for the reuse of grey water: hygiene, aesthetics, environmental tolerance, and economic feasibility. WHO recommends biological treatment and ultrafiltration in order to obtain high-quality water. However, the requirements that define "high water quality" are too sweeping and do not show the physical and chemical specifications. It is caused by a number of possible applications of reclaimed grey water that require various different quality standards. In 2006, WHO set microbiological criteria that reclaimed grey water should meet for its reuse for restricted and non-restricted agricultural irrigation. [7,8].

Numerous approaches, including low and high pressure-driven membrane techniques [9–11], coagulation [12,13], biological processes, and membrane bioreactors [3], have been proposed for the treatment of grey water. These technologies vary in both complexity and performance. For example, Li et al. recommended the employment of an aerobic treatment coupled with membrane filtration or an aerobic treatment with sand filtration, with disinfection as the last step for both systems [8]. Similarly, Ding et al. suggested applying a system that combines membrane filtration with biological treatment in a gravity-driven membrane filtration system. In this technique, the fouling layer attached to the membrane stabilizes a flux and improves the treatment effects due to the microbial activity of biofilm [14]. From these studies, it is clear that grey water requires both biological and physical treatment to meet non-potable reuse standards. Aerobic biological processes are effective to remove organics from grey water. Physical processes, particularly membrane filtration, are recommended for polishing effluents from the biological treatment step [9,15]. Ultrafiltration (UF) retains bacteria, suspension, colloids, natural organic matter, and partial micropollutants such as pharmaceuticals and personal care products. However, the efficiency of ultrafiltration in the removal of individual pollutants depends on membrane type and its properties such as molecular weight cut-off (MWCO). Membrane properties such as contact angle and zeta potential play an important role in the fouling behavior and hydraulic performance of the system [16]. Membrane fouling results in a permeability loss due to the increase in hydraulic resistances in the filtration system, especially in the case of porous polymer membranes. There are many studies reported in the literature that aimed at mitigating the fouling by selecting the most optimal operational parameters for ultrafiltration such as transmembrane pressure, velocity, temperature, or developing a cleaning method. [17–19]. On the other hand, they do not consider the effect of structure and surface properties of membrane i.e., MWCO and contact angle on fouling behavior. Another issue is that most of these studies were conducted for artificial grey water. While artificial grey water provides useful information for model development, it does not reflect real conditions, and the composition affects the process significantly.

The main novelty of this work shows that high-quality water can be obtained from grey water. For a need of this study, we established that high-quality water is water: (1) that fulfills criteria included in the Polish Regulation from the Minister of the Environment on the conditions to be met when discharging wastewater into water or soil and (2) with the basic quality parameters (smell, color, turbidity, TOC, ammonium, nitrate, chloride, conductivity, hardness, *E. coli*, and enterococci) like the standards for tap water. In order to gain this purpose, real grey water was treated using a sequential biological reactor (SBR) followed by ultrafiltration. Treatment efficiency was evaluated for SBR and ultrafiltration using extended physicochemical and microbiological analyses. Additionally, ultrafiltration was studied in terms of fouling behavior for three polymer membranes with different MWCO. In that context, the novelty of this work is also a determination of the effect of membrane type on the treatment of SBR effluent and fouling behavior.

## 2. Methodology

### 2.1. Grey Water Characteristic

Grey water was collected from a single-family household located in Silesia in Poland. This house is equipped with an installation to collect grey water from the shower, kitchen sink, and washing machine. The characteristic of the grey water is given in Table 1. Samples were taken directly from a storage tank and transported to the lab facilities, stored at 4 °C and analyzed within 48 h according to the methodology presented in Section 2.6.

**Table 1.** Characteristic of grey water, as taken.

| Parameter | Unit | Value | | |
|---|---|---|---|---|
| | | Min | Max | Average |
| pH | - | 7.47 | 7.88 | 7.6 |
| Conductivity | μS/cm | 735 | 772 | 758 |
| Color | mgPt/L | 182 | 228 | 201 |
| Turbidity | FTU | 26 | 31 | 28.2 |
| $Absorbance_{UV254}$ | 1/cm | 0.468 | 0.611 | 0.55 |
| COD | mg/L | 415 | 638 | 543 |
| $BOD_5$ | mg/L | 190 | 240 | 212 |
| $N\text{-}NO_3^-$ | mg/L | 0.4 | 0.8 | 0.6 |
| $P\text{-}PO_4^{3-}$ | mg/L | 3.8 | 4.6 | 4.15 |
| $N\text{-}NH_4^+$ | mg/L | 0 | 0.3 | 0.2 |
| $N_{tot}$ | mg/L | 9.4 | 11.3 | 10.7 |
| Surfactants nonionic | mg/L | 1.33 | 1.64 | 1.42 |
| Surfactants anionic | mg/L | 12.9 | 17.5 | 15.2 |
| TOC | mg/L | 51.2 | 59.3 | 56.3 |

### 2.2. Biological Treatment in SBR

The SBR technology used provides for the dosing of wastewater in alternately used anaerobic reactor oxygen conditions. This guarantees a high degree of removal of carbon, nitrogen, and phosphorus compounds. The operation of the reactor is simple, and during operation, modifications of individual separate phases of the technological cycles are possible. The biological process was carried out under laboratory conditions, using activated sludge taken from the municipal sewage treatment plant in Gliwice, Poland. The treatment of grey water was carried out in a laboratory SBR with a capacity of 3 L. During the treatment experiment, the excessive activated sludge was periodically removed from the SBR in order to keep its concentration at the level of 3.0 g/L. The solid retention time (SRT) was 20 days. The load of the sludge with the contaminants was equal to 0.1 g COD/$g_{DM}$d and the concentration of oxygen was at the level of 3 mg/L. The system was operated as the sequential biological reactor in one cycle per day. The length of particular operation stages was as follows: filling and mixing phase for 2 h, aeration phase for 21 h, and sedimentation and SBR effluent removal for 1 h. The aeration system consisted of aquarium cubes placed on the bottom of the tank with connected aeration pumps of the Tetratec APS 300 type (Tetra, Melle, Germany). In order to thoroughly mix the contents of the chamber, it was mixed by means of a magnetic stirrer with the possibility of speed regulation in the range from 50 to 1000 revolutions/min. The chamber was filled and emptied using Heidolph peristaltic pumps (Heidolph, Schwabach, Germany). The electronic weekly programmer controlled the start and end of the relevant SBR work phase.

### 2.3. SBR Effluent Treatemnt by Ultrafilltration–Ultrafiltration Run

Ultrafiltration is especially dedicated to drinking water production because colloids, particulates, macromolecules, and pathogens are removed in this process. It is also a suitable process to purify effluent after biological treatment as this effluent contains bacteria, particles of organic matter, and suspended solids. Ultrafiltration was carried out in the lab, using a scale cross-flow configuration equipped with a plate-and-frame membrane module SEPA CF-NP (GE Osmonics, Minnetonka, MN, USA) as seen in Figure 1. Three ultrafiltration membranes were used in separate processes. Membrane properties are presented in Table 2. Before each experiment, the clean water flux was determined using ultrapure water. The process was operated at a constant pressure of 5 bar and a constant temperature of 22 ± 1 °C and a constant velocity of 1 m/s with continuous dosing of SBR effluent to the feed tank. Each filtration run consisted of three cycles including 60 min of filtration followed by forward flushing with ultrapure water during 60 s. The volume of permeate was monitored in order to determine the permeability from the following equation:

$$L_p = \frac{V}{\Delta p \cdot A \cdot t} \quad (1)$$

where: $L_p$ is permeability ($L \cdot m^{-2} \cdot h^{-1} \cdot bar^{-1}$) in short (LMHB), $V$ is permeate volume (L), $A$ is membrane surface area ($m^2$), $t$ is permeate time collection (h), and $\Delta p$ is transmembrane pressure (bar).

**Table 2.** Properties of ultrafiltration membranes.

| Symbol | DSGM | V3 | BN |
|---|---|---|---|
| Manufacturer | GE Osmonics (Minnetonka, MN, USA) | Synder (Allison Parkway Vacaville, CA USA) | Synder |
| Polymer material | Polyamide-TFC | PVDF | PVDF |
| MWCO, Da | 8000 | 30,000 | 50,000 |
| Thickness *, µm | 150 | 220 | 200 |

* own measurements with an electronic micrometer.

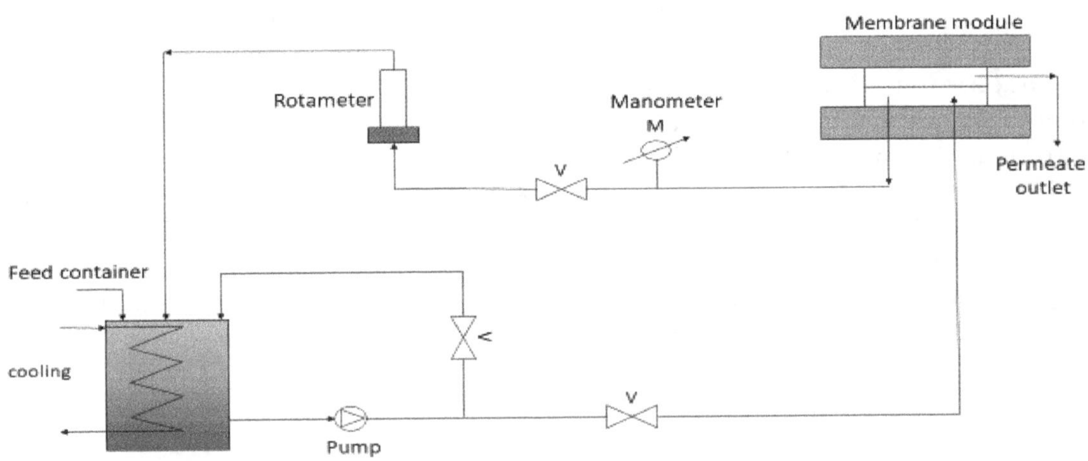


**Figure 1.** Schematic diagram of the cross-flow filtration set-up.

## 2.4. Membrane Fouling Characterization

In this study, hydraulic resistances were used to characterize the fouling behavior of ultrafiltration membranes treating SBR effluent. Hydraulic resistances of membrane and fouling layer were calculated using the resistance in the series model and Darcy's law using the correlations as shown below [20].

$$k_{tot} = k_m + k_f \quad (2)$$

$$k_f = k_{irr} + k_{rev} \quad (3)$$

$$J = \frac{\Delta p}{\mu \cdot k} \quad (4)$$

where: $k$ is hydraulic resistance, where subscripts $m$, $f$, $irr$, $rev$, and $tot$ are related to membrane, fouling, hydraulically irreversible fouling, hydraulically reversible fouling, and total ($m^{-1}$), respectively, $J$ is the flux ($m^3 \cdot m^{-2} \cdot s^{-1}$), $\Delta p$ is the transmembrane pressure ($kg \cdot s^{-2} \cdot m^{-1}$), and $\mu$ is the dynamic viscosity of water at given temperature ($kg \cdot m^{-1} \cdot s^{-1}$). Membrane resistance ($k_m$) was measured for the clean membrane with ultrapure water prior to feed water filtration. Hydraulically irreversible fouling was determined from the flux after forward flushing, while hydraulically reversible fouling was determined from the difference in fouling and irreversible resistances.

### *2.5. Membrane Characterization*

Zeta potential of membranes was determined by the electrokinetic analyzer SurPASS ™ 3 (Anton Paar, Graz, Austria). Measurements of the contact angle were performed using the goniometer PG-1 (Fibro System AB, Hägersten, Sweden) and the sessile drop method was applied. By syringe on top, a drop of ultrapure water was put on the dried membrane surface. Through an enlarged projection of the water drop on the gauge, the value of the contact angle was measured. For every type of membrane, 10 samples were measured and the average value was calculated.

### *2.6. Quality Analysis and Microbiological Assessment*

The treatment efficiency in SBR and ultrafiltration was evaluated by the monitoring of the typical quality parameters (color, turbidity, COD, $BOD_5$, TOC, phenolic index, absorbance of $UV_{254}$, nitrate ($N\text{-}NO_3^-$), phosphate ($P\text{-}PO_4^{3-}$), ammonium ($N\text{-}NH_4^+$), total nitrogen ($N_{tot}$), conductivity, pH, and anionic and non-ionic surfactants). Color and turbidity measurements were performed with a UV-Vis Spectroquant®Pharo 300 (Merck, Kenilworth, NJ, USA). Phenolic index, COD, nitrate, phosphate, ammonium, total nitrogen, and anionic and non-ionic surfactant concentrations were determined spectrophotometrically with Merck test kits (codes of the Merck test kits are given in Table S1). The absorbance was measured at 254 nm, using a UV-visible light (UV-Vis) Cecil 1000 (Analytik Jena AG company, Jena, Germany). TOC was measured using a TOC-L series analyzer (Shimadzu, Kioto, Prefektura Kioto, Japan). pH and conductivity were monitored by multifunctional analyzer CX-461 (Elmetron, Zabrze, Poland). The $BOD_5$ was determined by respirometric measurement using the OXI Top System WTW set (Xylem Analytics, Weilheim Germany). Microbiological analysis including *E. coli* and enterococci was conducted by an external accredited lab according to ISO methods PN-EN ISO 9308-1:2014-12/PN-EN ISO 9308-1/A1:2017; PN-EN ISO 7899-2:2004.

## 3. Results and Discussion

### *3.1. Grey Water Treatment in SBR*

#### 3.1.1. Reduction of COD, $BOD_5$, and TOC

The biological sewage treatment aimed to reduce biodegradable organic matter and remove the biogenic substances i.e., nitrogen and phosphorus. In the preliminary stage, the activated sludge was taken from the municipal wastewater plant, then it was adapted to a new influent (grey water). The results discussed in Figures 2–4 show the analysis of individual parameters after the adaptation process. Figure 2 shows the SBR system performance in relation to the concentration of organic compounds in purified grey water. Detailed results of physicochemical analyses are presented in Table S2 (in Supplementary file).

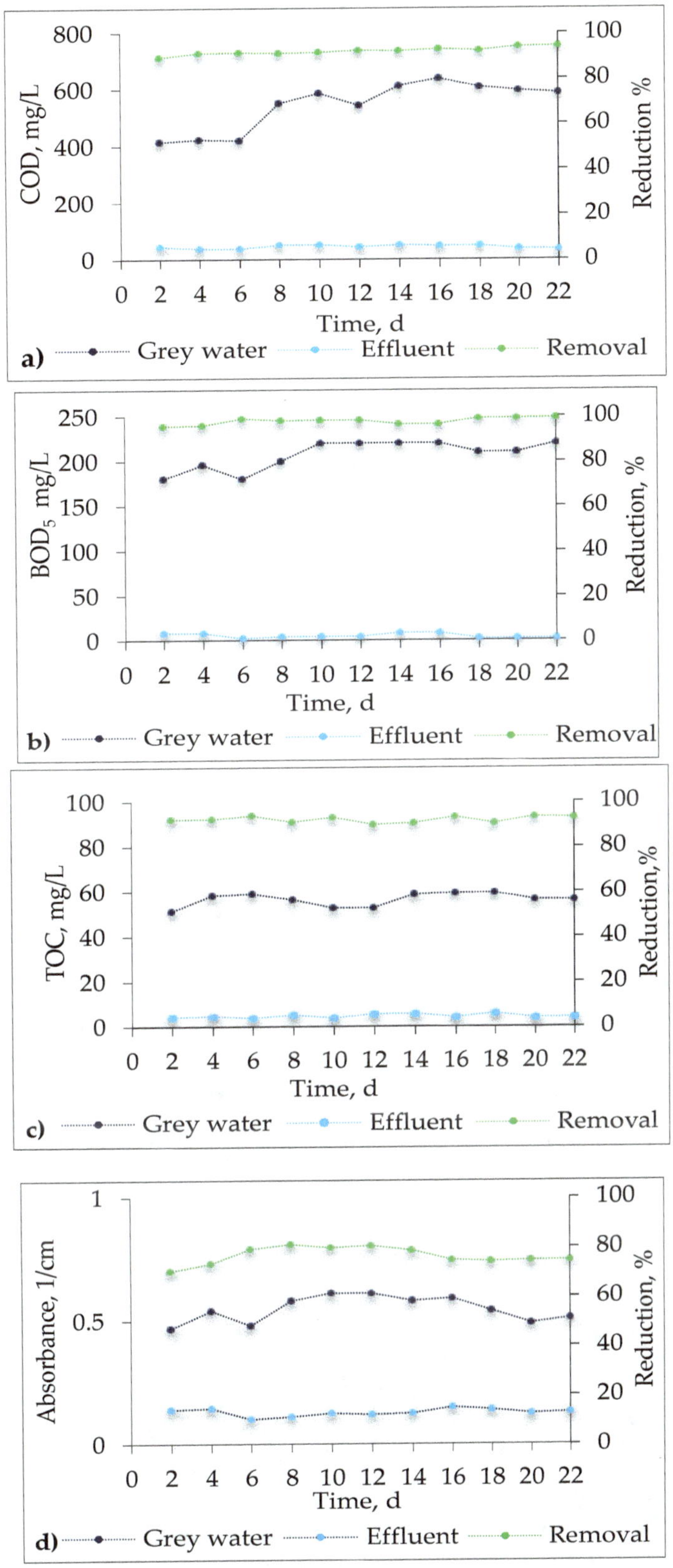


**Figure 2.** The performance of the sequential biological reactor (SBR) system with respect to the changes in the concentration of the organic compounds.

It was shown that the degree of pollution removal was high, and the values of the quality parameters in SBR effluent did not exceed the permissible values included in the currently in force in Poland, Regulation of the Minister of the Environment on the conditions to be met when discharging wastewater into water or soil [21]. Over the whole experimental period (22 days), the SBR system reduced COD by around 92%, from an average value of 543 mg/L (influent) to 44 mg/L (effluent) (Figure 2a). Fountoulakis et al. obtained a similar result in their studies [22]. More specifically, the removal of COD was approximately 87%. Such, a high removal was also obtained in studies on grey water purification in a membrane bioreactor operating in a flow-through system [3], where the COD reduction degree was 88%.

It was found that the easily biodegradable compounds expressed by $BOD_5$ were removed in 99% of cases, and the average concentration in the effluent was 4 mg/L (Figure 2b). Similarly, a high reduction was recorded for TOC and absorbance, i.e., 91% and 76%, respectively (Figure 2c,d). Anionic and nonionic surfactants were reduced by 97% and 100%, respectively (as seen in Table S1 in the Supplementary File). In the study [22], the removal efficiency of the anionic surfactants was about 80% in the submerged membrane bioreactor. This is due to the fact that non-ionic and anionic surfactants are readily biodegradable under aerobic conditions [23–25].

### 3.1.2. Removal of Biogenic Compounds

As seen in Figure 3, the concentration of nutrients in SBR effluent was low and did not exceed permissible values laid out in the Polish regulations [21].

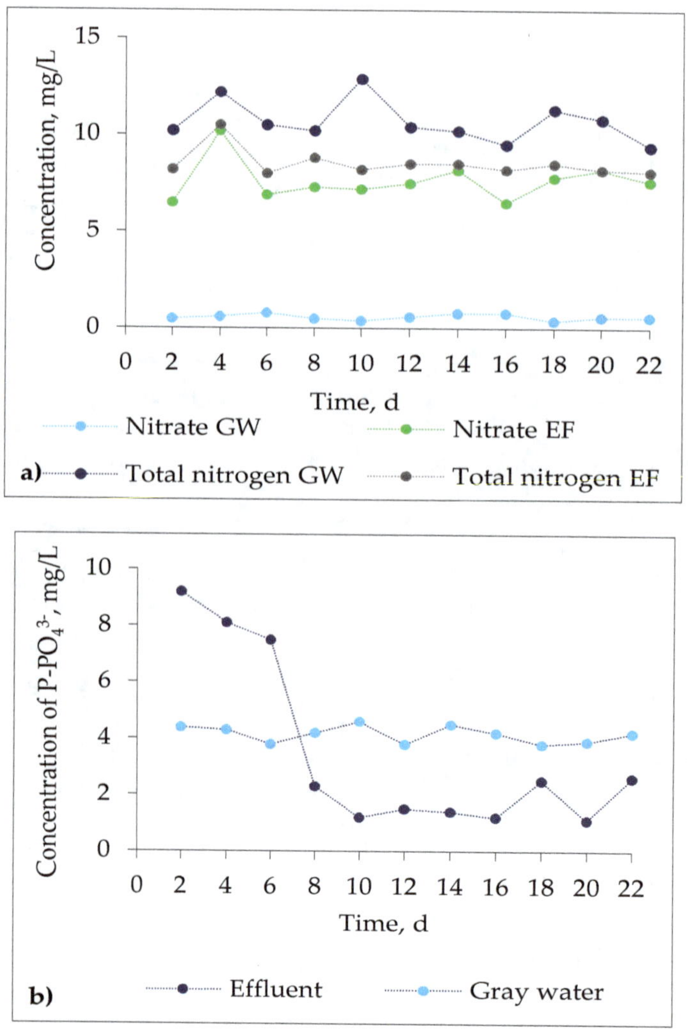


**Figure 3.** The performance of the SBR system with respect to the changes in the concentration of the nutrients. GW—grey water, EF—effluent.

Raw grey water was characterized by a minimum content of ammonium and nitrate nitrogen at the level of 0.2 mg/L and 0.8 mg/L, respectively. After biological treatment with the SBR system, the concentration of nitrate and nitrogen increased to an average value of 7.6 mg/L. Differences in the nitrate concentrations in the influent and effluent indicate that a large part of nitrogen in grey water was organically bound [16]. The average value of total Kjeldahl nitrogen in the influent was 10.6 mg/L. The removal degree of total nitrogen was at the average level of 20% and the concentration in the effluent was 8 mg/L. It should be emphasized that despite the lack of total denitrification, the concentrations of individual forms of nitrogen in SBR effluent did not exceed the maximum permissible values given that are currently in force in Poland, (Regulations of the Minister of the Environment) on the conditions to be met when discharging wastewater into water or soil [21]. At the initial stage of the grey water purification process, there was a problem with phosphorus removal. Its value exceeded several times the permissible value. It can be explained by the variable physicochemical nature of the SBR influent. The effective dephosphatation takes place when the influent contains an easily biodegradable COD fraction and the $COD/BOD_5$ ratio is 2 [26,27].

The second factor intensifying a release of phosphate under anaerobic conditions is the constant delivery of easily biodegradable organic compounds, e.g., volatile fatty acids and their salts. Their occurrence in grey water is likely to be changeable. Since grey water was collected in a holding tank over a longer time, it could undergo an acid fermentation during which volatile fatty acids may be formed. It was found that in the second week of the process, the degree of $P\text{-}PO_4^{3-}$ removal increased along with the increase of COD concentration of the inflowing grey water. In the following weeks of operation of the SBR reactor, the phosphate phosphorus concentration ranged from 1.1 to 2.6 $mgP\text{-}PO_4^{3-}/L$. Despite the increase in efficiency of phosphorus removal, its permissible concentration specified in the Regulation of the Minister of the Environment [21] ( $P_{tot}$ = 2 m g /L) was still exceeded. In order to improve the effectiveness of phosphate removal, the cycle of operation of SBR should be modified by means of changes in the duration of aerobic–anaerobic phases [28].

### 3.1.3. Removal of Color and Turbidity

Figures 4 and 5 show the treatment efficiency of the SBR system with respect to the changes in the color and the turbidity. From these results, it is clear that over the whole experiment, the efficiency of the SBR system in terms of color and turbidity was at a constant high level. During 22 days of continuous SBR system operation, the color of grey water decreased from an average value of 201 mgPt/L to 44 mgPt/L, which corresponds to an average removal of 78%. Meanwhile, the average grey water turbidity decreased from 28 FTU to 4 FTU corresponds to an average removal of 84%.

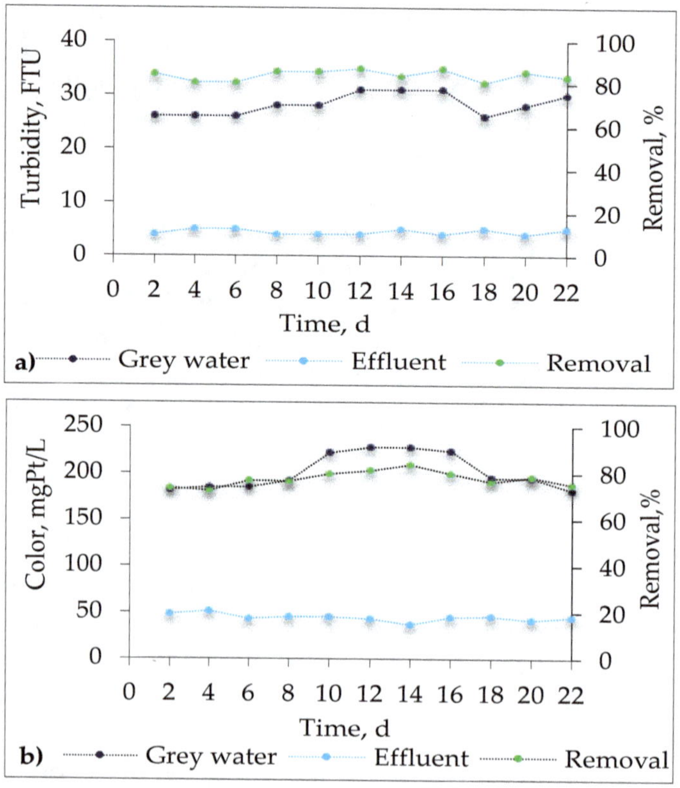


**Figure 4.** The removal performance of the SBR system with respect to the changes in the color and the turbidity.

**Figure 5.** Changes in the color and the turbidity of grey water under treatment with SBR followed by ultrafiltration.

### 3.2. *SBR Effluent Treatment by Ultrafiltration*

The treatment efficiency of SBR effluent in an ultrafiltration unit with different membranes is presented in Figure 6 and Table S3 (Supplementary file). As expected, for ultrafiltration, removal of organics expressed by the color, absorbance, COD, and TOC was very high. More specifically, the color, absorbance$_{UV254}$, COD, and TOC were reduced maximally by 73%, 91%, 84%, and 91%, respectively. Slightly lower elimination was observed for inorganic quality parameters. Conductivity, nitrate, phosphate, ammonium, and total nitrogen were reduced by 61%, 64%, 79%, 85%, and 59%. It is important to emphasize that negative charge of membranes at pH 7–8 (as seen in Figure 8) played an important role in the reduction of conductivity, phosphate, and nitrate ions in SBR effluent. The literature describes that negatively charged ultrafiltration membranes reject phosphate ions by 87%, as an effect of electrostatic repulsion [29]. Another important influencing factor on the rejection of ions and organics in the ultrafiltration unit treating wastewater/surface water can be hydrophobic/hydrophilic interactions between organics and ions in feed water [30]. For example, Shang et al. reported that high phosphate removal was attributed to the adsorption of phosphate on biopolymers in effluent and removal with these biopolymers [31].

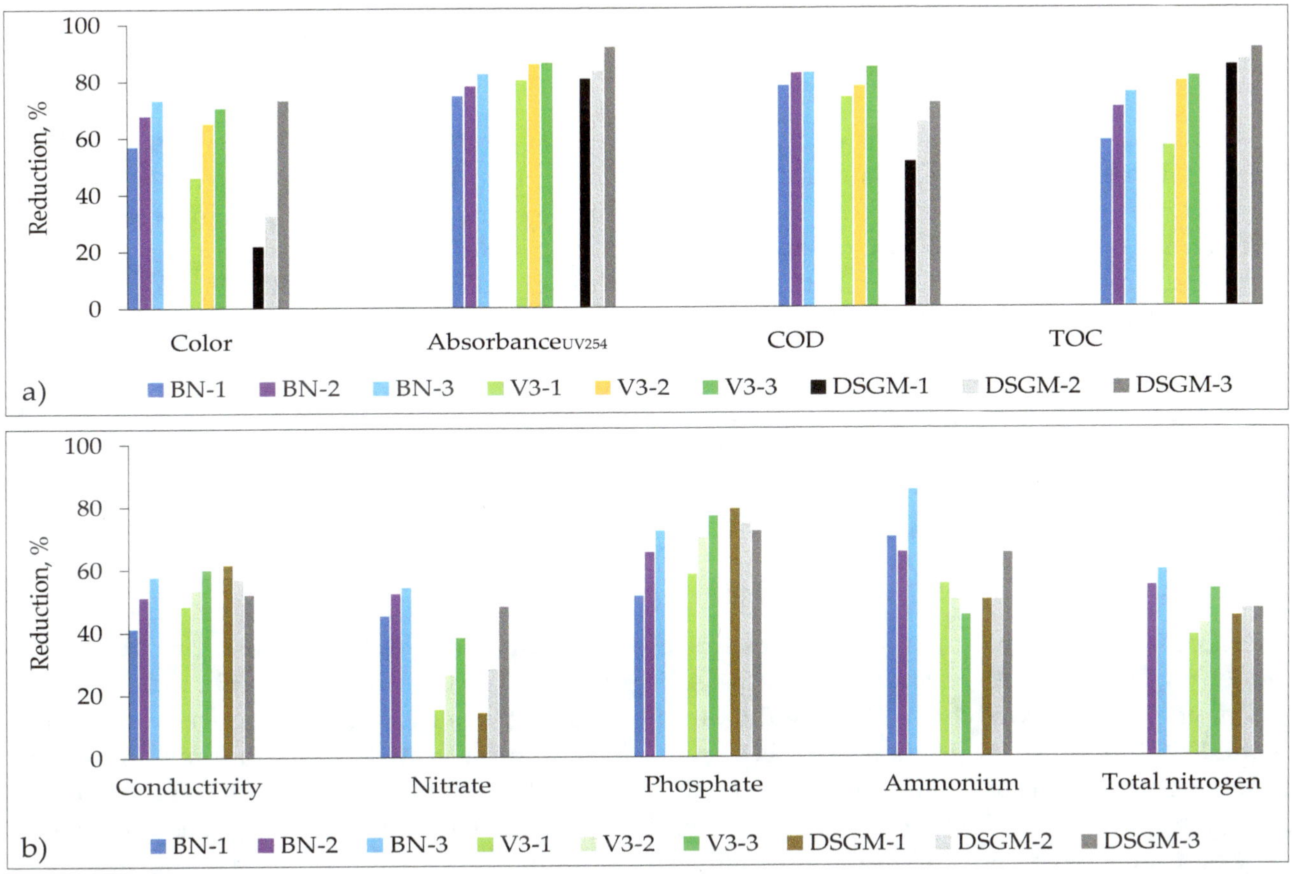


**Figure 6.** Reduction values of organic (**a**) and inorganic (**b**) parameters during the ultrafiltration of SBR effluent. Each bar corresponds to the filtration cycle 1–3.

Surprisingly, membrane DSGM with the lower value of MWCO provided permeate quality values very similar to the values for BN and V3 membranes with higher MWCO. This brings an important finding that the treatment efficiency was more affected by operational conditions of cross-flow ultrafiltration than by membrane properties. When good mixing and turbulent flow along the membrane surface are guaranteed, as it is in cross-flow, other factors affecting permeability and selectivity are less important [32]. Feed nature could also play an important role. From the quality parameters of SBR effluent, we can assume that SBR effluent did not contain a significant portion of low molecular weight organics that needed to be removed. Another reason could be the fouling layer that supported the retention of pollutants by BN and V3 membranes. Jerman et al. found that the cake or gel layer acts as an additional membrane in ultrafiltration and improves retention by up to 40% [33].

It was also found that treatment efficiency increased within cycles 1–3 for all membranes (Figure 6). It can be related to the formation of the cake layer on the membrane surface of BN and V3. This corresponds well with the permeability loss observed for these membranes. Many authors suggest that the cake layer decreases membrane permeability by the reduction of effective pore size and improves the retention of organics and ions [34–36]. However, in the case of the DSGM membrane, there was no fouling so the cake layer was probably not created. The reason for increasing retention over time for DSGM can be related to the concentration effect of feed components within cycles

1–3. In the literature, the influence of the feed concentration on both ions and organics removal can be found [37,38]. Muthumareeswaran et al. have found that a change of the concentration of feed components in the multicomponent system affects their retention due to the ionic interaction between the feed components (Columbic interaction), molar volume, and interaction between ions and membrane surface charge [39].

Importantly, physical, chemical, and microbiological specifications (smell, color, turbidity, TOC, ammonium, nitrate, chloride, conductivity, hardness, *E. coli*, and enterococci) of the permeate correspond well with typical tap water values of basic quality parameters [40].

### 3.3. Microbiological Quality of Grey Water

Pathogens, such as *Escherichia coli* and Enterococci, have been identified in grey water (Table 3). The concentration of these bacteria was above 100 CFU per 100 mL. During the biological purification process in the SBR system, the presence of *Escherichia coli* did not change, while Enterococci decreased to an average of 3 CFU per 100 mL of sample. Then, ultrafiltration for each UF membranes resulted in the complete removal of the determined pathogens [41]. High pathogen removal from the *E. coli* group was also noted in the SMBR system from Khalid Bani-Melhem et al. [3].

**Table 3.** Microbiological quality of grey water.

| Parameter | Unit | Grey Water | SBR Effluent | UF Permeate of BN/V3/DSGM |
|---|---|---|---|---|
| *Escherichia coli* | CFU/100 mL | >100 | >100 | 0 |
| Enterococci | CFU/100 mL | >100 | 3 | 0 |

### 3.4. Membrane Permeability and Fouling Behavior in Ultrafiltration

Hydraulic performance of membranes was evaluated by permeability loss as a function of time. As seen in Figure 7, the permeability decreased gradually for BN and V3 membranes, while for DSGM, permeability was constant along the ultrafiltration. In other words, fouling was not observed for the DSGM membrane. This finding is very important in the context of membrane lifetime and reducing operational cost. Similarly, Acero et al. reported that UF membranes (with higher MWCO) revealed higher fouling than NF membranes (with lower MWCO) [42]. It can be explained by different surface properties of given membranes, such as hydrophilicity/hydrophobicity and surface charge (zeta potential curve and isoelectric point). The contact angle and isoelectric points of clean and fouled membranes are listed in Table 4. Zeta potential curves for clean and fouled membranes are presented in Figure 8. Owing to the lowest contact angle and strong negative surface charge, the DSGM membrane had the best antifouling properties. On the contrary, the surface of BN and V3 membranes was much more hydrophobic, and thus more prone to adsorb the feed components. Some authors reported a high importance of membrane hydrophilicity and negative charge for fouling mitigation in ultrafiltration [43–45] treating water or wastewater.

In order to investigate this further, the hydraulically reversible and hydraulically irreversible resistances were calculated. As seen in Figure 9, the total fouling (sum of reversible and irreversible resistances) was higher for the V3 membrane. However, considering the proportion of reversible and irreversible resistances for these membranes, it is clear that fouling behavior for these membranes exhibited the opposite trend. More specifically, a significant proportion of the increase in resistance for the V3 membrane was hydraulically reversible, while hydraulically irreversible for the BN membrane. Reversibility/irreversibility of fouling depends on the surface properties of membranes. It is in good correspondence with changes of course of zeta potential curves and shift in isoelectric point for clean and fouled membranes. Membrane V3 with the lower negative charge was more sensitive to fouling overall than the BN membrane. However, for the BN membrane, we observed the higher shift in isoelectric point (from 3.01 to 3.55) than for V3 (from 5.1 to 5.25) indicating a persistent change in BN membrane properties caused by irreversible fouling.

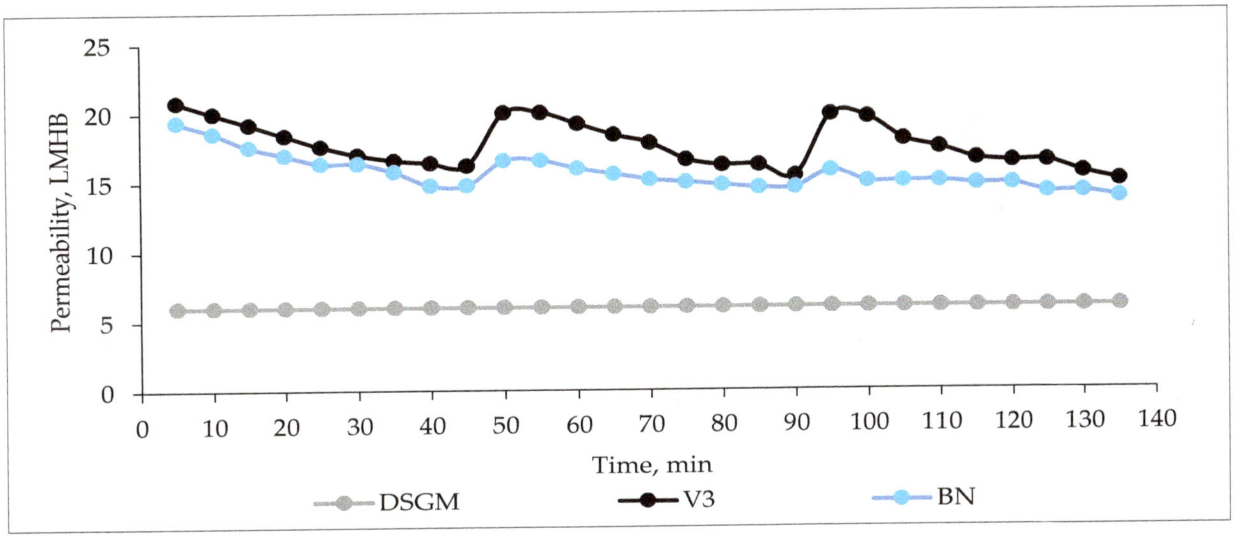


**Figure 7.** Permeability loss as a function of time of UF for different types of membranes. Ultrafiltration with DSGM membrane was performed without forward flushing due to constant permeability.

**Table 4.** Contact angle and isoelectric point for clean and fouled membranes.

| Membrane | Contact Angle | | Isoelectric Point | |
|---|---|---|---|---|
| | Clean Membrane | Fouled Membrane | Clean Membrane | Fouled Membrane |
| DSGM | 45.5 ± 3.2 | 46.25 ± 2.8 | - | - |
| V3 | 71.5 ± 1.9 | 73.5 ± 3.8 | 5.05 | 5.25 |
| BN | 75.0 ± 1.4 | 78.6 ± 1.2 | 3.01 | 3.55 |

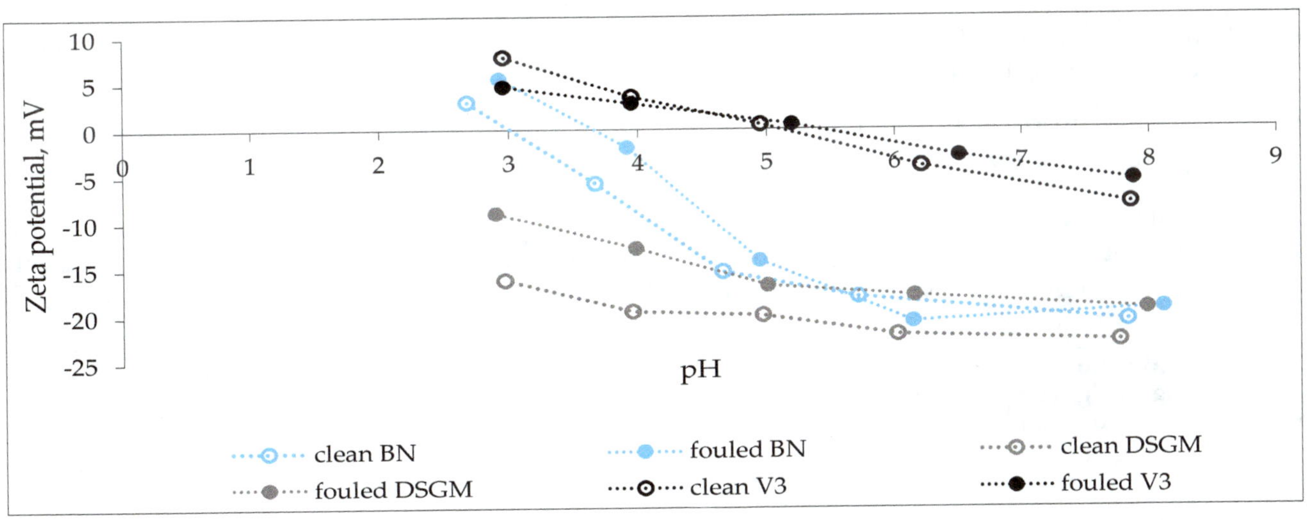


**Figure 8.** Zeta potential vs pH for clean and fouled membranes.

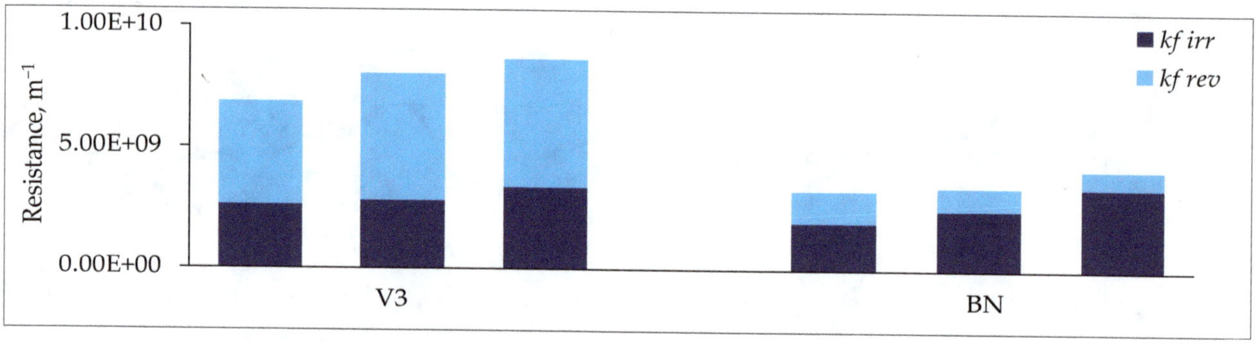


**Figure 9.** Irreversible and reversible fouling resistances of ultrafiltration for all membranes. Each bar corresponds to the filtration cycle 1–3. *kf irr* and *kf rev* are irreversible and reversible resistances, respectively.

## 4. Conclusions

This study proved a great performance of SBR followed by ultrafiltration to obtain high-quality water for non-potable purposes from real grey water. Purified grey water fulfilled criteria for wastewater discharged into water or soil as well as physical, chemical, and microbiological requirements for tap water. It is a good starting point to find new reuse application of reclaimed grey water.

The following conclusions can be reached from the experimental results.

1. The SBR system ensured the complete removal of nonionic surfactants and 97% removal of anionic surfactants from grey water.
2. Removal of phosphate in SBR was, on average, 42%, but its concentration still exceeded acceptable values. It could probably be related to the variable concentration of raw grey water. It was stated that in order to improve the phosphorus reduction, in subsequent studies, the cycle of operation of SBR should be modified by means of changes in the duration of aerobic–anaerobic phases.
3. In ultrafiltration treatment, there was an efficiency increase in filtration time due to the formation of a cake layer or concentration effect.
4. Treatment efficiency in ultrafiltration was mainly influenced by the cross-flow filtration mode and nature of feed water.
5. Membrane properties played an important role in fouling behavior, but did not greatly affect treatment efficiency.
6. Membrane DSGM did not show fouling in permeate flux monitoring. It is a consequence of the initial properties of this membrane such as the lowest contact angle and the most negative zeta potential when compared with the BN and V3 membranes, for which fouling was observed.

**Author Contributions:** Conceptualization G.K. and A.M. methodology, G.K. and A.M.; investigation, G.K. and A.M.; writing—original draft preparation, G.K. and A.M., Writing—review and editing, G.K. All authors have read and agreed to the published version of the manuscript.

## References

1. Use of Freshwater Resources. Available online: https://www.eea.europa.eu/data-and-maps/indicators/use-of-freshwater-resources-2/assessment-3 (accessed on 23 December 2019).
2. Xin, D.; Saige, W.; Bin, C. The blue, green and grey water consumption for crop production in Heilongjiang. *Energy Procedia* **2019**, *158*, 3908–3914.

3. Bani-Melhem, K.; Al-Qodah, Z.; Al-Shannag, M.; Qasaimeh, A.; Qtaishat, M.R.; Alkasrawi, M. On the performance of real grey water treatment using a submerged, Membrane bioreactor system. *J. Membr. Sci.* **2015**, *476*, 40–49. [CrossRef]
4. Martínez-Alcalá, I.; Pellicer-Martínez, F.; Fernández-López, C. Pharmaceutical grey water footprint: Accounting, influence of wastewater treatment plants and implications of the reuse. *Water Res.* **2018**, *135*, 278–287. [CrossRef] [PubMed]
5. Galvis, A.; Zambrano, D.A.; Van der Steen, N.P.; Gijzen, H.J. Evaluation of pollution prevention options in the municipal water cycle. *J. Clean. Prod.* **2014**, *66*, 599–609. [CrossRef]
6. Prajapati, B.; Jensen, M.B.; Jørgensen, N.O.G.; Petersen, N.B. Grey water treatment in stacked multi-layer reactors with passive aeration and particle trapping. *Water Res.* **2019**, *161*, 181–190. [CrossRef]
7. WHO. Guidelines for the Safe Use of Wastewater, Excreta and Greywater, Excreta and Greywater Use in Agriculture. 2006. Available online: https://apps.who.int/iris/bitstream/handle/10665/78265/9241546824_eng.pdf;jsessionid=47807504FED628D9A9BB71654F04A9E7?sequence=1 (accessed on 23 December 2019).
8. Li, F.; Wichmann, K.; Otterpohl, R. Review of the technological approaches for grey water treatment and reuses. *Sci. Total Environ.* **2009**, *407*, 3439–3449. [CrossRef]
9. Boddu, M.V.; Paul, T.; Page, M.A.; Byl, C.; Ward, L.; Ruan, J. Gray water recycle: Effect of pretreatment technologies on low pressure reverse osmosis treatment, *J. Environ. Chem. Eng.* **2016**, *4*, 4435–4443. [CrossRef]
10. Manouchehri, M.; Kargari, A. Water recovery from laundry wastewater by the cross flow microfiltration process: A strategy for water recycling in residential buildings. *J. Clean. Prod.* **2017**, *168*, 227–238. [CrossRef]
11. Ding, A.; Liang, H.; Li, G.; Derlon, N.; Szivak, I.; Morgenroth, E.; Pronk, W. Impact of aeration shear stress on permeate flux and fouling layer properties in a low pressure membrane bioreactor for the treatment of grey water. *Water Res.* **2016**, *510*, 382–390. [CrossRef]
12. Mohammadi, M.J.; Takdastan, A.; Jorfi, S.; Neisi, A.; Farhadi, M.; Yari, A.R.; Dobaradaran, S.; Khaniabadi, Y.O. Electrocoagulation process to Chemical and Biological Oxygen Demand treatment from carwash grey water in Ahvaz megacity, Iran. *Data Brief* **2017**, *11*, 634–639. [CrossRef]
13. Bani- Melhem, K.; Smith, E. Grey water treatment by a continuous process of an electrocoagulation unit and a submerged membrane bioreactor system. *Chem. Eng. J.* **2012**, *198–199*, 201–210. [CrossRef]
14. Ding, A.; Liang, H.; Li, H.; Szivak, I.; Traber, J.; Pronk, W. A low energy gravity driven membrane bioreactor system for grey water treatment: Permeability and removal performance of organics. *J. Membr. Sci.* **2017**, *542*, 408–417. [CrossRef]
15. Leong, J.; Oha, K.S.; Poh, P.E.; Chong, M.N. Prospects of hybrid rainwater-greywater decentralised system for water recycling and reuse: A review. *J. Clean. Prod.* **2017**, *142*, 3014–3027. [CrossRef]
16. Adamczak, M.; Kamińska, G.; Bohdziewicz, J. Application of waste polymers as basic material for ultrafiltration membranes preparation. *Proceedings* **2019**, *16*, 14. [CrossRef]
17. Drews, A. Membrane fouling in membrane bioreactors—Characterisation, contradictions, cause and cures. *J. Membr. Sci.* **2010**, *363*, 1–28. [CrossRef]
18. Duan, L.; Jiang, W.; Song, Y.; Xia, S.; Hermanowicz, S. The characteristics of extracellular polymeric substances and soluble microbial products in moving bed biofilm reactor-membrane bioreactor. *Bioresour. Technol.* **2013**, *148*, 436–442. [CrossRef]
19. Pinto, A.C.S.; Grossi, L.; Carvalho de Melo, R.A.; Macedo de Assis, T.; Ribeiro, V.M.; Santos Amaral, M.C.; de Souza Figueiredo, K.C. Carwash wastewater treatment by micro and ultrafiltration membranes: Effects of geometry, pore size, pressure difference and feed flow rate in transport properties. *J. Water Process Eng.* **2017**, *17*, 143–148. [CrossRef]
20. Crittenden, J.C.; Trussell, R.R.; Hand, D.W.; Howe, K.J.; Tchobanoglous, J. *Water Treatment: Principles and Design*, 3rd ed.; John Wiley & Sons: Hoboken, NJ, USA, 2012.
21. Dz.U.2014.poz.1800. Directive of the Minister of the Environment from 18th of November 2014. Available online: http://prawo.sejm.gov.pl/isap.nsf/download.xsp/WDU20140001800/O/D20141800.pdf (accessed on 3 January 2020).
22. Fountoulakis, M.S.; Markakis, N.; Petousi, I.; Manios, T. Single house on-site grey water treatment using a submerged membrane bioreactor for toilet flushing. *Sci. Total Environ.* **2016**, *551–552*, 706–711. [CrossRef]
23. Guang-Guo, Y. Fate, behavior and effects of surfactants and their degradation products in the environment. *Environ. Int.* **2006**, *32*, 417–431.

24. Ivanković, T.; Hrenović, J. Surfactants in the Environment. *Arh. Hig. Rada Toksikol.* **2010**, *61*, 95–110. [CrossRef]
25. Leal, H.; Temmink, H.; Zeeman, G.; Buisman, C.J.N. Comparison of Three Systems for Biological Greywater Treatment. *Water* **2010**, *2*, 155–169. [CrossRef]
26. Świerczynska, A.; Bohdziewicz, J. Determination of the most effective operating conditions of membrane bioreactor used to industrial wastewater treatment. *Environ. Prot. Eng.* **2015**, *41*, 41–51.
27. Hocaoglua, S.M.; Atasoya, E.; Babana, A.; Orhonb, D. Modeling biodegradation characteristics of grey water in membrane bioreactor. *J. Membr. Sci.* **2013**, *429*, 139–146. [CrossRef]
28. Świerczyńska, A.; Bohdziewicz, J.; Puszczało, E. Treatment of industrial wastewater in the sequential membrane bioreactor. *Ecol. Chem. Eng. S* **2016**, *23*, 285–295. [CrossRef]
29. Shang, R.; Verliefde, A.R.D.; Hu, J.; Zeng, Z.; Lu, J.; Kemperman, A.J.B.; Deng, H.; Nijmeijer, K.; Heijman, S.G.J.; Rietveld, L.C. Tight ceramic UF membrane as RO pre-treatment: The role of electrostatic interactions on phosphate rejection. *Water Res.* **2014**, *48*, 498–507. [CrossRef] [PubMed]
30. Peeva, P.D.; Million, N.; Ulbricht, M. Factors affecting the sieving behavior of anti-fouling thin-layer cross-linked hydrogel polyethersulfone composite ultrafiltration membranes. *J. Membr. Sci.* **2012**, *390–391*, 99–112. [CrossRef]
31. Shang, R.; Verliefde, A.R.D.; Hu, J.; Heijmann, S.G.J.; Rietveld, L.C. The impact of EfOM, NOM and cations on phosphate rejection by tight ceramic ultrafiltration. *Sep. Purif. Technol.* **2014**, *132*, 289–294. [CrossRef]
32. Broeckmann, A.; Busch, J.; Wintgens, T.; Marquardt, W. Modeling of pore blocking and cake layer formationin membrane filtration for wastewater treatment. *Desalination* **2006**, *189*, 97–109. [CrossRef]
33. Jermann, D.; Pronk, W.; Boller, M.; Schafer, A.I. The role of NOM fouling for the retention of estradiol and ibuprofen during ultrafiltration. *J. Membr. Sci.* **2009**, *329*, 75–84. [CrossRef]
34. Bohdziewicz, J.; Kamińska, G.; Pawlyta, M.; Łukowiec, D. Comparison of effectiveness of advanced treatment of municipal wastewater by sorption and nanofiltration. Separate processes and integrated systems. *Environ Prot. Eng.* **2015**, *41*, 119–132.
35. García-Martín, N.; Silva, V.; Carmona, F.J.; Palacio, L.; Hernández, A.; Prádanos, P. Pore size analysis from retention of neutral solutes through nanofiltration membranes. The contribution of concentration–polarization. *Desalination* **2014**, *344*, 1–11. [CrossRef]
36. Meng, S.; Zhanf, M.; Yao, M.; Qiu, Z.; Hong, Y.; Lan, W.; Xia, H.; Jin, X. Membrane Fouling and Performance of Flat Ceramic Membranes in the Application of Drinking Water Purification. *Water* **2019**, *11*, 2606. [CrossRef]
37. Kramer, F.C.; Shang, R.; Rietveld, L.C.; Heijman, S.J.G. Influence of pH, multivalent counter ions, and membrane fouling on phosphate retention during ceramic nanofiltration. *Sep. Purif. Technol.* **2019**, *227*. [CrossRef]
38. Arsuaga, J.M.; López-Muñoz, M.J.; Aguado, J.; Sotto, A. Temperature, pH and concentration effects on retention and transport of organic pollutants across thin-film composite nanofiltration membranes. *Desalination* **2008**, *221*, 253–258. [CrossRef]
39. Muthumareeswaran, M.R.; Agarwa, G.P. Feed concentration and pH effect on arsenate and phosphate rejection via polyacrylonitrile ultrafiltration membrane. *J. Membr. Sci.* **2014**, *468*, 11–19. [CrossRef]
40. Acero, J.L.; Benitez, F.J.; Teva, F.; Leal, A.I. Retention of emerging micropollutants from UP water and a municipal secondary effluent by ultrafiltration and nanofiltration. *Chem. Eng. J.* **2010**, *163*, 264–272. [CrossRef]
41. Haas, R.; Opitz, R.; Grischek, T.; Otter, P. The AquaNES Project: Coupling Riverbank Filtration and Ultrafiltration in Drinking Water Treatment. *Water* **2019**, *11*, 18. [CrossRef]
42. Świerczyńska, A.; Bohdziewicz, J.; Kamińska, G.; Wojciechowski, K. Influence of the type of membrane-forming polymer on the membrane fouling. *Environ. Prot. Eng.* **2016**, *42*, 197–210.
43. Ghiasi, S.; Behboudi, A.; Mohammadi, T.; Khanlari, S. Effect of surface charge and roughness on ultrafiltration membranes performance and polyelectrolyte nanofiltration layer assembly, Colloids and Surfaces A. *Physicochem. Eng. Asp.* **2019**, *580*. [CrossRef]
44. Kumari, P.; Modi, A.; Bellare, J. Enhanced flux and antifouling property on municipal wastewater of polyethersulfone hollow fiber membranes by embedding carboxylated multi-walled carbon nanotubes and a vitamin E derivative. *Sep. Purif. Technol.* **2020**, *235*. [CrossRef]

# Permissions

All chapters in this book were first published in MDPI; hereby published with permission under the Creative Commons Attribution License or equivalent. Every chapter published in this book has been scrutinized by our experts. Their significance has been extensively debated. The topics covered herein carry significant findings which will fuel the growth of the discipline. They may even be implemented as practical applications or may be referred to as a beginning point for another development.

The contributors of this book come from diverse backgrounds, making this book a truly international effort. This book will bring forth new frontiers with its revolutionizing research information and detailed analysis of the nascent developments around the world.

We would like to thank all the contributing authors for lending their expertise to make the book truly unique. They have played a crucial role in the development of this book. Without their invaluable contributions this book wouldn't have been possible. They have made vital efforts to compile up to date information on the varied aspects of this subject to make this book a valuable addition to the collection of many professionals and students.

This book was conceptualized with the vision of imparting up-to-date information and advanced data in this field. To ensure the same, a matchless editorial board was set up. Every individual on the board went through rigorous rounds of assessment to prove their worth. After which they invested a large part of their time researching and compiling the most relevant data for our readers.

The editorial board has been involved in producing this book since its inception. They have spent rigorous hours researching and exploring the diverse topics which have resulted in the successful publishing of this book. They have passed on their knowledge of decades through this book. To expedite this challenging task, the publisher supported the team at every step. A small team of assistant editors was also appointed to further simplify the editing procedure and attain best results for the readers.

Apart from the editorial board, the designing team has also invested a significant amount of their time in understanding the subject and creating the most relevant covers. They scrutinized every image to scout for the most suitable representation of the subject and create an appropriate cover for the book.

The publishing team has been an ardent support to the editorial, designing and production team. Their endless efforts to recruit the best for this project, has resulted in the accomplishment of this book. They are a veteran in the field of academics and their pool of knowledge is as vast as their experience in printing. Their expertise and guidance has proved useful at every step. Their uncompromising quality standards have made this book an exceptional effort. Their encouragement from time to time has been an inspiration for everyone.

The publisher and the editorial board hope that this book will prove to be a valuable piece of knowledge for researchers, students, practitioners and scholars across the globe.

# List of Contributors

**Andrea G. Capodaglio**
Department of Civil Engineering and Architecture, University of Pavia, 27100 Pavia, Italy

**Gustaf Olsson**
Department of Biomedical Engineering, Division of Industrial Electrical Engineering and Automation, Lund University, Lund Sweden

**Sofia Dias C and Marisa R. Almeida**
CIIMAR—Interdisciplinary Centre of Marine and Environmental Research of the University of Porto, Terminal de Cruzeiros do Porto de Leixões, Avenida General Norton de Matos, 4450-208 Matosinhos, Portugal
Chemistry and Biochemistry Department, Faculty of Sciences, University of Porto, Rua do Campo Alegre 687, 4150-171 Porto, Portugal

**Pedro Rodrigues**
Chemistry and Biochemistry Department, Faculty of Sciences, University of Porto, Rua do Campo Alegre 687, 4150-171 Porto, Portugal

**Ana P. Mucha**
CIIMAR—Interdisciplinary Centre of Marine and Environmental Research of the University of Porto, Terminal de Cruzeiros do Porto de Leixões, Avenida General Norton de Matos, 4450-208 Matosinhos, Portugal
Biology Department, Faculty of Sciences, University of Porto, Rua do Campo Alegre 790, 4150-171 Porto, Portugal

**Rute Duarte Crespo**
Biology Department, Faculty of Sciences, University of Porto, Rua do Campo Alegre 790, 4150-171 Porto, Portugal

**José M. Angosto, María J. Roca and José A. Fernández-López**
Department of Chemical and Environmental Engineering, Technical University of Cartagena, Paseo Alfonso XIII, 52, 30203 Cartagena, Murcia, Spain

**Daniele Cecconet and Arianna Callegari**
Department of Civil Engineering and Architecture, University of Pavia, 27100 Pavia, Italy

**Jakub Raček and Petr Hlavínek**
AdMaS Research Centre, Faculty of Civil Engineering, Brno University of Technology, 61200 Brno, Czech Republic

**Qun Xiang**
The United Graduate School of Agricultural Sciences, Ehime University, 3-5-7 Tarumi, Matsuyama, Ehime 790-8566, Japan

**Shuji Fukahori**
Paper Industry Innovation Center of Ehime University, 127 Mendoricho Otsu, Shikokuchuo, Ehime 799-0113, Japan

**Naoyuki Yamashita and Hiroaki Tanaka**
Research Center for Environmental Quality Management, Kyoto University, 1-2 Yumihama, Otsu, Shiga 520-0811, Japan

**Taku Fujiwara**
Research and Education Faculty, National Sciences Cluster, Agriculture Unit, Kochi University, 200 Monobe Otsu, Nankoku, Kochi 783-8502, Japan

**Samar Hadroug**
Wastewaters and Environment Laboratory, Water Research and Technologies Center, Soliman 8020, Tunisia
National Agricultural Institute of Tunisia, University of Carthage, Tunis 1082, Tunisia

**Salah Jellali**
Center for Environmental Studies and Research, Sultan Qaboos University, Al-Khoud 123, Muscat, Oman

**James J. Leahy, Marzena Kwapinska and Witold Kwapinski**
Department of Chemical Sciences, Bernal Institute, University of Limerick, V94 T9PX Limerick, Ireland

**Mejdi Jeguirim**
CNRS, IS2M UMR 7361, University of Haute-Alsace, University of Strasbourg, F-68100 Mulhouse, France

**Helmi Hamdi**
Center for Sustainable Development, College of Arts and Sciences, Qatar University, Doha

**Juanhong Li and Xiwu Lu**
School of Energy and Environment, Southeast University, Nanjing 210096, China

**Sílvia Costa and Bárbara Brown**
Câmara Municipal de Lisboa, Direcção Municipal do Ambiente, Estrutura Verde, Clima e Energia, Laboratório de Bromatologia e Águas, Avenida Cidade do Porto S/N-1700-111 Lisboa, Portugal

**Pedro Teixeira**
Câmara Municipal de Lisboa, Direcção Municipal do Ambiente, Estrutura Verde, Clima e Energia, Laboratório de Bromatologia e Águas, Avenida Cidade do Porto S/N-1700-111 Lisboa, Portugal
Centro de Estudos do Ambiente e do Mar (CESAM), Faculdade de Ciências da Universidade de Lisboa, Campo Grande, 1749-016 Lisboa, Portugal
Departamento de Saúde Ambiental, Instituto Nacional de Saúde Doutor Ricardo Jorge, Avenida Padre Cruz, 1649-016 Lisboa, Portugal

**Raquel Rodrigues and Elisabete Valério**
Departamento de Saúde Ambiental, Instituto Nacional de Saúde Doutor Ricardo Jorge, Avenida Padre Cruz, 1649-016 Lisboa, Portugal

**Susana Silva**
Departamento de Epidemiologia, Instituto Nacional de Saúde Doutor Ricardo Jorge, Avenida Padre Cruz, 1649-016 Lisboa, Portugal

**Amanda L. Ciosek and Grace K. Luk**
Department of Civil Engineering, Faculty of Engineering and Architectural Science, Ryerson University, Toronto, ON M5B 2K3, Canada

**Aida Kesraoui and Mongi Seffen**
Laboratory of Energy and Materials (LabEM): LR11ES34, Higher School of Science and Technology of Hammam Sousse, University of Sousse, 4011 Hammam Sousse, Tunisia

**Roberto Boada and Manuel Valiente**
GTS-UAB Research Group, Department of Chemistry, Faculty of Science, Universitat Autònoma Barcelona, 08193 Bellaterra, Spain

**Amina Othmani**
Laboratory of Energy and Materials (LabEM): LR11ES34, Higher School of Science and Technology of Hammam Sousse, University of Sousse, 4011 Hammam Sousse, Tunisia
Faculty of Sciences of Monastir (Monastir University), 5000 Monastir, Tunisia

**Motasem N. Saidan**
Department of Chemical Engineering, School of Engineering, The University of Jordan, Amman 11942, Jordan

**Mohammad Al-Addous**
Department of Energy Engineering, School of Natural Resources Engineering and Management, German Jordanian University, Amman 11180, Jordan

**Radwan A. Al-Weshah**
Civil Engineering Department, School of Engineering, The University of Jordan, Amman 11942, Jordan

**Ibrahim Obada**
Ministry of Water and Irrigation, Amman 11181, Jordan

**Malek Alkasrawi**
Department of PS & Chemical Engineering, University of Wisconsin, Stevens Point, WI 54481, USA

**Nesrine Barbana**
Department of Environmental Technology, Technische Universität Berlin, 10623 Berlin, Germany

**Muhammad B. Asif, Faisal I. Hai and Long D. Nghiem**
Strategic Water Infrastructure Lab, School of Civil, Mining and Environmental Engineering, University of Wollongong, Wollongong NSW 2522, Australia

**William E. Price**
StrategicWater Infrastructure Lab, School of Chemistry, University of Wollongong,Wollongong NSW 2522, Australia

**Jinguo Kang**
StrategicWater Infrastructure Lab, School of Chemistry, University of Wollongong,Wollongong NSW 2522, Australia
Strategic Water Infrastructure Lab, School of Civil, Mining and Environmental Engineering, University of Wollongong, Wollongong NSW 2522, Australia

**Jason P. van de Merwe and Frederic D. L. Leusch**
Australian Rivers Institute and Griffith School of Environment, Griffith University, Gold Coast QLD 4222, Australia

**Kazuo Yamamoto**
Environmental Science Centre, Department of Urban Engineering, University of Tokyo, Tokyo 113-0033, Japan

**Lin Li, Jingwei Hou, Yun Ye and Vicki Chen**
The United Nations Educational, Scientific and Cultural Organization (UNESCO) Centre for Membrane Science and Technology, School of Chemical Engineering, University of New South Wales, Sydney 2052, Australia

**Jaleh Mansouri**
The United Nations Educational, Scientific and Cultural Organization (UNESCO) Centre for Membrane Science and Technology, School of Chemical Engineering, University of New South Wales, Sydney 2052, Australia
Cooperative Research Centre for Polymers, Notting Hill 3168, Australia

**Yatao Zhang**
School of Chemical Engineering and Energy, Zhengzhou University, Zhengzhou 450001, China

**Gabriela Kamińska and Anna Marszałek**
Department of Water and Wastewater Engineering, Silesian University of Technology, Konarskiego 18, 44-100 Gliwice, Poland

# Index

www.ingramcontent.com/pod-product-compliance
Lightning Source LLC
LaVergne TN
LVHW081026210926
848485LV00004B/956